*International
Process Safety Management
Conference and Workshop*

*International
Process Safety Management
Conference and Workshop*

September 22–24, 1993
Hyatt Regency Hotel
(Embarcadero Center)
San Francisco, California

SPONSORED BY

**Center for Chemical Process Safety
of the American Institute of Chemical Engineers**

Health & Safety Executive, U.K.

U.S. Environmental Protection Agency

European Federation of Chemical Engineering

Society of Chemical Engineers, Japan

Copyright © 1993
American Institute of Chemical Engineers
345 East 47th Street
New York, New York 10017

Library of Congress Cataloging-in Publication Data
International Process Safety Management Conference and Workshop (1993
  : San Francisco, Calif.)
    International Process Safety Management Conference and Workshop :
September 22–24, 1993, Hyatt Regency Hotel (Embarcadero Center), San
Francisco, California / Sponsored by Center for Chemical Process
Safety . . . [et al.].
        p.   cm.
    Includes bibliographical references.
    ISBN 0–8169–0589–4 : $100.00
    1. Chemical engineering—Safety measures—Congresses.
I. American Institute of Chemical Engineers. Center for Chemical
Process Safety. II. Title.
TP149.555   1993
660′ .2804′068—dc 20   93–25606
                                                        CIP

Contents

Process Safety Management Program Considerations to Reduce Future Risk Management Plan Efforts

Bert C. Morris
Daniel J. Statile
Hazard Management, Westinghouse Electric Corporation

INTRODUCTION

Recent Legislation, including the Clean Air Act Amendments (CAAA) of 1990 have authorized the introduction of two required programs for accident prevention that overlap in execution, specifically, Process Safety Management of Highly Hazardous Chemicals (PSM), promulgated by the Occupational Safety and Health Administration (OSHA), and Risk Management Plans (RMP), to be promulgated by the Environmental Protection Agency (EPA). This paper identifies aspects of Risk Management Plans that should be included in the establishment and execution of the Process Safety Management program. Additionally, some recommendations are made on the combination of total quality management and the process safety management function.

Title III of the 1990 Clean Air Act Amendments regarding the need to develop RMP's addresses the "Prevention of Accidental Releases of Hazardous Air Pollutants". The objective of these RMP's is to identify, prevent, and minimize the consequences from accidental releases of certain regulated hazardous materials. While the requirements of these plans have not been developed fully, some general guidelines have been identified. Based on this information, the material required in the RMP has much in common with the requirements developed under the recently promulgated PSM regulations. Therefore, by coordinating it's efforts for both of these programs, a company may realize both economic savings, through the reduction of possible duplication of efforts and quality improvements, by interrelating the two programs.

The PSM program is a performance-based program that addresses many aspects of the management of a process, including design and operating information, procedures, training, accident investigation, mechanical integrity, safe work practices, process hazard reviews and emergency response planning. The PSM guidelines identify over 100 highly hazardous chemicals (with threshold quantities) that require the development of a PSM program. Of the PSM focus areas, the two most relevant to the planned RMP requirements are process hazard analyses (PHA) and emergency response planning (ERP). Coordination of efforts would be most beneficial in these areas.

ESTABLISHING A PROCESS SAFETY MANAGEMENT PROGRAM

One of the objectives of this paper is to provoke the reader to consider the requirements of the future RMP when establishing each element of a PSM program. A key objective in the establishment of a process safety management system is to mitigate risks associated with the operation of the facility. A discussion of the elements of the RMP is provided along with recommendations for inclusion into certain PSM elements.

The PSM directive from federal and state regulators identifies the need to organize and implement programs and culture that minimize risks to workers, assets, the general public and the environment. Although there exists a multitude of "correct" management systems, it is generally accepted that management is responsible for establishing the written and unwritten policies and procedures which shape the conduct and outcome of the business. It is the RMP's goal to assemble these into a format that allows the agency to examine how the facility is being run, ascertaining the commitment level of the management team in dealing with risk management. Most of the basis for today's speculation on the content of risk management plans is based in the California Risk Management and Prevention Program regulation, in addition to the details provided in the CAAA paragraph creating federally mandated risk management plans. Based on the requirements of the California RMPP, the program must include a clear description of the management organization, policies, and process safety management programs designed to mitigate the likelihood of accidental releases.

The following describes the requirements of the California RMPP; the CAAA RMP may require similar elements.

1)   Facility Description

A brief overview of the facility including; address, geographic location, roadways, schools, residential areas, day cares, hospitals, etc.

A description of the site history, products manufactured, maps, and processes which use, store, and manufacture, acutely hazardous materials (EPA list). Written safety information identifying hazards, equipment, and technology employed.

2)   Facility Organization and Responsibilities

An overview of the management structure and responsibilities.

3)   Accident History

A listing of the incidents and near misses over the past three years. A compilation of accident investigation reports. A listing of the incident investigation procedure and responsibilities.

4)   Equipment History

An equipment list, including age, condition, maintenance and test schedules, safety procedures for startup, pressure protection device settings, equipment replacement schedules, and system modification and changes.

5)   Design and Operations

A system overview, including description and operation of the system, shutdown and lockout procedures to maintain safety, standard operating procedures, plant startup procedures, emergency shutdown procedures, normal shutdown procedures, codes and standards, and precautions for handling of materials.

6)   Detection and Monitoring

Accident or release detection methods, automatic shutdown systems, over-pressure protection, and automatic detection systems.

7)   Hazard Identification and Recommended Actions

This portion should include HAZOP results, including recommendations, off-site consequence analysis, vulnerability analysis, and seismic analysis.

8)   Auditing and Inspecting

This section should include inspection audit frequency and procedures, special emergency preparedness safety procedures, equipment inspection and replacement, environmental activities with acutely hazardous materials, safety meetings schedules with the administering agency, mechanical integrity program, pre-startup safety reviews, and any other inspections.

9)   Recordkeeping

Administrative procedures for records storage and maintenance. These records should include plant audits and inspections conducted and the results of the audits and inspections, accident records, lessons learned and implemented, and maintenance and testing results.

10)   Emergency Response and Training

Written plans and procedures, encompassing emergency response training and evacuation and communications plans and procedures.

11)   Training

Operator training programs and procedures.

12)   Certification

A statement by the certifying entity stating that all contained information is true and complete.

13)   Supporting Technical Information

Compilation of related HAZOPs, consequence analysis, seismic assessments, and other technical information created in performance of the RMP.

Table 1 provides a cross-correlation between PSM and RMP for convenient reference. In the next few paragraphs a discussion of the various PSM elements and arguments for content in setting up the PSM program are made.

FACILITY ORGANIZATION, RESPONSIBILITIES, AND ORGANIZATIONAL BEHAVIOR

The bottom line in the establishment of process safety management is to organize the day-to-day operation and culture of the facility to protect itself from the rigors of today's competitive environment. The imposed regulations require the establishment of practices that, when collectively implemented, protect the workers, the general public, and the environment from the various process hazards including accidental releases.

TABLE 1 RMP AND PSM CROSS REFERENCE												
RMP ELEMENTS \ PSM ELEMENTS	Process Safety Information	Process Hazards Analysis	Operating Procedures	Training	Contractor Safety	Pre-startup Safety Review	Mechanical Integrity	Hot Work Permits	Management of Change	Incident Investigations	Emergency Planning & Response	Compliance Safety Audits	Employee Participation
Facility Organization and Responsibilities						●			●				●
Facility Description	●												
Accident History										●			
Equipment History	●						●						
Design and Operations	●		●			●	●		●				
Detection and Monitoring		●					●						
Recommended Actions		●											
Auditing and Inspections						●	●					●	
Record Keeping	●					●	●	●					
Emergency Response and Training											●		
Training				●	●				●				
Certification													
Technical Information Documents		●											

With the intent of the regulations in mind, establishing programs that address the PSM elements without first assessing the skills of the management team and providing that team with the tools necessary to execute PSM is an exercise in vain. Experience shows that facilities that have excellent management skills have bred a culture that already include PSM in the established policies and programs. Therefore, as the first step towards compliance with PSM, each facility should evaluate it's ability to deal with several key management skills necessary to foster PSM. These key elements include, but are not limited to;

- **Organization and Planning** The facility must be organized to maximize safety management and does it execute the plans developed in a timely manner. Each manager must have the skills necessary to organize his/her workload to accomplish the facilities mission in a timely manner.

- **Problem Solving** The staff must demonstrate the skills necessary to resolve technical and managerial problems.

- **Judgement** The staff must demonstrate good judgement. The culture must encourage good judgements to be made.

- **Delegation** Management must empower the employees.

- **Leadership** Management must provide the leadership to the employees that facilitates critical communication. Employees must feel that management is "in touch" with their concerns and the critical facility issues.

Each facility must assess it's management environment and provide training in deficient areas. Without these basic management skills, the facility may expend monies and manpower to buy PSM compliance but never realize the maximum yield from it's investment. There are numerous competent consultants available to establish an assessment system for a particular facility. As a company begins to evaluate it's facilities true safety culture they should consider making a management behavioral assessment.

EMPLOYEE PARTICIPATION

We began the discussion on PSM by discussing organizational behavior. Nowhere in PSM's basis is this more relevant than in establishing a safety program, specifically an open employee participation program. As indicated in the PSM compliance directive CPL 2-2.45A, the OSHA inspector will query employees to ascertain management's commitment to addressing employee concerns. These concerns may be communicated verbally, arise in the process hazards analysis, or be identified during the pre-startup safety review. If the company culture does not address these concerns in a genuine manner, the owner or operator is at risk of being in non-compliance. You must ask, does the facility culture allow employees to speak freely? Is management committed to responding to employee concerns? Does management express a genuine concern for addressing identified issues? If you answer anything other than yes, the management team has a cultural hurdle to overcome. It is recommended that help be obtained from outside your immediate facility to aid in pinpointing problem areas and develop an aggressive program to place the management team on the correct course.

ESTABLISHING CHECKS AND BALANCES

An important aspect to the management structure, and an essential element of the Process Safety Management program is the establishment of checks and balances. One component of this is accomplished through the "management of change" element, which essentially creates a check and balance system to test the validity and safety of each system alteration. The risk management plan should take full credit for the existence of the "management of change" program and highlight the fact that this minimizes the potential for accidental release of possibly hazardous chemicals by ensuring a thorough review.

Many elaborate studies have been conducted for developing checks and balances necessary to ensure quality. There is no magic mold for achieving quality, however, most would agree that establishing a team environment is pivotal for making progress. The management's ability to have change accepted and lasting is contingent upon its ability to delegate responsibility down to the operating level. For small facilities, the team environment demonstrates an increased quality factor that generally cannot be matched by large operations. For large facilities, pressures of the operating environment may lead to attempts to circumvent bottlenecks, sometimes at the sacrifice of safety and quality. Although neither process safety management nor risk management plans explicitly identify a quality assurance function, the creation of the function is imperative to minimizing risk and maximizing quality. Many corporations are undertaking the Total Quality concept. The adoption of these principles and the appointment of an independent quality manager can in itself serve as an important element in the risk management program.

The quality manager's role in the PSM program is that of quality assurance. The quality manager's role may include, but is not limited to:

- Auditing of all aspects of process safety management and risk management programs.

- Establishing and monitoring code-related design practices.

- Development of total quality management programs and monitoring the progress of these programs.

- Providing a check on the entire management system to ensure no elements are circumvented.

- Primary responsibility for administrating management of change. Please note that this does not mean the quality manager is responsible for executing management of change, but rather, is the responsible procedure sponsor, and is given the job of ensuring adherence to the procedure.

OTHER CONSIDERATIONS

When establishing the PSM program consider the following:

- Determine who is responsible for each operating, maintenance, and emergency procedure, the procedure's review cycle, how to retire a procedure, how procedures are to be amended, and finally how long procedures are to be kept prior to elimination.

- When establishing the various elements of mechanical integrity, try to organize the information with the RMP in mind, including,

  - Developing a recordkeeping system for each identified piece of equipment that includes tracking all work performed on that equipment.
  - Developing a method for determining the retirement of equipment.
  - Determining the status of equipment.

- Consider developing a description of the facilities detection, shutdown, and over-pressure protection systems, on a level that can be clearly understood by laymen. This is extremely valuable information when performing a process hazards analysis.

RISK MANAGEMENT PLAN - A LIVING DOCUMENT

As with process safety management policies and procedures, it is recommended that facilities make the risk management plan a living document. A responsible management team examines each change to the facility and "engineers" contingency plans.

When compiling the risk management plan, it is necessary for the facility to demonstrate the checks and balances, identifying all aspects of the operation of the facility that will preclude an accidental release of highly hazardous chemicals. Most important to the risk management plan, and subsequently, its foundation, is the execution of a process hazards analysis. The next few paragraphs highlight some aspects of performing PHA's that should be considered.

LINKING PROCESS HAZARD ANALYSIS AND RISK MANAGEMENT PLANS

As previously indicated, certain aspects of the PSM program elements, specifically process hazard analysis and emergency response planning, share many common requirements with the projected RMP requirements. Therefore, pre-planning and execution of the PSM elements with an awareness of the RMP requirements can result in reduced compliance efforts and improved consistency between these interrelated programs.

One of the elements of the PSM program requires a facility to perform a process hazard review of units which involve the use, storage, handling, or on-site movement of one of the 131 materials identified as "highly hazardous chemicals" in the OSHA Regulation 29CFR1910.119. In addition, this requirement also applies to facilities which handle flammable liquids or vapors in excess of 10,000 lbs, with certain exceptions on use and storage conditions. These reviews can include one or more of the following methodologies:

- What if,
- Checklist,
- What if/Checklist,
- Hazard and Operability Study, HAZOP,
- Failure Modes and Effects Analysis,
- Fault Tree Analysis,
- An appropriate equivalent method.

The objective of these process hazard reviews is to review the design, operation and maintenance of a unit to identify potential process related hazards. Integral to these reviews is the examination of the facility's safety culture and work environment, and an evaluation of the administrative and management controls in effect to reduce the likelihood of a potential hazard. At a minimum, OSHA requires that the hazard reviews address the following areas:

- The hazards of the process. Identifying the potential sources and types of hazards that exist in a facility.
- The identification of any previous incident which had a likely potential for catastrophic consequences in the workplace. The review should analyze the causes and potential consequences of any previous incident

and determine what actions were taken or need to be taken to prevent a reoccurrence of that incident.

- Engineering and administrative controls applicable to the hazards. The review should determine any available controls which would act to prevent or mitigate the hazard. These controls might include detection and mitigation systems, and operating procedures.
- Consequences of failure of engineering and administrative controls. The review should qualitatively determine the potential effects if the identified control systems fail to perform their functions.
- A review of the facility location should include unit spacing, escape routes, potential blast distances, and atmospheric dispersion considerations.
- A human factors review, including operator/equipment interfaces, clarity of process information, and operator scheduling requirements.
- A qualitative evaluation of the range of possible safety and health effects on employees, resulting from the failure of the controls.

In addition, the PSM program requires the development of a site-wide emergency response plan, in accordance with 29CFR1910.38. OSHA also specifically requires that the plan include procedures for handling small releases of the regulated materials.

Since OSHA is mandated to protect worker safety, the focus of the PSM program is on the "on-site" employer process safety environment. There has been no attempt to identify or evaluate the off-site effects to the surrounding community or environment. Also, the general nature of the PSM is a qualitative rather than quantitative assessment of the potential hazards.

The CAAA of 1990 includes provisions to help reduce or prevent the accidental release of certain hazardous chemicals. These provisions are directed toward minimizing the effects off accidental releases of hazardous materials to the surrounding community, i.e. population and environment, of accidental releases of hazardous substances. The CAAA requires that by November 1992, the EPA develop an initial list of at least 100 Extremely Hazardous Substances, along with threshold quantities, which will require the development of accidental release programs, including RMP. The CAAA includes a base list of 16 mandated chemicals including chlorine, ammonia and hydrogen fluoride. By November 1993, the EPA must promulgate regulations to prevent, detect, and respond to the accidental release of these extremely hazardous substances. These regulations would take effect three years after promulgation, and are to cover all aspects of the use and storage of these substances. On January 19, 1993, the EPA published (40CFR68) an initial proposed list of hazardous substances to be regulated under the accidental release prevention provisions. The list includes 100 toxic substances, 62 flammable substances, and commercial explosives. While there are many common substances on the PSM list and the new CAAA list, not all PSM regulated materials are on the CAAA list, and vice versa. Additionally, even for chemicals that are common to both lists, the threshold quantities for application of the regulations may be different. Therefore, it is possible that a facility may be required to prepare either a PSM program, a CAAA RMP program or both.

For the affected facilities, the required RMP will be expected to address the prevention of accidental releases, the identification of the consequences of such releases, and provisions for emergency response to minimize these consequences. Among the expected requirements of the RMP, are the following:

> A Hazard Assessment including
> - A five year release history, describing previous releases of highly hazardous chemicals, their causes, and preventative measures taken.
> - A "what if" or similar analysis, identifying possible accidents and potential release quantities, describing possible equipment failures and human errors that would result in an accidental release.
> - An evaluation of the worst case accident release scenario, including an estimate of release quantity and duration.
> - An estimate of downwind exposures and affected populations (vulnerability analysis), including determination of the concentrations of concern.

- A program for preventing accidental releases highly hazardous materials, describing the engineering and administrative controls to minimized releases.

- An emergency response program to address the required actions to be taken in the event of an accidental release. This program would include both on- and off-site notification, required emergency medical care, and necessary employee training.

A comparison of the PSM and RMP guidelines is provided in Table 2. This comparison indicates many common concerns, including,

> - the need to perform a hazard assessment to identify the possible accidental release scenarios,
> - the need to review the previous release history, with a focus on determining the existing controls or systems which would prevent a reoccurrence,
> - an evaluation of the consequences of identified releases,
> - the development of emergency response plans to address these potential releases.

Table 2
Comparison of Related PSM and RMP Requirements

Process Safety Management	Risk Management Plans
Process Hazard Analysis	**Hazard Assessment**
· Identify Process Hazards · Previous Incidents with a likely Potential for Catastrophic Consequences · Engineering and Administrative Controls · Consequences of Failure of Controls · Qualitative Evaluation of Safety and Health Effects of Failure of Controls · Facility Siting · Human Factors	· Five Year Release History · A Program for Preventing Accidental Releases · "What if" or Similar Analysis Identifying Possible Accidents and Release Quantities · Evaluation of the "Worst Case" Accident Release Scenario
Emergency Response Plan	**Emergency Response Program**
· Site-wide Plan consistent with 29CFR1910.38 · Include Procedures for Small Releases	· Required Actions in the Event of Accidental Release · On- and Off-Site Notification · Required Emergency Medical Care · Required Employee Training

The requirements of the programs differ, since the PSM program is centered on a "qualitative" review of all potential accidents: accidental releases, fire, etc., with the focus being on-site worker protection, while the RMP program is directed toward the "quantitative" evaluation of possible off-site effects of accidental releases as they affect the surrounding community and environment. However, a primary theme of both programs is to review the facility design and operation to reduce the number and consequences of accidental releases of hazardous materials. For facilities which must prepare both programs, these programs can and should be complementary. The future off-site RMP information should be built upon the on-site hazard information currently being developed in the PSM program.

A review of Table 2 indicates that much of the information (such as the release history, identification of process hazards, and accident scenarios, developed during the PSM PHA) may be directly useful in developing the RMP. Additionally, other PHA information (such as the qualitative consequence analysis and on-site emergency response plan) may serve as the basis for developing the more detailed, quantitative, off-site analysis required by the RMP.

RECOMMENDED ADDITIONS TO PROCESS HAZARDS ANALYSIS

Since the majority of affected facilities have selected the Hazard and Operability Study technique for their process hazards analysis, the following are considerations for inclusion into the HAZOP programs that can provide direct value to the RMP.

- The HAZOP team leader should develop a short form that identifies chemicals found in the facilities. This short form should include maximum and minimum temperatures and pressures, maximum quantity of materials to be found, and any reactions these materials may have with chemicals found in close proximity.

- The HAZOP team should make a genuine effort to determine release potentials, and should discuss potential impacts on-site and off-site. Using the short form, a maximum release quantity can be estimated and discussed. Documentation of plausible events should be logged for future reference for starting the quantitative portion of the RMP.

- The HAZOP team should review any existing dispersion analyses to aid in determining on-site impacts. When reviewing this data, potential off-site impacts can be discussed.

  It should be noted that if dispersion analyses are not available, they should be recommended as an action item in the HAZOP.

The HAZOP team should discuss emergency situations when discussing siting issues, addressing escape routes, fire fighting scenarios, and safety systems. Elements of the Emergency Response Plan can be discussed to provide a clear view of potential pitfalls for the most likely release scenarios identified during the study.

SUMMARY

The graduation from process safety management to the risk management plan should be a painless effort. It is a management imperative to assess and improve the facilities planning process. Planning today for the future will help alleviate the cost burden for changes brought about by the risk management plan.

Based on the overlap in the engineering analyses required by these regulations, and since the PSM work precedes the RMP programs, a facility should review the RMP requirements prior to performing the relevant PSM analyses to ensure that the requirements of both programs will be met.

REFERENCES

"RMPP, Guidance for the Preparation of a Risk Management and Prevention Program", Prepared by the California Office of Emergency Services, Hazardous Material Division.

29CFR119.1910 Process Safety Management of Highly Hazardous Chemicals, Volume 57, No. 36, February 24, 1992

Integrating Process Safety Management and Risk Management Plans - Minimizing Implementation Efforts, Morris, B.C., Statile, D.J., PETRO-SAFE'93, January 26, 1993, Houston, TX

Planning Guidelines for Acute Risk Management: The Canadian Chemical Industry Experience

Graham D. Creedy
The Canadian Chemical Producers' Association

1. INTRODUCTION AND BACKGROUND

Acute risk management plans for hazardous installations are not specifically required by legislation in Canada. For members of The Canadian Chemical Producers' Association (CCPA), however, such plans are an implied requirement under the association's Responsible Care® initiative.[1]

With the approaching target date for Responsible Care implementation, the CCPA faced increasing requests for guidance to members on how to satisfy this requirement. Typical questions were:

- Does the requirement apply to all facilities, or only to those which are likely to pose a serious hazard?
- If the latter, what criteria determine who is included or excluded?
- What should the plan address?
- What techniques should be used for assessment of risk, and how far should the analysis go?
- Where the plan identifies a need for measures to reduce or control the risk, how far must these be put into effect before the the company can claim that the implementation target is met?
- How much should risk management be extended to other handlers of members' materials: carriers, warehouse and terminal operators, distributors, customers and suppliers?
- How much should the requirement take into account a member company's size and current resources (ability to perform and act on a risk assessment)?

This paper describes the approach taken by the CCPA in answering the above and other questions, and in providing an indication of the desired content and time frame for implementation of this step under the first phase of Responsible Care. It presents an approach which balances the need for appropriate management of risks against the challenge this poses to many operators of hazardous facilities, who may be so overwhelmed that they do not know where to start.

First, however, it is useful to review the underlying aim of Responsible Care and the action steps that this implies.[2]

2. UNDERLYING PRINCIPLES

Responsible Care is a voluntary initiative based on an ethic, or philosophical belief, rather than on specific, detailed obligations. This ethic grew out of public concern over the hazards of chemicals and the way the chemical industry was run, and its aim is to dispel that concern by addressing its root cause. This means that we - the chemical industry - must do two things:

- improve the level of safety in all aspects of chemical operations, to eliminate the basis for concern, and
- communicate this to the public, and to governments, and demonstrate that the industry is sensitive and responsive to public concern.

The goal is to achieve a protected, informed society, and to do so by managing chemicals responsibly and demonstrating this to others. This in turn clearly calls for an ability to assess and control the risks which may be associated with chemicals and chemical operations.

3. THE BASIC APPROACH

The ethical foundation of Responsible Care implies that CCPA members should be concerned not only with whether a given behaviour is legal, but whether it is *right*. The codes of practice and additional explanatory material in the supporting implementation guides give an indication of what is required, but often lead to the question of how far the process should be carried, at least to meet the goal established by member companies under their commitment schedule.

The codes of practice, for example, refer to the need for assessment of risk, but leave open the choice of methods and the extent to which they should be applied. The degree of control expected for risks once they have been identified is also left to the judgement of members, leading to requests for more and yet more clarification of what should be done.

The CCPA and its member companies do not have the resources to produce a master document describing in detail exactly what should be done in every situation - and, as mentioned above, this is not the point. As with any ethic, much is left to individual judgement - in this case the judgement of the company concerned.

There is, however, a basic approach which can guide a company in reaching this decision first on the assessment and then on the control of risks. This approach is valid regardless of the size and nature of the company, since the viewpoint is from outside the industry rather than within. It consists of three questions:

(a)   What could go wrong?

This refers to potential hazards, including both acute hazards such as a fire or spill and chronic hazards such as long term exposure to chemicals in the atmosphere, water supply or products with which one is in frequent contact. The emphasis is on the word *could*, i.e. it is not the same as "what is expected to go wrong" which involves an assessment of risk.[a] This emphasis on hazard identification without the usual next step of attempting to quantify the likelihood of occurrence is intentional, although it does not appear to follow the logic of most risk assessment texts. It arose because questions asked, and the ensuing discussions, revealed that people often need help in breaking out of a mindset about the risks they may be incurring. Focusing only on what is expected can all too easily lead to overlooking hazards whose significance is not obvious. A devil's advocate is useful at this stage, preferably with experience and knowledge to foresee all the possibilities.

(b)   How could this affect others?

The public, the environment, customers, employees, etc. are some examples, but there may be more. Could they or their families be killed or injured, evacuated from their homes or places of work? Could they suffer impairment of health or wellbeing which may perhaps not even be discovered until years after the damage is done? What about livestock or personal property, or future generations?

---

[a]   A hazard is an event with undesired consequences, while risk refers to those consequences multiplied by the probability of occurrence. The aim is first to eliminate or reduce the hazard, then bring the residual risk to an acceptable level by suitable management controls.

(c)   How might these potential effects be perceived?

Perceived risk is often far more significant than actual risk and can have serious and real consequences, as can be seen from the PCB incident at St. Basile, Quebec. It is perceived risk rather than actual risk which counts, regardless of what we might wish, and the implications of this must be acknowledged and understood. Perceived risk can rapidly be transformed from a dormant unease to near panic by an apparently minor triggering event while the actual risk remains unchanged - for example when public sensitivity to a hazard is aroused by media attention or release of a new report. Panic may then lead not only to real psychological or even physical harm to some individuals, but to public outrage if it appears that the company was aware of the situation all along but did nothing until it was too late.

We cannot give an absolute guarantee to the public that no incidents or harmful effects will ever occur, although goals such as "zero incidents" or "zero discharge" may still be worthwhile. We should, however, be able to justify to the public that, even if an incident should occur, we had acted responsibly by taking all reasonable steps to prevent it.

4. SEVEN KEY STEPS

The process of managing risk by appropriate reduction and control is clearly needed along with risk assessment, or there would be little point in doing the assessment. Requests were also received for guidance on how far the risk reduction and control process should be carried to meet the expectations of the initial implementation phase of Responsible Care.

The following seven-step process was developed to help companies understand what is expected, but without attempting to specify in rigorous detail the action or performance expected. The steps are given here with the accompanying explanatory notes. They are, in roughly logical order:

(a)   Be aware of all laws and regulations applicable to company operations and Responsible Care.

These represent a *minimum* standard, though one which falls considerably short of what is expected under Responsible Care. This step can be done by one person or several, or even contracted out - but it is obviously essential.

(b) Review *all* aspects of company operations to determine where those laws, regulations and Responsible Care apply.

The aim here is to identify where the potential exists for actual or perceived hazards (to the public, customers and users of the company's products, to those transporting, storing or distributing them or disposing of associated wastes, to employees, etc. and to the environment) or irresponsible behaviour under both normal and abnormal conditions. The approach described in section 3 above will help in complying with the spirit of Responsible Care rather than just the written requirements of the codes of practice.

This is the most critical step, and the one where it is easiest to make a mistake. Major incidents often occur because of some event or combination of events whose significance was not recognized until it was too late. This is why the approach of reviewing everything and eliminating items is more effective than trying to list hazards as they come to mind.

(c) Bring operations to a standard where they are *capable* of current compliance with laws, regulations and Responsible Care by elimination or reduction of hazards.

Typically a company realizes that some improvements are needed to eliminate or reduce major hazards now, while others are worthwhile but can be done later.

Any capital projects, equipment modification, phase-out of products, etc. deemed essential for proper hazard control must be completed under the initial implementation phase, or the situation must be rendered safe by operating controls such as reducing the quantity of material or throughput, additional supervision, etc. Other changes may be scheduled after initial implementation, but the following steps *must be based on the situation as it exists and not on future plans.*

(d) Ensure that systems are in place to establish and maintain appropriate human resources at all times for responsible operation under anticipated conditions.

This includes the number of people, their basic qualifications and training requirements. Coverage should provide for absence or impairment due to vacations, sickness, training, travel, attendance at meetings, etc. If the company has determined that a given minimum resource pool is needed for safe and responsible operation, this pool must be present and available for duty.

(e)   Ensure that systems are in place to control residual hazards within acceptable limits under all anticipated conditions, and to control hazards during and after any change to existing operations.

Once the equipment is inherently capable of safe operation and the necessary trained people exist, hazards can be controlled to reduce the probability of undesired events or effects and hence the risk to the appropriate level. The acceptable limits of safe operation should be clearly defined, as should the responsibilities for control and authority for any necessary action.

(f)   Ensure that systems are in place for communication and feedback of appropriate information on the above with employees and other affected parties.

Communication and feedback are actually involved throughout the process described so far, but are specifically stated here since their absence is the usual reason for failure of the preceding step.

(g)   Ensure that management systems are in place to monitor the above.

This step includes auditing (to verify that what should be done is actually being done); evaluation (to make sure that it is effective in achieving the desired aim, and to improve effectiveness); feedback on system performance to senior management, and keeping employees, etc. aware of executive commitment. The code milestones specifically require this step as an annual review after implementation has been completed. However, it is useful to introduce it as an ongoing process during the later stages of implementation, to confirm that implementation is on track and that the objectives of the original plan are indeed valid.

Note: The seven steps described above provide the foundation for effective control of acute and chronic hazards arising from future actions. They are key steps, but do not cover everything that is expected under the spirit of Responsible Care as they do not necessarily address the effects of past and present actions. Companies should therefore be guided by the comments in section 3 above, be sensitive to public concern, and endeavour to go beyond the status quo in interpreting what is appropriate in their particular circumstances.

The application of the approach described so far can now be seen by reviewing the typical questions received from the association's members and described in the introduction. What follows is an overview, since a full description of the approach is beyond the scope of this paper. In practice dialogue is used to contrast examples of obviously responsible and less-than-acceptable behaviour and indicate how limits can be established. Once this process is grasped, companies are then able to examine situations by themselves and by discussion with others to establish and then narrow the limits to arrive at an understanding of what constitutes responsible conduct.

In this paper, the question of acute risk is examined primarily in terms of external effects (e.g., on the public) and then on effects within the company (e.g., on employees and others on the plant site).

5. APPLICATION OF THE APPROACH

(a) Does the requirement apply to all facilities, or only to those which are likely to pose a serious hazard?

Some form of acute risk management is expected for all CAER sites operated by CCPA members. A CAER site is defined as any site operated by a CCPA member where the potential exists for a perceived hazmat incident. This includes warehouses, low risk processing plants, etc., but excludes sales offices where since the type of incident which could result is familiar to responders and the public, unless significant chemical samples are present. Nevertheless members are encouraged to list all facilities, then exclude those which are exempt so that these can be periodically reexamined to ensure that any assumptions remain valid.

(b) What criteria determine who is included or excluded?

(c) What should the plan address?

These two questions are dealt with together since, although the principles are the same for risk management at high and low risk facilities, it is unrealistic to expect the same actions from, say, a warehouse handling plastic resin as from an ethylene plant. Municipal emergency plans, for example, are not required by law in most provinces of Canada, and the minimum acceptable level of

preparedness in the community and the degree of persistence by the site in
attaining it are likely to be quite different for the two cases just mentioned.

The following criteria are therefore suggested in the guidelines:

For the lowest category - warehouses or plants with low hazard materials
only - the company should look for the existence of a general community
emergency plan and some organizational structure to put it into effect. The
guidelines do not call for further action at this stage if the community is not
willing to participate, although the company is encouraged to monitor the
situation and do what it can to advance the CAER message.

Where moderately hazardous materials in typical (i.e. not stockpiled)
quantities or small quantities of high hazard materials are present, however, the
company should expect the community to be prepared and trained to respond to
potential incidents. The community plan should be an official document (i.e.
approved by council) covering all appropriate agencies with responsibilities and
authorities clearly defined. It should meet provincial guidelines where these
exist and have been given at least a table top exercise within the last three years.

Where there are high hazard materials in more than small quantities, large
quantities of moderately hazardous materials or significant amounts of materials
likely to be politically sensitive because of public concern, the site should form
an active link with the community emergency planning committee, and the
community plan should be comprehensive including provision for public alerting
and communication during an emergency, evacuation, welfare, etc. At this
stage the plan should be on file with the province and have been updated and
comprehensively tested within the last three years. The site should participate in
community risk assessment once a generally accepted method becomes available.

(d) What techniques should be used for assessment of risk, and how far should the analysis go?

Assistance here was at first limited to a simple list of some of the standard
reference works, but it later became clear that more direct guidance was needed
because of users' unfamiliarity with the subject. The challenge was to provide a
process to enable recognition of the nature and scale of the more important
hazards and the effects of reduction/control measures, but without making it
sound so technical that people became apprehensive about their ability to
understand and apply it.

The advice therefore became more specific, while still being presented as
an example, i.e. companies are free to select an alternative or modified
approach if they feel it is more appropriate, so long as they are sufficiently
thorough.

The suggested process for acute site risk assessment then became:

(i)   Break the site into areas for hazard review using the plot plan.

(ii)   Screen to determine the scope of the review for each area according to the potential hazards present. Companies with limited technical resources are advised to use small teams of people with the skills to identify what could go wrong, and seek any special expertise the team deems necessary for further investigation.

(iii) Apply the steps outlined in section 4 above to ensure that the risk from all identified hazards is satisfactorily controlled for each area. In those areas with materials of low hazard only this may consist of basic safety, fire protection, etc. plus steps to prevent entry of unexpected materials by feedstock supply errors, improper piping connections or contamination,etc.

Where materials of moderate or higher hazard are present, use the Dow Fire and Explosion Index (FEI)[3] and/or Chemical Exposure Index (CEI)[4] to determine the approximate zones of exposure to effects from fire, explosion and/or toxic release. If any of the conditions in table 1 are met, examine the critical process systems and equipment in detail. Piping and associated equipment handling or containing high hazard materials should be checked to confirm proper design, drawings should be updated to correspond to the actual situation, and the systems should be examined using formal HAZOP studies, failure modes/effects analysis and/or other suitable methods given in the CCPS *Guidelines for Hazard Evaluation Procedures.*[5]

TABLE 1. Suggested criteria for further study of hazards.

- the FEI is 130 or higher[b];
- the plot of the CEI against the sum of General and Special Process Hazards from the FEI falls into the "further study required" zone in figure 1;
- for toxic hazards, LC_{50} is exceeded for 30 minutes;
- for explosion hazards, overpressure is 7 KPa or higher;
- for fire hazards, radiation to people is 4.7 kWm^{-2} or higher or radiation to buildings is 12.6 kWm^{-2} or higher; and/or
- the potential impact on personnel or the environment is otherwise significant.

[b] For new facilities an FEI of 110 is suggested as the point beyond which further study is recommended.

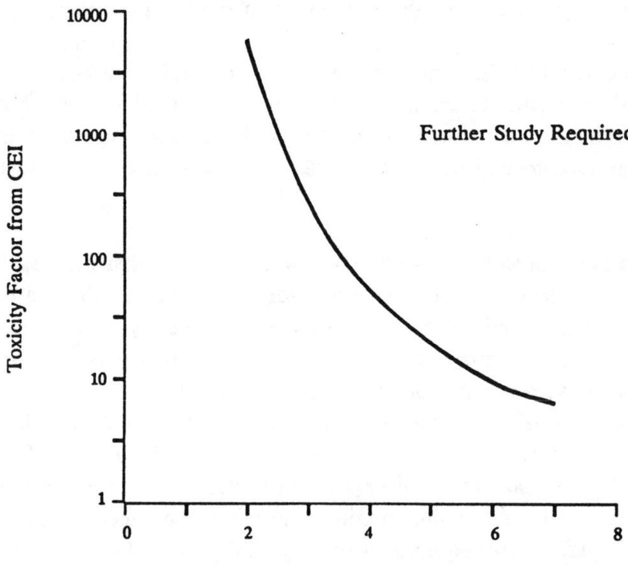

Figure 1. Toxic Hazards Detailed Review Decision

(e)   **Where the plan identifies a need for measures to reduce or control the risk, how far must these be put into effect before the the company can claim that the implementation target is met?**

This is answered by step (c) of section 4 above. Any major deficiencies should be corrected by equipment modification where practical, but in any case safe limits should be established for all appropriate parameters and a combination of engineering/supervisory systems put into place to control operation of processes within those limits.

In theory the risk can be quantified and steps taken to reduce individual risk to a given level -- e.g., for offsite exposure the buffer zone guidelines of the Major Industrial Accidents Council of Canada (MIACC) can be used.[6] Rigorous quantitative risk analysis, however, is not practical for most of the companies for whom these guidelines are intended, but the process described does give a rough indication of what is expected.

A short list of key tools for further reference has also been issued, and members of the CCPA's risk group and process safety subcommittee are available for over-the-phone advice and guidance to the association's members.

(f) How much should risk management be extended to other handlers of members' materials: carriers, warehouse and terminal operators, distributors, customers and suppliers?

This again depends on the situation, and the same basic approach of section 3 above can be used as a guide. A full description is beyond the scope of this paper, but members are advised to consider the hazardous nature of the material and the quantity shipped, and cover the following: communication of hazard information; safe storage and handling; policy on responsible behaviour and its application; emergency plans; employee protection; community protection including site security; provisions for disposal of hazardous waste. A self-assessment tool has been developed to aid in this process.

For the higher hazard categories the situation should be reviewed by a team of code experts on a case-by-case basis. The team should examine the potential consequences which could occur with each customer or other party, both while under the party's control and also the implications further down the distribution chain if the material is in turn shipped out to others.

Appropriate action plans for risk control should then be developed with customers, etc. on a case-by-case priority basis.

(g) How much should the requirement take into account a member company's size and current resources (ability to perform and act on a risk assessment)?

The extent of risk management should be guided by the need for protection of the public and other stakeholders and the questions of section 3, which do not refer to company size or resources.

There may be differences in the rate at which companies are able to implement Responsible Care, but each is expected to comply with the conditions of all codes of practice before claiming that it has met the implementation target. The target time frame for completion of initial implementation (as opposed to subsequent continuous improvement) was three years. Full realization of the extent of the Responsible Care commitment as implementation progressed, together with the current recession, has led to some slippage, although implementation was reported at 89 percent complete by the target date of December 1992. Members have agreed that the standards should not be relaxed, and networking is intensifying as they strive to meet the remaining requirements during 1993.

The message is clear -- regardless of company size or resources, risks must be assessed and brought under satisfactory control if the association's members are to meet the high standards they have established under Responsible Care.

The author acknowledges the assistance of the CCPA's process safety subcommittee in applying the above approach to site risk assessment. Special thanks are due to Peter Hughes of Novacor Chemicals Ltd. for the technical guidelines on acute risk assessement which appear in this paper.

*   Responsible Care is a registered trademark of The Canadian Chemical Producers' Association.

REFERENCES

1   Responsible Care: A Total Commitment. Guiding principles and codes of practice, The Canadian Chemical Producers' Association, September 1991.

2   The approach described in this paper was issued to CCPA members via several internal communications, notably:
- Guidelines for completion of Responsible Care implementation (Phase I), February 20, 1992.
- Risk Assessment Guidelines, July 1992.
- Manufacturing code of practice Site Acute Risk Assessment implementation aid, November 1992.

3   Dow Fire and Explosion Index Hazard Classification Guide, 6th. edition, 1987, is avialable from the A.I.Ch.E.

4   This item was kindly made available by Dow for use by CCPA members. For a general description of the CEI, see reference 5 below.

5   Guidelines for Hazard Evaluation procedures, second edition with worked examples, Center for Chemical Process Safety, 1992.

6   MIACC recommended buffer zone guidelines for land use planning propose the following classification for land use based on calculation of individual risk:

From risk source to 1 in 10,000 contour: no other land uses than the source facility

1 in 10,000 to 1 in 100,000: uses involving continuous access and the

presence of limited numbers of people but easy evacuation, e.g., open space, warehouses, manufacturing plants

1 in 100,000 to 1 in 1,000,000: uses involving continuous access but easy evacuation, e.g., commercial uses, low-density residential areas, offices

Beyond the 1 in 1,000,000 contour: all other land uses without restriction including institutional uses, high-density residential areas, etc.

These guidelines assume an emergency plan is in effect. If this is not the case the contours should be increased by a factor of 10.

Risk Management: A Business Risk or a Risky Business?

Keith Cassidy
Head of Major Hazards Assessment Unit, Health & Safety Executive, U.K.

'There are three kinds of risk manager: those who quantify risks, and those who don't.'

ANON

1. <u>INTRODUCTION</u>

The management of risk is now recognised as central to the effective and efficient operation of industry and commerce and is widely practiced.

Risk management has economic, political and human dimensions, which in all cases involve pivotal judgements relating to the acceptability or (as appropriate) tolerability of the criteria which underpin the executive decisions and actions in the management process.

How robust are the techniques which are used to arrive at such judgements? And how can existing variations in tolerability criteria be explained or justified? The developing methodologies contain many uncertainties (for example, selection of failure cases from a range of possibilities; failure possibilites for each case; scale of consequences and modelling uncertainties; model validation; parameter values of the models used; uncertainties in enhancing and mitigating factors.)

 How far do these uncertainties affect the validity of risk management decisions? And how sensitive are these decisions to aspects of uncertainty?

How far do the influences affecting public perception of the nature, type and magnitude of any risks affect the nature of risk

management? (For example issues such as voluntary vs
involuntary exposure, natural vs man made risks, perceptions
of personal control, familiarity, perceptions of benefit or
disbenefit, the nature of the hazard, the nature of the threat, the
special vulnerability of 'sensitive' groups, public perceptions of
comparators, reversibility of effects all may be felt to influence
significantly the decision making process.)

These issues will be addressed specifically in the context of the
control of risks from major chemical hazards, and risk
associated with some mass transportation systems, but with
wider reference as appropriate. And will be accompanied by
illustrative case studies showing the variability of decision in
terms of the influences of perception of tolerability and of
assessment uncertainties, as part of presentation of the paper at
the Conference.

2. RISK MANAGEMENT

There is no generally accepted definition of risk management.
Indeed perceptions of the meaning of the term, and the nature
of its role, will vary according to the viewpoint of the observer
or practitioner. But whatever the definition or perception, risk
management is considered to be a 'good thing', a worthwhile
objective, an attractive goal.

There are many models of the process, some more tangible
than others; but all embrace the following common factors:

- organisational characteristics
- external factors, and
- fundamental safety/risk management principles

It is a three stage process, where first stage controls aim at the
control of inputs, to minimise the hazards entering the
organisation. The second stage control involves the regulation

of the activity, to eliminate or minimise risks inside the organisation, by the creation and management of a supportive organisational culture. Finally, tertiary control relates to outputs, to minimise risks outside the organisation arising from work activities, products, and services. This matrix of control allows the very necessary setting of performance standards at all stages of the operation, particularly in those areas where the potential for control lies within the organisation. (1)

A simple, global model of this risk management model is shown at Figure 1. It shows a combination of external factors, fundamental principles, and organisational characteristics (usually unique to a specific operation) as they impinge upon the management process. A structured version of this model is being developed as part of a current European Community Collaborative Research Project. (2)

The validity of such models is emphasised by common causal characteristics in a range of major disasters. Such characteristics include:

- poor horizontal and vertical communication
- managerial failure to address 'lifecycle' implications
- impaired or restricted organisational learning
- lack of formal hazard/risk identification
- resourcing and allocation of responsibilities

The model has both local and global application of course; and it is one which is time dependent and time sensitive. In the specific context of the control of major chemical hazards (with which this paper is concerned) it can be applied at varying levels. Essentially, the question is one of integrated risk assessment and management:-

> assessment - in the sense that ideally all environmental, health, and safety risks within a defined region should be systematically identified, analysed, and assessed in such a way that rational choices can be made about which risks should be reduced, weighing the social and economic costs of such risks,

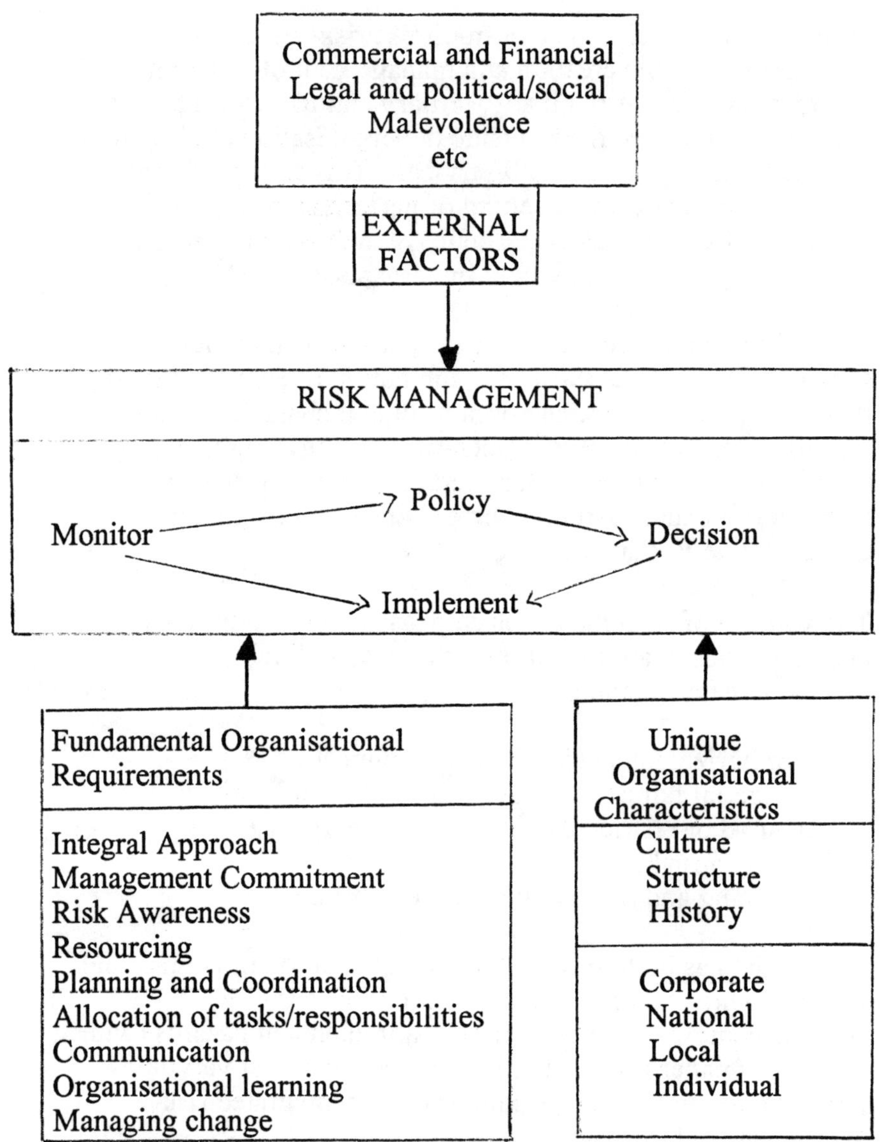

<u>Figure 1</u>: A developing global risk management model

the benefits of risk reduction and associated costs, and formulating the basis of integrated environmental and safety management;

management -   in the context that all the options of risk
management: locational, preventative,
mitigating, protective and institutional
should be explored in a holistic way and
complementarily so that the resources
committed in the safety management
process are optimised.

Such integrated assessment and management approaches
necessitate consideration of all relevant risks, but the level of
detail in the various facets may be varied, depending upon
predetermined priorities (which may be determined by some of
the social and economic factors referred to above.) They
require adequate communication and co-ordination between the
various parties involved in the risk management process -
government, industry and the community. As well as
conventional sources of chemical risk - viz the continuous
emissions to air, water and land, and the accidental releases,
both referred to earlier, - they will also involve

- transportation systems, as a source of
continuous, episodic, and accidental
emissions;
- the interaction of natural hazard sources;
- large scale agricultural activities (both in terms
of fertilizers, and various biocide contributors),
including the effects of irrigation needs;
- industrialization and urbanization, and
associated infrastructures, including waste
generation and disposal.

Of course, the ultimate development of these concepts leads
inexorably to an overall life cycle analysis approach - the
so-called 'cradle to grave' analysis. It is a procedure which has
been performed for some time in technical and economic
studies, but is only now being systematically used in
environmental impact studies, where it can be interpreted as an
enlarged, more comprehensive approach to the problem of
'pollution' in its most basic sense.

The principles of life cycle analysis are simple. Such analysis
of a product starts with depletion of natural resources which are
used to produce the raw materials required. During all the
production steps environmental impact occurs as emissions
(either continuous or episodic), and waste of one kind or
another is generated. The use of the product may often also
bring pollution; and the product will eventually end up as
waste, (which to some extent may be recycled for a time, but
which eventually results in some kind of environmental
impact.) The 'major hazard' impact, which is being considered
in several papers at the Conference, is a particularly significant
manifestation of a challenge to the environment (either natural
or human) from the effects of human activities. The overall
principle is as follows:-

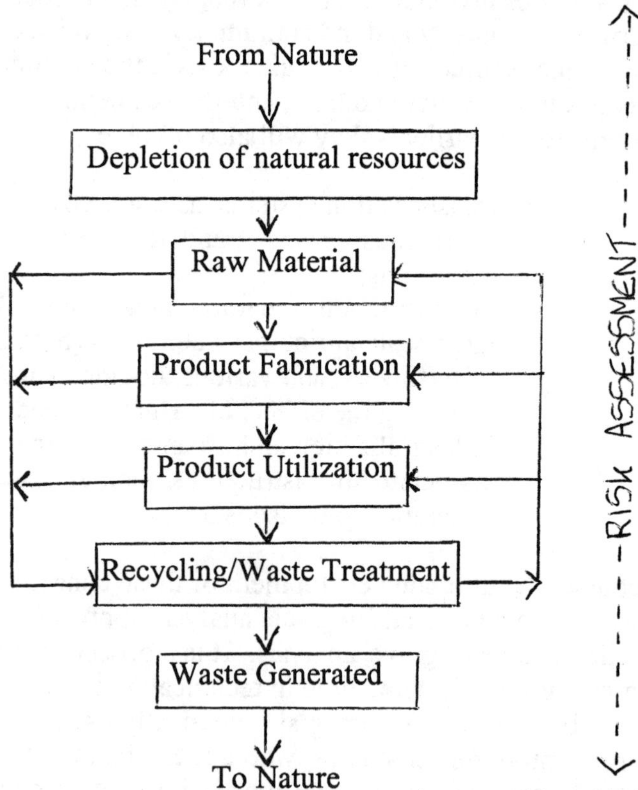

<u>Figure 2 - The life-cycle process</u>

such life cycle analysis informs the process of integrated risk management, and makes it possible to take decisions on such matters as replacing products having a heavy impact on the environment with products having a lesser impact. Similarly, the benefits of a specific product can be balanced against environmental impact in a narrow or a wider sense. It is a classical cost/benefit approach, applied at either the micro or the macro scale. Risk assessment of one kind or another, is pivotal in these risk management decision processes. (3)

3. Risk Management of Major Chemical Hazards - The Nature of the Problem

The issues involved may be very complex, and the scale of the concomitant problems enormous, but it is possible to characterise the overall process into a simple, coherent control strategy, based on the following principles

- IDENTIFICATION - the recognition and location of any potential problem;

- ASSESSMENT - the bounding and dimensioning of any potential problem;

- CONTROL - the limiting of the scale of any potential problem, by prevention or avoidance;

- MITIGATION - the amelioration of the residual elements of the potential problem.

This is a strategy first applied, in the UK, to the control of major chemical hazards as a result of the recommendations of the UK Advisory Committee on Major Hazards (4) , but it is an overall approach which is universally applicable.

Measures used to parameterise, or to limit, the component
elements may vary between hazards and risks, between
different components of the
overall environment, or between different economic and
cultural systems.

But the underpinning logic of the approach remains. It is
briefly, a taxonomy comprising overall environment, or
between different economic and cultural systems.

Is is briefly, a taxonomy comprising

- the 'SOURCE'

- the 'SOURCE TERM'

- the 'DISPERSION'

- the 'DOSE'

- the 'IMPACT'

that is to say (for chemical hazards)

> - the nature of the release: routine planned releases as
>   part of normal plant operations (eg chimney
>   emissions); expected unplanned releases as part of
>   normal plant operations (eg pump gland leakages);
>   and unplanned releases from abnormal plant
>   operations or accidents
>
> - the size and duration of the release;
> - the environmental pathways the release may follow;
> - the resulting environmental concentrations;
> - the effects of these concentrations on the component
>   parts of the environment - man, flora, fauna etc.

and will involve some consideration of the associated
uncertainties. All this will lead to judgements as to

'acceptability' - or preferably 'tolerability' - of the effects, and this will involve
the derivation and application of relevant criteria. These issues
 are considered below.

4. Risk Management: The Critical Questions

All the forgoing concepts can be distilled down into a very
simple question set, at the core of a risk management approach.
Essentially, these core questions are:

WHAT   IF ?

WHAT   THEN ?

THEN   WHAT ?

SO   WHAT ?

'What if?' requires a combination of technical expertise,
experience, and a degree of imaginative insight. 'What then?'
and 'Then What?' are essentially the techniques and practices of
risk assessment. 'So What?" is the area of judgement, informed
but not constrained by the earlier inputs. It is a decision
process, often rigorous, which involves:-

a) quantification of likely risk with an understanding of the
inherent uncertainties involved in the assessment process

b) reference to the likely benefits generated and the political
and economic considerations associated with it

c) judgements as to tolerability or acceptability for groups
directly or indirectly affected; and

d) sometimes, decisions as to further reductions in risk taking
cost (including effort, and available technology) into account.

It is, in short, a process which is essentially economic and
political, technically informed.

It is these aspects of the 'So What?' question I propose to address next; but first a small digression.

5. <u>The Nature and Definition of 'Risk'</u>

Central to all these issues is the basic question which any risk manager must address: 'How much of what kind of what risk of what to whom/what?' Here there are some emerging, agreed definitions, at least in principle - the versions following are those proposed by the UK Institution of Chemical Engineers (5) :-

hazard: a physical situation with a potential for human injury, damage to the environment, or some combination of these

risk: the likelihood of a specified undersized event occurring within a specified period or in specified circumstances (and may be expressed either as frequency, or probability)

individual : the frequency at which an individual may be
risk   expected to sustain a given level of harm from the realization of specified hazards

Societal: the relationship between the frequency and the
risk   number of people suffering from a specified level of harm in a given population from the realization of specified hazards.

The definitions above are expressed in essentially social terms; but they are equally applicable in principle to financial or economic contexts, or to combinations of these, again with increases in the uncertainties inherent in the process. A particular problem in the management of risks associated with the large scale events associated with major chemical hazards is that the assumptions which underpin normal risk management decisions (eg the Heinrich Triangle approach for accident ratio risks, - similar principles apply to financial extrapolations) may not apply in the context of low probability, high consequence events, even when contingency effects are excluded.

6. <u>General Principles Risk Criteria</u>

Whatever the assessed level of risk, the decision maker is ultimately faced with the 'So What?' question, where increasingly quantitive probabilistic criteria are being applied by regulators or enforcers. Where such target criteria are applied (often in a 3-Zone situation, the middle zone embracing the 'ALARA' or 'ALARP' principle) a standard framework is emerging. This framework reflects well established approaches in international risk control, particularly related to advanced, high technology activities, such as those of the nuclear industry, the latter being expressed in the 1977 Report of the International Commission on Radiological Protection. These ICRP recommendations embody the interrelated principles of justification of practice, optimisation of radiation protection, and individual dose limits. No practice or activity involving exposure to radiation was to be adopted unless its introduction produced a positive net benefit in a society, this benefit to be maximised by the 'as low as reasonably achievable (ALARA)' principle, and inequitable distribution at the level of the individual avoided by dose limits for that individual.

Such principles apply, of course, to the control of most risks. In an ideal world, any hazardous activity would not impose risks which were disproportionate to the benefits (such benefits can form a wide spectrum, and inevitably involve economic as well as other, often less tangible, social value parameters), and any such risks would be equitably distributed amongst society in proportion to the benefits received. In practice, of course, such distribution is not possible, and the principles of distribution described above are applied in a much more general way, involving tests to ensure:-

a) whether a given risk is so great, or the outcome so unacceptable that it must be refused altogether; or

b) whether the risk is or has been made so small that no further precaution is necessary; or

c) if a risk falls between these two states, that it has been reduced to the lowest level practicable, bearing in mind the benefits flowing from its acceptance, and taking into account the costs of any further reduction.

These principles combine with other generally accepted tenets:-

d) that risks should never be imposed unnecessarily; and

e) that no individual or community should bear an unfair proportion of any risk.

Such value judgements involve very complex social processes. Hazards and risks are viewed quite differently, depending on the origins of the hazards and the nature of the risks they present. Natural hazards seem to be 'accepted' more readily than those which are man made; and hazards which presage catastrophe appear less 'acceptable' than those presenting a lower level, continuous risk. A relatively well established hierarchy of 'tolerability' has emerged, which involves issues such as:-

* voluntary vs involuntary exposure
* 'natural' vs man-made risks
* perceptions of personal control
* familiarity
* perceptions of benefit or disbenefit
* the nature of the hazard or consequence
* the nature of the threat
* the special vulnerability of 'sensitive' groups
* public perception of the extent and type of risk
* perceptions of comparators
* the reversibility of effects

It is a decision hierarchy which turns on the confidence of those exposed to the risks in those authorities and bodies who create and control the risks - government, the regulatory authorities, plant operators, 'experts', and the emergency services. Priority questions include:-

- does the public believe that all views and interests have been considered in the decision making process, or has there been some 'dealing'?

- does the public have confidence in the effectiveness and independence of the regulatory authorities?

- is there a consistent and credible consensus of scientific opinion about the project that the public can trust, or do the 'experts' disagree amongst themselves?

- what is known about the quality of the project and plant management?

- are the emergency and medical services able to cope with any event, in the short or long term?

It is, in effect, a combination of physical and social detriments, in which some major elements are not quantifiable in any meaningful way.

7. Some Individual Risk Criteria

In recent UK publications (eg 6) it has been generally agreed that the risk of <u>death</u> of one in a thousand per annum is about the most that is ordinarily accepted under modern conditions for workers in the UK - that is to say, this figure represents the current upper bound of 'tolerability'. The report further suggested that a maximum tolerable risk to any member of the public from any large scale industrial hazards should be a decade, or order of magnitude, lower than this, at $1 \times 10^{-4} \mathrm{yr}^{-1}$, both levels being subject, of course, to further control ALARP. A lower level, which might be 'broadly acceptable' - but not altogether negligible - was proposed at about $1 \times 10^{-6} \mathrm{yr}^{-1}$. In general, there seem to have been a wide (but not total) measure of agreement that these criteria comprise a reasonable representation of current public expectation - although it may be that the numerical difference between the 'at work' and 'imposed' risk situations, in terms of duration of exposure, is not as clear cut generally appreciated. In effect, a more valid comparison of the two suggested criteria (on the basis of approximately 2000 hrs per year at work) is :-

$$\text{imposed risk (continuous)} \quad 1 \times 10^{-8}/\text{hr}^{-1}$$

$$\text{voluntary risk (working)} \quad 5 \times 10^{-7}/\text{hr}^{-1}$$

but some modification of the 'continuous' assumption is also arguable to reflect absences. The 'negligible' levels do, in fact, appear to have some historical confirmation or validation in that recent statistics produce an annual risk of death for workers in some industries to be in the $1 \times 10^{-4}/10^{-5}\text{yr}^{-1}$ region, with less 'industrial' operational environments/offices, shops, etc, producing risks below $1 \times 10^{-5}\text{yr}^{-1}$.

In a recent report HSC has addressed the risks in the UK from the transport of major hazard quantities of hazardous substances (7). In general the individual risk tolerability criteria above were endorsed in this Report, which also recognised the validity of some of the criticisms, made at the Hinkley Point Inquiry, of the $1 \times 10^{-4}\text{yr}^{-1}$ criterion for members of the public. Also endorsed (in the context of their specific applicability) were the discussion criteria advanced by HSE for developments in the vicinity of hazardous chemical installations (8). These criteria form part of a 3 dimensional matrix involving risk criteria related to zones of risk, and to categories of development which are intended to reflect in some respects public expectations implicity in the criteria described above; and some elements of the public perception of vulnerability of 'sensitive' populations who might be exposed.

Essentially, the HSE criteria are as shown in Table 1 below.

TABLE 1 - Decision Matrix for HSE Land-Use Planning Decisions

Development Category/Type	Inner Zone	Middle Zone	Outer Zone
B (Commerce/ industry)	Acceptable	Acceptable	Acceptable
C(Commerce/ leisure)	Maybe/Not	Maybe/Not	Acceptable
A (Housing)	Not Acceptable	May be/Not	Normally acceptable
D (Sensitive development)	Not Acceptable	Not acceptable	Maybe/Not

The zones described above are set at the following level of probability of receiving a 'dangerous dose' (see below) or worse:-

Inner Zone   - > $1 \times 10^{-5} \text{yr}^{-1}$

Middle Zone   - > $1 \times 10^{-6} \text{yr}^{-1}$ and $< 1 \times 10^{-5} \text{yr}^{-1}$

Outer Zone   - > $3 \times 10^{-7} \text{yr}^{-1}$ and $< 1 \times 10^{-6} \text{yr}^{-1}$

The 'dangerous dose', (of heat, overpressure, or toxic gas) is set at a level at which all the following effects may be expected:-

>   severe distress to almost everyone
>   a substantial fraction requiring medical attention
>   some serious injuries, requiring prolonged treatment
>   some highly susceptible people may be killed.

The uncertainties inherent in the 'dangerous dose' approach may compound the problems of risk management using the criterion. Nevertheless,of course, individual risk criteria can be used comparatively, as well as in 'absolute' tolerability contexts. This is particularly valuable in ALARP judgements, as well as CBA.

It is perhaps a consequence of the sensitive nature or risk criteria that few are overtly published. The above UK criteria appear to fall within the range of criteria which are in the public domain. These latter include the following:-

TABLE 2 - A Selection of Published Risk Criteria

Authority	Intolerable Risk (per year)	Negligible Risk (per year)
VROM (new plants)	10^{-6}	10^{-8}
VROM (existing or combined new plants)	10^{-5}	10^{-8}
EPA, W Australia (new plants)	10^{-5}	10^{-6}
UK HSE (nuclear ps)	10^{-4}	10^{-6}
HSE/UK (new housing)	10^{-5}	10^{-6}
Hong Kong (new plants)	10^{-5}	-
NSW (new plants and housing)	10^{-6}	-

The application of individual risk criteria in 'novel' situations has also been investigated. In particular, a recent study for the HSE of the magnitude of risks to fairground workers, and to members of the public using such fairground rides used the above criteria, and also comparator activity to assess the need for further controls. Similarly, HSE is proposing a numerical criterion for aspects of offshore installations, which has an individual risk aspect. However, the dominant aspect of the latter is one of societal risk, and it is to this aspect I now turn.

8. <u>Some Societal Risk Criteria</u>

There are two distinct types of societal risk - the case where a degree of physical harm is done to the whole of national and international society, resulting in a more or less randomly distributed proportion of deaths (eg release of radioactivity); or a case of localised physical harm (eg most major chemical hazards) but where the social effects may be very widespread. Both types are usually expressed either in the form of F/N 'curves' showing the cumulative frequency F of accidents involving N or more fatalities, or as an average annual predicted frequency (the expectation value) of some harm - usually death. Comparison of multiple death risks is greatly facilitated by this approach.

Societal risks to workers are sometimes calculated separately from those to members of the public, with less certainty about the status of workers on adjacent installations than is the case with individual risk. The arguments for and against distinguishing between worker and other criteria are also applicable to societal risk, but it is not at all clear whether the public perception distinguishes this (except in contexts where there is no direct public potential, eg offshore and mining accidents).

In the UK, only limited use of societal risk criteria has been publicly debated. In the Hinkley Point 'C' Report of the Public Inquiry, Michael Barnes QC suggested a criteria of $1 \times 10^{-5}\text{yr}^{-1}$ of a nuclear accident leading to 100 or more deaths ($1 \times 10^{-4}\text{yr}^{-1}$ for the Uk as a whole, implying $2 \times 10^{-5}\text{yr}^{-1}$ for each of a programme of 20 such installations). Barnes' comments do not relate to the total national SR for existing reactors in the UK.

The 'Risk Criteria' (8) document recognised and accepted the need to allow, in risk criteria, for the 'disaster' aspect of major hazards, but debated at length the great inherent difficulties in the problem. Partly as a result of these difficulties, HSE decided not to suggest, at the time, specific societal risk criteria, but rather to apply qualitative judgements, using some surrogates based on its individual risk criteria, on the societal risk significance of various sizes and types of development. The following equivalencies were suggested:-

TABLE 3 - Land Use Planning Societal Risk 'Equivalences'

Housing	Retail	Leisure	Holiday/Hotel Accommodation
10 Houses	100 People	100 People	25 People
30 Houses	300 people	300 People	75 People

Within any type of development, those which are very large (shops over 5000m^2 floor area), which contain populations which are vulnerable (handicapped, outdoor) or where evacuation may be difficult, are given special consideration. In general, HSE would advise against where the individual risk was greater than $10^{-6}yr^{-1}$ and in many cases (eg large theme parks, tower blocks,hospitals) would wish to see risks below $3 \times 10^{-7}yr^{-1}$ (again, of dangerous dose).   Such SR criteria are essentially local, although there may be indistinct links between local SR and national SR in such a context.

The ACDS Transportation study (7) is a recent landmark in the development of societal risk criteria for technological risks, where quantitative benchmarks had to be derived to allow testing of the need for, and the effect of improvements in control. Briefly, the criteria were derived and applied as follows:-

(a)   the starting point was the second 'Canvey Island' risk assessment where, after a searching process of technical and sociopolitical assessment, including Public Inquiries and UK Parliamentary Debate, the risks were deemed to be just below the borderline of tolerability. This allowed a benchmark to be set for the upper limit of societal risk to a local community, using an F/N curve with a slope - 1 through

$F = 2 \times 10^{-4}yr^{-1}$ and $N = 500$ (fatalities);

(b)   a negligible level was set with a frequency 3 orders or magnitude below this 'Canvey' related line, and with same slope;

(c)   these lines were used as an initial benchmark for judgements against 3 seaports individually assessed;

(d)   for ports where the tonnage of hazardous substances might be substantially less than at Canvey, the frequency would be adjusted downwards pro rata to tonnage per year to a 'scrutiny' level (a level above which particular effort may be needed to reduce the risks). This scrutiny level is not as severe in its effects as a tolerbility level, but it is well up the ALARP region;

(e)   in addition to local seaports limits and scrutiny levels, a 'national scrutiny' level was proposed. This was derived in the same way as the local scrutiny line (ie pro-rata to tonnage) but for the total national throughput of hazardous substances through seaports;

(f)   for road and rail, one test was to use the 'Canvey' based limit for any particular major route or national trade for a particular substance. This would ensure no population near such a route was exposed to a higher societal risk than the Canvey community;

(g)   the road and rail risks were also (separately) aggregated to produce national SRs. These were compared with the national scrutiny level for ports, and with the 'Canvey' line. Both the comparisons were applied to test the likely significance of the risks and ALARP of safety precautions, accepting the very loose links with the initial benchmark. The uncertainties

involved in the reading across were fully
acknowledged. In addition, some element of
'aversion' was recognised to exist in a slope
of -1, but no specific measures were taken
to model any aspects of aversion in the
disposition of the slope.
It was, however, recognised that substantially
steeper slopes might contain elements of
extreme aversion.

As stated above, there are few cases where explicity societal
risk criteria exist in the public domain. Examples include,
however :-

TABLE 4 - Some Societal Risk Criteria

Authority	F/N Curve Slope	Intolerable Intercept with N = 1	Negligible Intercept with N = 1	Limit on N
VROM (new plants)	-2	10^{-3}	10^{-5}	Not Below 10
Hong Kong (new plants)	-1	10^{-3}	10^{-5}	1,000
HSE (existing ports, road & rail in one locality)	-1	10^{-1}	10^{-4}	Includes Workers

In an offshore context HSE is proposing a numerical risk
criterion for offshore installations safety cases, as a result of a
recommendation in the Cullen Report on the Piper Alpha
disaster, that a quantified assessement be carried out of the risk
of breach of the 'temporary safe refuge'. (Lord Cullen was of
the view that criteria should be set by installation operators and
subject to scrutiny by HSE - unlike the Norwegian approach in
the North Sea - but that the TSR criterion was exceptionally to
be proposed by HSE).

The HSE proposals suggest that the frequency of breach should be less than 1 x 10-3yr-1. Extrapolating this benchmark to the installation count in each North Sea sector, the resultant criterion is in excess of the 'Canvey' derived ACDS benchmarks. However, this is 'worker' exposure and the direct read across may not be appropriate.

9. The Increasing Use Of QRA Criteria

Risk assessment techniques are supra national, but criteria are very specifically sensitive to the social and cultural conditions in which they are to be applied. Tolerability criteria, whether they are to be used for social judgement or as part of a licensing or an approval regime, are essentially components of the society in which they are located. Only rarely are quantitative criteria placed in the public domain or, more importantly, subject to public debate. Table 5 identifies patterns believed to be current national practices in the issue of operating licenses (or similar activities) for land based non-nuclear installations.

There is some movement towards a more consistent approach for the methodology of quantified risk assessment and analysis, and to its application. Development of criteria will necessarily accompany such progress, and may eventually lead to consistency in the 'tolerability' arena, and in the application of those criteria. Until then, the diversity, complexity and uncertainty involved in the criteria and their use will remain an obstacle to structured risk management, given the pivotal role of criteria therein, and given the sensitivity of decision making to the criteria.

10. Risk Assessment Methodologies

I now return to the other core questions the risk manager may need to address in the other aspects of his procedures - the

TABLE 5

Country	Classifi-cation of installa-tion	Hazard Indentifi-cation	Partial Quantifi-cation	Probabi-listic Assess-ment	Quantita-tive Objectives
Belgium	yes	Mandatory	Sometimes	No	No
Canada*	No	No	Rare	Some	No
Denmark	Yes	Mandatory	Rare	Some	Some cases
Finland	Yes	Mandatory	Sometimes	Some	Some cases
France	Yes	Mandatory	Frequent	Rare	Occasional
Germany	Yes	Mandatory	Sometimes	No	Occasional
Greece	Yes	Mandatory	N/K	No	No
Italy	Yes	Mandatory	N/K	No	No
Japan	Yes	Mandatory	Rare	Rare	No
Luxem-bourg	Yes	Mandatory	N/K	No	No
Nether-lands	Yes	Mandatory	Frequent	Yes	Yes
NSW	Yes	Yes	Yes	Yes	Yes
N Zealand	Yes	N/K	N/K	Some	No
Norway	Yes	Mandatory	Sometimes	Some	Some Cases
Sweden	Yes	Mandatory	N/K	No	No
Switzer-land	Yes	Mandatory	Sometimes	Some	Some Cases
UK	Yes	Mandatory	Frequent	Some	Some Cases
US*	Yes	Yes	Yes	Some	No

* At national (Federal) level

'what if, What then, then what?' dimension - the areas of identification, assessment, control and mitigation outlined above. There is substantial commonality and consensus in approaches here, although again baselines and criteria may be different, but these are determinants applied generally at a late stage of the decision making process. These common components are:-

- in <u>identification:</u> the use of a substance/threshold approach (and the importance of the search for a hazard or risk equivalence system);

- in <u>assessment</u>: the classical approaches to consequence and probabilistic assessment. These include

(i)   Comparative methods, such as
- Process/system checklists
- Safety Audit/Review
- Relative Ranking (eg Dow and Mond Indices)
- Preliminary Hazard Analysis

(ii)   Fundamental Methods, such as
- Hazard and Operability Studies
- 'What if' Analysis
- Failure Mode and Effect Analysis
- Failure Mode, Effect, and Criticality Analysis

(iii)   Logic Diagram Methods
- Fault Tree Analysis
- Event Tree Analysis
- Cause Consequence Analysis
- Human Reliability Analysis
- System Success Trees

- in <u>control</u>:   the application and enforcement of technical, operational and legal standards;

:   information and descriptive packages (eg the 'safety report' approach);

:   descriptive and analytical system justifications (the 'safety case' approach);

:   licensing approvals or other ways of granting permission.

- in mitigation:	on site emergency planning;
:	off site emergency planning;
:	information to those who may be affected by the risks;
:	controls over incompatible land use;
:	siting controls for risk sources.

It is an approach which (in the context of major chemical risks)

(a)   identifies the nature and scale of use of dangerous substances;

(b)   places the use in its geographical and social context;

(c)   identifies the type, consequence and relative likelihood of potential harm;

(d)   identifies the control regime and systems (both hardware and peopleware) to obviate that potential for harm;

(e)   justifies the adequacy of the level of control (in the context of any tolerability criteria) [6] and

(f)   demonstrates the broadly acceptable levels of any residual risk (again in the context of tolerability criteria).

It is a regime which (with adaptation) is equally applicable to environmental and wider applications.

Again however, there are many uncertainties in all aspects of the above, which bear directly on risk management decisions. These even at the basic level, include

the selection of failure cases from the range of possibilities
failure probabilities for each failure case
scale of release rates and duration
conversion of a failure case to a source term for use in further calculation
the validity of the dispersion model
meteorological inputs
topographical inputs

human and environmental response to toxic, pressure,
or thermal burdens
ameliorating factors
ignition factors
and in particular
parameter values in many of the mathematical models.

Ofcourse, there are techniques to compensate for such
uncertainites. Much effort is being expended on 'fuzzy logic'
approaches; sensitivity testing, always an essential tool, is very
useful here. But the criticality of such approaches
demonstrates the vulnerability and impression of the models
used, and the further sensitivities of risk management decisions
taken on the basis of QRA outputs. The critical nature of this
uncertainty can be seen by the substantial raft of research work
in this area, summarised in (9), which is continuing apace.
Areas in which considerable (in some cases increasing) effort is
being deployed to reduce critical uncertainities include, in the
major chemical hazards field alone:-

a)　　**In technologies for prevention and restoration**

　　(i)　　alternative, lower hazard operations and
　　　　　systems;
　　(ii)　　early warning systems (including systems for
　　　　　hazard detection and identification);
　　(iii)　　inspection and maintenance technologies;
　　(iv)　　design and construction of containment systems;
　　(v)　　reliability technology;
　　(vi)　　hazard characterisation procedures;
　　(vii)　　mitigative technologies;
　　(viii)　preventative logistics;
　　(ix)　　decontamination techniques.

b)　　**In chemical and physical hazard phenomena**

　　(i)　　multiphase and time variant source terms;
　　(ii)　　the interaction of shock waves and flame fronts
　　　　　with obstructive geometries;
　　(iii)　　short, medium and long term effects from the

               short term exposure to high concentrations of
               chemical substances;

(iv)   short, medium and long term effects from the longer term exposure to low concentrations of chemical substances;

(v)   uncertainty modelling;

(vi)   model validation.

c)   In Risk Management

(i)   assessment and audit methodologies;

(ii)   software and hardware failure rate databases for validation;

(iii)   trend identification in low probability areas;

(iv)   procedures for modelling and validating
- communication processes
- decision processes
- emergency planning procedures

(v)   organisational crisis management;

(vi)   the effects of group culture;

(vii)   regional and national variability;

(viii)   approached to risk tolerability.

Even if all this work is delivered, it brings with it additional problems. Uncertainties associated with phenomenology or hardware can be parameterised and defined, but much of current effort is being deployed in aspects of 'peopleware'. Not just in 'human factors' - however important that may be, and for too long interpreted in too limited a way as 'operator error'; but in the more critical fields of management and organisational culture and control, where uncertainties and imprecisions are notoriously difficult to manage, particularly in a context where they are to be set against long standing or long term risks. And when translated from the human to the wider environment, even greater uncertainties exist. These include (10) additional uncertainties in, for example, some of the areas outlined above:-

- in hazard analysis, identification and classification of ecotoxic substances (including intermediates which may be formed under fire or other conditions) validation of data

       acquisition of comprehensive (or failing that) representative ecotoxicity data for agreed surrogate species of fauna and flora, with justification of the surrogates.

- in dispersion modelling, identification and development of suitable dispersion models and their validation

       identification and where appropriate qualification of significant factors in dispersion of ecotoxic substances (and intermediates) through the atmosphere, and transport processes on land and water

- in consequence modelling, assessment methods for prediciting the scale of affected areas, and via transport mechanisms and bio-and chemico transformation, prediction of effects

       methods of determining population distribution of flora and fauna in affected areas and of prediciting dose levels and severity

       development of severity critieria based on industry/fatality dose thresholds of any affected species within the ecosystem.

       identification of significant factors governing severity

       criteria for 'tolerability'

       improved data on ecotoxicity, chemical and biological persistence, secondary effects of contamination, and environmental 'values'.

- in mitigation measures, strategies for dealing with environmental emergencies, including measures minimising impacts, decontamination methods and prophylactic responses (eg depopulation)

Resolution of any or all these areas will reduce, but will not remove uncertainties, and may introduce new, and less predictable, uncertainties.

11. __The Wider Context__

In any judgemental process on acceptability or tolerability, criteria are a priori needs. It would appear to be mainly their form of expression, their transparency, and their sensitivity to varied social expectations and contexts, where the major difficulties lie. Society has learned to tolerate (to live with) the use of tolerability criteria for nuclear installations, for aircraft reliability, for seaboard protection, and in many other areas, including personal health (eg immunisation), other mass transportation systems, etc. It is perhaps significant that recent debate is tending to centre on the value judgements which enter the overall equation, rather than on the probabilistic or consequence elements, and their uncertainties.

Economic and social development judgements present decision makers with a series of complex issues to address: they embrace infrastructure needs associated with industrialisation as well as population and land-use planning. Decisions have to be taken in an existing industrial and social context, which may already contain significant hazards and risks; and risk managment decisions need to address the life cycle implications of innovation across varying timescales and wider ecological considerations. Prioritization of effort may be an important prerequisite, whilst decisions which may have such wide ranging implications require a decision making process which provides community insight not only into hazards and risks which are assessed but also into the assessment process itself. Whilst such insight may not bring consensus, at least it brings structure to more informed debate.

However, there is a further problem. Whilst most assessment methodology uncertainties can be accounted for by differences in underlying assumptions, data, and model structures, there remains (best, and most thoroughly illustrated by structured benchmark exercises) a set or residual uncertainties which cannot be so explained (11). Indeed, it would appear possible that simple sensitivity testing (often the fall-back of the assessor and the decision maker) may misleadingly indicate consistency where no such consistency is present. In such cases, it is important that the individual components of uncertainty are identified, addressed and tested, as well as their combined contribution. Any statistically significant variation in any one of the risk input factors must be viewed as a significant variation in the final risk product. For this, and other reasons, these uncertainties must go to form an integral part of the overall risk analysis process, however far they are reduced; essentially, they may be meaningful in themselves.

12. Conclusions

In conclusion, there are many inconsistencies and uncertainties in the methodologies and value judgements which form the quintessential components of risk management. Examples readily illustrate the critical nature of such uncertainties in the outcome of the decision making process. That cost effective further effort be deployed in reducing such uncertainties is self evident for more targetted and successful risk management. But the remaining uncertainties may not only be part of the risk managers dilemma, they may well serve to focus on the critical components of that dilemma, and its resolution. Such unexplained random uncertainty must be treated statistically, and reported in the results, or at the very least this further dimension uncertainty in estimation must be reflected. In particular, when addressing 'absolute' levels of risk, decision makers need to be aware that the complexity of any assessment model, and the concomitant uncertainities, will influence (sometimes very strongly) the types of decisions which can be addressed.

Note.
The views expressed in this paper are those of the author; and except where the context expressly indicates, should not be assumed to be those of the UK Health and Safety Executive, or other(s).

References

1. Successful health and safety management 1992 HSE/HMSO

2. Management at Risk - Development of a structured

framework for risk management assessment, using a system
approach to identify and resolve safety related organisational
and management factors
                    CEC Contract No CT90-0104 (Reports in
                                                    preparation)

3. 'Continuous and episodic risks'. Ellis, AF & Podgorny, G.
International Conference on Risk Assessment, QEII
Conference Centre, London October 1992

4. The National and International Framework of Legislation
for Major Industrial Accident Hazards. Cassidy K. I Chem E
Symposium
Series No 110   ISBN   0-85295-225-2

5. Nonenclature for hazard and risk assessment in the process
industries
I Chem E 1992   ISBN   0 85295 297 X

6. The tolerability of risk from nuclear power stations
HSE/HMSO 1992   ISBN   0-11-886368-1

7. Major Hazard Aspects of the Transport of Dangerous
Substances 1991
HSC/HMSO   ISBN   0 11 885699 5

8. Risk Criteria for land-use planning in the vicinity of major
industrial hazards 1989
HSE/HMSO   ISBN -0-11-885491-7

9. Research to improve the quality of hazard and risk
assessment for major chemical hazards. Nussey C. I Chem E
Research Sysposium Dublin   October 1992

10. CIMAH and the Environment. Cassidy K. 1990 I Chem E
Symposium
Series No 117   ISBN   0 85295 252 X

11. Assessing the risks of transporting dangerous goods.
Proceedings of an International Consensus Conference
Toronto, 1992. Ed Saccomanno FF, Cassidy K. and Craig L
Institute of Risk Research, University of Waterloo, Canada (in
preparation)

Management Creep: Is It a Real Risk in the Process Industries?

R. Ward

Lecturer, School of Mechanical Engineering, University of Technology, Sydney, Australia

INTRODUCTION

The occurence of the engineering-metallurgical phenomenon known as 'creep' is reasonably well understood today; it may be described as the non-elastic deformation of a component, taking place at a stress below the elastic limit (Baker and Wiedemann, 1988). It is time-related, so that a comparitively low stress applied for a sufficient time results in deformation.

Materials have been known to fail quite unexpectedly and surprisingly under these conditions of comparitively low stress levels. The reasons behind the imposition of the stress is simply that the duty conditions caused it, and the designers were ignorant that low-stress failure could occur.

Can we find an analogy to suit management situations? In particular those in the process industries?

This paper will explore such a possibility and present two examples of what may be termed *management creeping towards disaster*.

A CASE OF COMPONENT CREEP

We may begin by reviewing an engineering example of creep, from which some components failed.

The behaviour of the exit pipes of an ammonia reforming furnace, in the factory in which I worked in the 1960s, is an illustration of creep in the engineering sense. These pipes connected the reformer tubes to the exit header, were shaped to allow for temperature distortion, and were designed for the stresses of their service conditions - - - or so it was thought. On inspection at the end of a few years' operation they were found to be distorted. Careful review of the calculations uncovered that the design calculations had omitted the self-weight of the pipes, which had increased the stresses to a level which would have still been acceptable at atmospheric temperature. But at 900°C the combination of stress and time caused the pipes to sag.

Large plate glass windows are said to exhibit the same behaviour. Although of constant thickness when manufactured many years later they have been found to be decidedly thicker at the bottom than at the top. Again, the self-weight is the factor which applies sufficient stress to cause that deflection.

So, given enough time for it to occur, creep is a function of the material properties, operating conditions, and stress.

TWO EXAMPLES OF MANAGEMENT CREEP

As an opening, we may consider two cases which illustrate 'management creep'. Later we will analyse them.

The first is from a company operating a continuous process, our oldest oil refinery. The refinery opened about 1930 with an output of around 30,000 bpd; today it delivers close to 200,000 bpd. Of course, it isn't the same refinery in detail, it has been built and rebuilt time after time. It is therefore rather like the bushman's axe which he'd had for thirty years, and apart from three new heads and four new handles it was as good as it always had been.

More than once, in the sixty years of development, that refinery has been tweaked up to get another few percent, by debottlenecking here and there. Each of those steps has been a very minor addition or modification, nothing of any consequence in itself, but the *sum* of those small steps in each stage amounted to a major increase.

The owner-company must be applauded for the decision, taken several times, to stop the incremental process and thoroughly rebuild a section. If the succession of small increases had gone on for too long without management intervention and appropriate action - - - who knows? Perhaps that would have been 'creeping towards disaster'.

The second example comes from one of my terms of employment, with a company producing resin and other paint base materials by batch process.

The company went into business making alkyd resins by the fusion process, which uses an excess of phthalic anhydride and drives off the water of reaction by simply heating the reacting mixture. This is a relatively simple process run at atmospheric pressure, admittedly at about 300ºC, but has very little process risk as it is effectively endothermic.

In time the company prospered and went to producing alkyds by the azeotrope process, which removes the water of reaction by using a solvent as a carrier, separates the solvent by decanting, and refluxes the solvent. All this occurs at the same temperature, more or less, and at atmospheric pressure or vaccuum.

But there is an increase in risk because a highly flamable solvent is used, at that high temperature, in the vapour phase.

The next development was into vynil acetate emulsions, again at atmospheric pressure and at low temperature (up to 90ºC).

But now they had an exothermic process, which can run away, as I have observed on several occasions.

They also went into polyurethanes, handling drums of TDI with no precautions whatever.

Shortly before I left the firm the Technical Director initiated development of a new product which was a co-polymer emulsion of vynil acetate and vynil chloride. This was also an exothermic process, but somewhat riskier because it used a more reactive (and more toxic) material and because the reaction was somewhat more vigorous.

The only major change in all these stages of development was to heat the resin reactors indirectly by hot oil instead of by means which provided a source of ignition, a very obvious and necessary change. In each of the other steps existing plant was used. There was no operator training apart from what was available on the job. There were no changes in the management structure or system, or to the plant infrastructure.

One may wonder what would have been the *next* step.

That, I believe, is an example of management 'creeping towards disaster'. I am happy to report that none of the possible disasters eventuated, the family which owned the company sold it to overseas interests, and it has been cleaned up extensively.

ANALYSIS

We may ask: what were the factors which motivated these companies to 'creep towards disaster'?

I suggest there are two factors which have the potential for causing that. One is *market pull*, the other is *research push*.

The first, the oil refinery which grew to some six times its original size, is an example of the influence of *market pull*. Australian demand for petroleum products increased rapidly through the three decades after that first refinery was built, and production rose to meet that demand. In the post-World War 2 period several more refineries were built, all in response to *market pull.*

The second case, of the small company in the resin business, was mainly of *research push.* Although there was some connection with responding to perceiving a market opening the impression given by the management of the time was that they had found out how to make this new emulsion and were going to put it on the market.

The conclusion drawn from this analysis is that *either* of these motivating factors can be dangerous. The knowledge that there are potential buyers out there, or that some ingenious new product has been discovered, may draw or direct a company into a high-risk area of operations.

A VERY DIFFERENT EXAMPLE

Another company in our area has produced polyethylene for very many years. Thirty years ago the possibility of producing polypropylene was floated, and I asked a member of staff why the proposal didn't go ahead when the overseas parent company had several plants.

The answer was that every polypropylene plant in the world, at that time, had had a major fire, caused by ignition of the solvent used as a carrier for the polymer, and the Australian company wasn't going to join *that* international set.

The *market pull* existed, but the company refused to be dragged along by it. The company waited for the correct *research push.* Finally a new type of plant was designed using nitrogen gas as a polymer carrier, and a polypropylene plant was built.

Although I was on the commissioning team for the plant (about 1980) I know of nothing official which would confirm what I have drawn from this example, but I believe what was said to me in conversation years before fits in with the later decision to build the plant.

ANALOGY

The analogy between 'engineering creep' and 'management creep' may be made by considering what causes the former: material properties, operating conditions, and stress, given sufficient time.

If we examine a management system we can find direct analogies for these, which allow us to explain 'management creep' in the same terms as 'engineering creep'.

MANAGEMENT MATERIAL PROPERTIES

In this context we will consider that the principal 'material' used in management, used by managers, is *people*.

That brings, immediately, a range of problems, because there are very wide variations between individual people, and all have different reactions or responses to work-situations they experience.

The differences and the different reactions are illustrated by the classifications given by tests such as the well-known 16PF and Humm-Wadsworth.

The latter has the advantage of classifying people in two modes, the unstressed and the stressed, and even though it uses only a small number of standardised behaviour patterns the mixture of these between the two modes provides a much larger total range of personalities of people.

Another and related 'management material' can be made from the individuals who make up the above basic management material; those people can be grouped together to work in *organisations*.

We have a need to operate in groups, rather than individually, because a group can achieve results much greater in scope and magnitude that the individuals can working alone.

Thus, organisations have existed for a long time and there's no doubt they'll continue well into the future.

Organisations are strange animals. One may be taken aback at the behaviour of quite normal people as individuals, but their behaviour in an organisation can be absolutely astonishing. Individuals by themselves can be a bit odd, but when they are put together into an organisation the resultant entity can be quite weird (Ward, 1987).

The result of that, from a management viewpoint, is that organisations are a mixed blessing. Certainly, they *can* be more effective than individuals, *if* the group functions as it should, but that requires at least some effort by a leader, and there is always the risk that the group will charge off in its own direction instead of following the leader's lead.

Such a directional error can be caused by the basic materials, the individuals, mixed in to form the group. As when mixing metals to form an alloy, one wrong component in the mixture can completely change the characterisitics of the final alloy, so that its response to stress is not what was expected.

The analogy is beginning to take shape: some of our management materials will react to stress more readily and unexpectedly than others.

OPERATING CONDITIONS

Externally, management 'operating conditions' are affected very largely by the economic conditions in the environment, as illustrated by the Hunt Organisation Model (Hunt, 1972), which sets out four principal organisational variables: the formal structure, the informal structure, the technical system, and the individual variable.

In the process industries external pressures from the environment, such as higher interest rates, more stringent government regulations, or a shortage of labour or materials, act on the technical system which responds by introducing equipment changes. This in turn acts on the formal structure which is strengthened to cope with the new technology, and that change decreases the strength of the informal structure and decreases job satisfactions among the employees.

Other internal pressures come from what I have termed *market pull* and *research push*.

The market place is a very compelling factor in a business organisation's life. It is outside the organisation, part of the external environment, but influences the internal workings of a business either by presenting opportunity for a new or existing product (when the business perceives a gap), or by providing buyers for a newly developed product.

Drucker has gone so far as to state that the only reason for the existence of a business is to create a market (Drucker, 1955), which applies literally to the latter; in the case of the former the business may be said to be creating an *increase* in the market, not far off the same as the initial creation of a market.

Thus, the presence of an established, expanding, market or the opportunity to develop a new market will *pull* a firm into a new product by external influences acting on the internal system.

The difference between this and the other examples offered as external pressures is that the others are *imposed* on the organisation and are accepted reluctantly; *market pull* is accepted without question, even gladly.

What I have termed *research push* comes from inside the company, and I believe is initiated by human curiousity. Not all humans have this quality to a high level, and it appears to vary with age, education, and no doubt many other factors such as the people with whom any individual associates.

But in some companies, with research teams of highly inquisitive people, well equipped and funded, curiousity flourishes. They make a discovery of some new product. Someone else sees a market for it. But the output process receives its impetus from the original *research push.*

The end result of market pull or research push is the same as from pure external pressure: the technical system responds, the formal system tightens, and the informal system and the job satisfaction are reduced.

MANAGEMENT STRESS

As with physical systems, stress exists in management systems. However, it's curious that many of the well-known management writers appear not to have mentioned it; a scan through some respected 'pure management' works (Drucker, 1979; Hicks and Gullett, 1981; Koontz, O'Donnell and Weihrich, 1984) found no mention of stress in their indexes.

Time of writing may be a reason for omission. Or the reason may simply be the writer's perception of the content of 'management'; several writers *have* considered stress in management (Schuler, 1981; Kets de Vries, 1981; Wallace and Szilagyi, 1982; Milton, Entrekin and Stening, 1984; Stoner, Collins and Yetton, 1985) and agree on some points.

If we look back before 1980, in general we find only the social scientists show interest in stress; as an example of what *can* be found, de Vries cited social science references back to the early 1960s.

Indeed, use of the word 'stress' in a psychological context is credited to a social scientist, Hans Selye, in 1936 (de Vries, 1981; Lawless, 1979).

Not all stress is bad. Although too much stress can be destructive, a certain, optimal, level of stress is a good stimulant (Stoner et al, 1985).

Most of the management writers deal with stress as it affects the individual (DuBrin, 1974).

The general agreement is that stress is caused by factors such as not knowing what is expected, conflicting demands, too much to do in time available, inadequate skills, poor relationships (up, down, and laterally in the organisation), no participation in decisions, working across organisation barriers, and being responsible for others' work (Milton et al, 1984).

High stress level has been linked with low job satisfaction, which has been linked with poor mental health. While these do not necessarily follow-through a cause-and-effect cycle they probably have a common cause (Milton et al, 1984).

It's now generally recognised that stress is one of the human factors which must be considered in industry.

Which section of an organisation contains the 'human components' which are stressed most heavily?

Of all levels of management I suggest *middle*-management suffers this most, and the above list contains the items which one typically encounters in middle-management.

ORGANISATION STRESS

The stress suffered by an individual member affects the group as a whole by reducing the output of that person. From this I suggest that work stress is contagious, what one member suffers can be picked up by others.

That, however, is still *individual* stress, even though it has spread beyond the one person. What can we say about *organisation* stress?

Here I am going to hypothesise, and suggest that *organisations* (particularly in the process industries) can suffer from stress caused by the three factors previously defined: *external pressure, market pull*, and *research push*.

COMPANY CULTURE

Another factor places some determining limits on reaction to these organisation stressors; this is *corporate* or *company culture.*

The nature of the company organisation, its ability to deal with external influences such as from the government, its specific interest in the market place, its research orientation, are all aspects of *corporate culture* (Deal and Kennedy, 1982).

Organisation cultures differ in many other ways, and the one relevant to this present line of thought is how it reacts to the prospect of taking a risk, that is, whether the culture allows or prevents taking risks.

This comes, to some extent, back to considering individuals, and there's no doubt some people are more inclined to take risks than others.

As remarked earlier, one serious organisation stressor is the market-place. Another organisation stressor is the inclination of many managers to accept a challenge, if for no other reason that 'because it's there'. Such managers are members of the risk-taking club.

Not all managers are like that. Some are risk-averse. Enough of either type, risk-averse or risk-taking, in an organisation makes the organisation as a whole one or the other.

The use of the word 'club' has an interesting connatation, too, for if the organisation is, for example, strongly and conservatively risk-averse it will reject a newcomer who has the opposite characteristics because 'he is not one of us'. He is not able to be 'one of the club'.

The opposite is, of course, also true: a conservative person will be rejected by a risk-taking group.

The combination of *market pull*, or *research push* and a risk-taking culture in management is a major factor which can lead to an organisation taking on new products and processes. This is usually an incrimental matter, progressing step by step.

It's believed possible, therefore, for a firm to take a series of such incrimental steps, each comparitively small, each leading to the next, without realising that one day there can be a step which begins deterioration of overall process safety.

CONCLUSION

'Management creep' in the process industries is a poorly recognised phenomenon, caused by the variability of people in the management and managed system, by the external and internal pressures due to the operating conditions, and by the stresses on individuals and the organisation from those pressures.

It appears reasonable to suggest that although a combination of those factors is needed to precipitate a disaster, the presence of those factors will *not necessarily* lead to a disaster. This would be in keeping with general risk philosophy: the presence of a *potential* for disaster only indicates a *probability*, not a *certainty*.

Further research is recommended into the effect of these factors on management systems, on individual managers in such systems, and on what enables an organisation to resist or modify the effects of those factors.

REFERENCES

Baker, E. and Wiedemann, H.G.R. (eds) 1988. **Materials Engineering: A Design Oriented Appoach.** University of Technology, Sydney.

Deal, T. E. and Kennedy, A. A. 1982. **Corporate Cultures.** Addison-Wesley Publishing Company, Inc., Reading, Mass.

Drucker, P. 1955. **The Practice of Management.** William Heinemann Ltd., London.

Drucker, P. 1979. **Management.** (7th printing, 1983.) Pan Books (Heinemann), London.

DuBrin, A. J. 1974. **Fundamentals of Organizational Behavior.** Pergamon Press Inc, New York.

Hicks, H. G. and Gullett, C. R. 1981. **Management.** 4th Edition. McGraw Hill Inc., Tokyo, Japan.

Hunt, J. W. 1972. **The Restless Organisation.** John Wiley and Sons Australasia Pty. Ltd., Sydney.

Kets de Vries, M. F. R. 1981. *Organizational Stress: A Call for Management Action,* in **Contemporary Readings in Organizational Behavior,** 3rd Edition, Luthans, F. and Thompson, K. R., eds. McGraw Hill Book Company, New York.

Koontz, H., O'Donnell, C. and Weirich, H. 1984. **Management.** 8th Edition. McGraw Hill Inc., Tokyo, Japan.

Lawless, D. 1979. **Organizational Behavior.** Prentice-Hall, Inc., Englewood Cliffs, New Jersey.

Milton, C. R., Entrekin, L. and Stening, B. W. 1984. **Organizational Behaviour in Australia.** Prentice-Hall of Australia Pty. Ltd., Sydney.

Schuler, R. S. 1981. *Effective Use of Communication to Minimize Employee Stress,* in **Contemporary Readings in Organizational Behavior,** 3rd Edition, Luthans, F. and Thompson, K. R., eds. McGraw Hill Book Company, New York.

Stoner, J. A. F., Collins, R. R. and Yetton, P. W. 1985. **Management in Australia.** Prentice-Hall of Australia Pty. Ltd., Sydney.

Wallace, M. J. and Szilagyi, A. D. 1982. **Managing Behavior in Organizations.** Scott Foreman and Company, Glenview, Illinois.

Ward, R. 1987. **The Engineering of Management.** University of Technology, Sydney.

Mixing Management, Mystery, and Safety: An Innovative Seminar

R. Ward

Lecturer, School of Mechanical Engineering, University of Technology, Sydney, Australia

INTRODUCTION

Education is a difficult business, both for the student and for the teacher. In undergraduate programs there's students who work hard and others who float along. Relatively few are truly critical of the process in which they exist.

Postgraduate students tend to be more demanding, and also more critical of what they receive.

When it comes to public seminars, for which those attending (or their employers) hand out hard cash for a few days on a specialised topic area, the attendees usually have strong expectations of getting value-for-money. That's understandable. And they also expect not to be bored.

Looking at a proposal for a seminar on a well-known and well-worked over topic commercially, which must be at least part of the presenter's viewpoint, there's the need, therefore, to offer something that doesn't bring up the 'oh, not *another* one!' reaction.

Knowing that, the task of devising a 'safety seminar' or 'safety workshop' to be run at a profit, in a community where there's been several in the last couple of years, presents quite a few problems.

Nevertheless, at UTS we've succeeded in developing a two-day seminar-workshop program which presents safety in a management context (and vice versa) as a mixture of talk-sessions and workshops.

The feature of this which makes it innovative and successful is its continuity through a sequence of interlocking problems set in a fictitious company, and the presence of an overall 'mystery problem' which must be solved as well as the individual problems, for a successful conclusion.

This paper outlines the rationale and content of the package and the background of the development.

The content is experiental in nature, covering what has been developed and used in a teaching activity, hence it refers to little as bibliographical material, for example, to the collections of cases used.

However, as the package is scheduled for presention after the date for submission of papers to this conference the results cannot be included here.

A HYPOTHESIS OF MANAGEMENT BEHAVIOUR

Observation of many managers, including myself, has convinced me that the immediate day-to-day problems enjoyed or suffered by a majority of managers prevent them from looking at wider issues, particularly if those every-day problems are related to the manager's original speciality.

To take the class of managers closest to myself, engineering managers in particular, we must admit to enjoying working on technical problems to such an extent that we at least neglect and at worst ignore people and hierachy problems.

Why that is so is arguable. I believe it's a mixture of the basic personality characteristics of the 'typical' engineer and the nature of the fact-based training engineers receive.

That was certainly the case in my generation and the one following, and although the newer generations may be progressively different my observations of the behaviour and performance of these student engineers (all of whom have had work experience by the time they reach my subject) confirms that the technical orientation of the 'typical' engineer persists and focusses attention on the trained speciality. Some evidence of that tendency will be outlined later.

It seems curious, however, that one can observe similar behaviour in non-engineers who are promoted to higher levels in a hierachy; their attention can also become fixed in 'what matters' to the organisation, whether it be production quantities, profit making, or cost reduction.

While these are necessary and laudable motives for a manager they have been seen to reduce, at least, or even blot out, at worst, matters which are outside those factors.

Thus, it appears that one might reason *generally* that managers tend to focus on immediate problems and tend to prefer work in which they were originally trained. Furthermore, today's 'original training' rarely includes safety as a prominent item.

From these observations my hypothesis was that safety, as a management item, can be included in that exclusion zone.

Having arrived at that tentative conclusion there was, inevitably, a desire to test it.

A SIMULATION OF MANAGEMENT BEHAVIOUR

The hypothesis was first tested on four groups of undergraduate students (a total population exceeding two hundred), many of whom had work experience in a junior supervising role, by presenting them with a weekly series of assignments, a series of cases which were basically technical management (that is, engineering management) problems with accidents occurring in parallel.

Satisfaction with the result was limited: there was some satisfaction because some students responded as anticipated; limited because the test population was not a truly representative sample of management people. Also, the one-week gap between their receiving the next problem in the sequence was believed to have two negative effects which could work together: the concentration on the management problems was lost, and the week interval gave time for them to think aound the overall problem.

It was clear that a more intensive test was required.

But first, we should go back to the beginning to outline how this was started and how the content of the problems was developed.

THE BEGINNING AND THE CONTENT

The idea had several beginnings, but looking back it's clear that one of the most significant starting points was trying to give undergraduate mechanical engineering students a very basic management subject which would interest them sufficiently to encourage them to learn something from the content.

That was in 1986-87, when a text designed specially for undergraduate engineers was put together (Ward, 1987). At the same time the first efforts were made to write case studies specifically for the same audience, and for that specific audience even those first cases were better for that purpose than many of the ones which may be found in many of the management texts.

The principal factors which made these especially suitable for the specific audience were:

> selection a factory within a manufacturing company as the overall scenario, a setting to which most of the students could relate,

use of engineering problems as a background to the management problems,

use of management problems which were typical of those in a technology-based business (both sets of problems based, generally, on my own work-experience),

introduction of characters with whom the students could identify (also based, mainly and shamelessly, on many of the people with whom I've worked, and in spite of the years between us *some* students thought they recognised characters as people with whom *they* worked), and

having a continuing series of cases through the semester, progressing to some form of management climax which the students had to resolve.

Those early years of development were reported four years ago (Ward, 1989).

One of the elements which was voted strongly as a factor in the success of these cases was the progressive development of the scenario. The general sequence in each case series was to begin with an introduction to the characters, then show them making some decisions (through which the students also had to work). By the time the fifth week was reached most of the class 'knew' the people they were reading about and could follow *why* certain events took place, due to characters' reactions.

Another factor which has made these more than somewhat successful is a 'central' or 'unifying' 'theme' in each set, which ties the set together. The theme of the very first set was justification of a project, the second was on organisational conflict and ran on several levels in the company. The third and fourth were about a series of accidents. The most recent four have been two about building and commissioning the project justified in the first set, and two on the first year of operations after commissioning.

(Needless to say, the chronology of writing does not follow the internal chronology of the overall series.)

Finally, there's the 'game' factor, which arises from following a plot which is set out before the semester starts. I, the writer, am the invisible hand which causes events to come out as they do. I know the twists and turns of the plot, what surprises are in store for the reader after lulling him or her into complacency. By halfway through a series some students show they are able to roll with the punches, most appear to be taking a battering.

This has been intentional, as a further simulation of the real world. The future is to us, generally, a mystery, we can anticipate future events as *possibilities*, not as certainties, and even though a good manager will prepare (mentally, at least) for the future, he will expect the unexpected at each step taken, each decision made.

However, the earlier cases were poor as readable fiction. One of the improvement needs was character consistency, so that the students could, as suggested above, almost predict reactions. Several actions since then have improved the readability, such as teaching a literature subject at UTS, teaching Creative Writing at a local Community Art Centre, and entering writing competitions. By now the cases are fairly readable short stories, and there are eight sets of ten series of cases. The only limitation is size: each case is restricted to six pages, about two thousand words. Three sets have been used twice, well-spaced in time so that students in the second group were as unprepared as those in the first.

THE POST-GRADUATE CLASS

A similar scenario was used for some assignments with a post-graduate class, in the Occupational Health and Safety Post-Graduate Diploma. This was not as successful as with the undergraduate engineers, and the reasons appeared to be that the students were older, all part-time, all too busy in their work-environment to enjoy 'playing games', and about half (being ex-nurses) lacked the technical orientation.

However, that experience provided some more understanding of using a 'game-scenario' for class problems.

REACTION TO THE ACCIDENT SERIES

Now to the two-semester series which included accidents.

When I say the series was about 'a series of accidents' I refer only to the background scenery.

The *basic* problem in each, the problem which each student was expected to analyse and about which each student should make a decision, was a *management* problem.

The *technology* (engineering) problem was only there as background, 'to add verisimilitude to an otherwise unconvincing narrative'.

The *accident* just happened to occur, in each case, in some relationship to the technology or the management problems. Some accidents were process-related, some were ridiculous events which should never occur (but do, in the real world, too).

The process accidents were easy to arrange, as all the cases were set in an ammonia-and-products factory, modelled on the one in which I worked for six years (with the exception of those in the first series, which was took place in the company's Head Office). So were the ridiculous ones. I wrote ten of these week by week in one semester, then nine in the semester following, finishing with an unexplained fatality.

I was determined from the beginning to have a fatality, partly for dramatic effect and partly because I remembered looking around the Ammonia Factory where I worked and thinking of how many ways to kill a person there were all around me, if anyone really *wanted* to do that. (In one of the 'chapters' that thought was expressed by one of my characters.)

At this point I must drop a word about how the fatality occurred, and how it was presented. The body was that of an operations auditor sent from Head Office at the beginning of the second semester, and being a most obnoxious person (as well as unwanted, in the minds of the people in the factory) was the obvious one to eliminate. But as in any good murder story (for that was how it was developed) the writer gave a clue to 'who done it' - - - but the clue was hidden in the series two semesters before, and although this was mentioned to the class and a good hint was dropped in the final section of narrative *no-one* put the finger on the murderer.

The reaction of the majority of the students through the semester was not surprising. That majority made a fair attempt at solving the management problems, and some touched on the technical-engineering problems. They showed no expectation of the unexpected.

What *was* surprising was that a few worked quite hard on the management aspects of the *accidents*, trying to work out how and why they were happening, making recommendations for action, and expressing concern about the future. This was only a few, perhaps six out of six dozen, but they gave enough to show they were able to see beyond the immediate day-to-day problems. Thus, the few who paid attention to the *whole* scenario were exceptional.

The few who predicted 'if this goes on someone is going to get killed' were *exceptionally* exceptional.

WHY NOT *SELL* THE SERIES?

The idea of 'selling the series' in seminar or workshop form was, one might say, almost inevitable, for two reasons: one was simply to get a wider audience for the work, the other was to get a better sample on which to test the hypothesis.

The accident series seemed to provide the background of management generally, and of management of safety, *and* of management of the unexpected. The aim would be to follow through as with the undergraduate classes, by seeming to concentrate on *management* problems with the accidents thrown in almost as asides, in an overall scenario in which the management of the factory had so much to worry about that they would begin by treating the accidents as 'part of the scenery'.

In time, by halfway through, the perceptive ones should be beginning to worry about the accidents and be trying to do something to stop them (which would be impossible, due to the invisible hand outside), or at least analyse them.

It should therefore be suitable for engineers, production staff, OHS staff, and probably others such as some consulting engineers and some insurance people.

To make it work as it should, as a laboratory simulation exercise, the audience would be split into groups of about five. Each person would be indentified by the name of one of the principal characters, and would be expected to act out that role.

We decided on a two-day program because it is easier to get people away from work for two days than for three; there is also the inescapable fact that two days cost less than three.

But most important from the hypothesis-testing viewpoint, having them there for an intensive two days would allow those running the workshop to mis-direct them into concentrating on the *management* issues, to see how many would attempt to cover both those and the accident series.

This is, of course, precisely what a stage magician does: the audience is told to concentrate on something and while they do that he sneaks the rabbit into his hat.

However, that shorter time-span introduced a content problem, with only (say) sixteen hours it would be impossible to fit in alternating talk-sessions and syndicate sessions which would use the nineteen incidents in the undergraduate assignments.

So the number of incidents was weeded down to ten, making a tight program but enough to allow for alternating half hour talks and half hour discussions. And the 'mystery factor' was maintained by ending with the same fatality under the same unexplained and suspicious circumstances.

But the need to 'play fair' with the audience would be stronger for several reasons; for example, they would be paying for attending and would want to have a good chance of solving the overall problem. Also, *we* would have had *our* fun misleading them into concentrating on operations management problems when they should have had their eyes more widely open, so *they* should have the information-means as well as the opportunity to have fun by being able to work out 'who done it'.

That produced the idea of 'pre-reading material', to be sent to all those registering, at least a week before the workshop, with a strong recommendation to read it so all could launch into the problems quickly.

This pre-reading material was a condensation of all the events in the fictitious company up to the time of the first syndicate problem; in spite of being as brief as possible that added up to just under nine pages of A-4 text and one page as a diagram (the company's overall organisation chart).

That number of pages is not surprising, when one considers it's a synopsis of nearly six hundred pages of original text and diagrams. To reduce it to nine pages, including the clue to 'who done it', was a reasonable feat.

INNOVATION?

Any number of seminars and/or workshops are offered and run every year, even among the relatively small Australian industrial community. What makes this one different, innovative?

The innovative features are those which worked so well in the undergraduate classes: the industrial setting, the background of technical management problems, the short story format of the problems with a unifying theme, recognisable characters in the narrative, the continuity from one problem to another, and a climax situation to provide final learning-by-shock-tactics.

SUMMARY

The hypothesis that managers concentrate on day-to-day issues and tend to ignore a peripheral matter, such as safety, has been tested on undergraduate classes with limited success.

The limited success is believed to be because the undergraduates were not a suitable sample of managers, and the test material was not presented intensively.

The hypothesis was also tested on a postgraduate class. This also gave limited results, believed to be for similar reasons, plus some related to that class in particular.

At this time of writing the package, including the pre-reading background, and an advertising brochure for the two-day workshop have been prepared.

The present proposal is to use the material in a two-day intensive program within the next four months, after which the results may be reported definitively.

REFERENCES

Ward, R.   1987.   **The Engineering of Management.**   University of Technology, Sydney.

Ward, R.   1989.   *Adventures in Engineers' Management*, in **Proceedings of the First Conference of the Australasian Association for Engineering Education.** Sydney.

The series of case studies mentioned (in chronological order of the narrative) are:

A Plant for Appropriate Technology.
A Project in Ammonia.
A Commissioning in Ammonia.
A Continuing in Ammonia.
A Shutdown in Ammonia.
A Conflict in Ammonia.
A Hazard in Ammonia.
A Murder in Ammonia.

In 1993 "A Conflict in Ammonia" will be rewritten, expanded into two parts, the first on *destructive* conflict, the second on *constructive* conflict.

A Review of the Role of Cost–Benefit Analysis as an Input to the Management of Safety

John Council
Safety Technology Department, Lloyd's Register, U.K.

ABSTRACT

This paper reviews the development of cost benefit analysis and approaches to economic evaluation of expressions of risk in a safety management context. Reference will be made to the use of cost benefit analysis in the nuclear industry and its application in the transport and chemical process industries. The paper goes on to discuss the advantages and limitations of cost benefit analysis as an input to safety management.

INTRODUCTION

There is increasing public concern regarding the safety of industrial installations that process, or handle, potentially dangerous substances such as radioactive materials and hazardous chemicals. This public concern is reflected in the introduction of legislation requiring stringent controls for hazardous industrial installations. As a result, developers and operators of hazardous sites have a legal duty to demonstrate that they are protecting employees' welfare and the health and safety of the public.

In response to demands from the public, customers and government regulatory agencies, companies have developed more sophisticated ways to improve industrial safety. These developments have been accompanied by the introduction of safety management systems. These consist of comprehensive sets of policies, procedures, and practices designed to ensure that barriers to major incidents are in place and are effective. Generally, the systems comprise explicit sets of arrangements for planning, organising, implementing and controlling industrial safety.

While the characteristics of a management framework are generic, and apply to the management of any industrial safety activity, the system elements are not universal. These elements are determined by the type of organisation and the nature of the industrial hazard. Nevertheless, an element common to all industrial safety management systems is risk.

Risk management has been defined (Reference 1) as "the process whereby decisions are made to accept a known, or assessed, risk and the implementation of actions to reduce the consequences or probability of occurrence". On this basis, risk management is concerned with the systematic identification, analysis, the evaluation and control of potential risks.

Evaluation of risks is undertaken to establish acceptability, with regard to relevant criteria and determine requirements for risk reduction and mitigation. Frequently, the acceptability of risk is established by management judgement and experience. There are many inputs and factors involved in risk evaluation decisions, not all of them being explicit. Cost Benefit Analysis (CBA) has been used in industry as one of the decision tools for considering the costs of project options, safety related risks, and potential benefits.

The evaluation of risk and the use of CBA, has been developed in several industries. An outline of cost benefit methodologies and a review of various applications is provided in this paper. For this review, economic costs are quoted in US Dollars, based on a sterling exchange rate of $1.43 (as 02/03/93).

RISK EVALUATION AND COST BENEFIT ANALYSIS

The main conditions that apply when evaluating industrial risk involve establishing whether:

- the risk is so great that it is unacceptable and must be refused altogether; or

- the risk is so small as to be insignificant, and is generally acceptable; or

- the level of risk falls between the unacceptable and the insignificant levels

A concept that has been introduced by the UK Health and Safety Executive (Reference 2) as an input to establishing acceptable risk is "tolerability". In this approach, a level of risk, falling between what is judged to be unacceptable and that which is insignificant, is regarded as tolerable. The tolerable risk boundaries are defined by many considerations and society's preferences. Several inputs are therefore considered in arriving at a consensus of what constitutes a tolerable level of risk. These include:

- guide-lines from the appropriate Government Agency and Regulatory Authority

- discussions and agreements with the parties affected by proposed developments

- industry standards

- international discussions and agreements

- legal requirements

A zone that has been defined as "tolerable" can be regarded to be justifiable if it can be established that a risk has been reduced to a level where to continue to reduce the risk would cause an unwarranted drain on resources.

The balance between risk and cost varies with risk, increasing as the risk rises towards intolerable levels. In the UK there is a requirement to demonstrate that risk has been reduced to a level As Low As Reasonably Practicable (ALARP). Reasonably practicable being taken to imply that risk should be reduced until there is a gross disproportion between the expenditure involved and the corresponding reduction in risk.

One technique that has been applied when deciding whether a particular health risk has been reduced to the optimum and therefore a justifiable level, is CBA.

COST BENEFIT ANALYSIS

Benefits

For risk management purposes, tolerable levels of risk are justified by comparing the benefits of further reduction in risk with the measures required to undertake the reduction. In this role, CBA consists of evaluating the adverse consequences associated with a given risk band and those costs that would arise in reducing the consequences.

The adverse consequences are called a detriment, and can be evaluations of predicted health effects arising from injurious effects of harmful substances, thermal radiation levels, explosion overpressures. The detriment can also be expressed as environmental damage, the economic loss of plant and property or the social and political impact of an accident. The so-called "benefit" is expressed as a reduction in the detriment and is often quantified in monetary terms to enable cost comparisons to be made.

Health Costs

It is in the assignment of safety related benefits where difficulties arise, particularly with regard to estimating the economic costs of health consequences. At this stage, questions on the evaluation of the health costs can raise a corollary regarding the value of human life. Approaches have therefore been developed to establish health detriment costs on the premise that such estimates of the value of a life saved are distinct from valuations of human life.

The costs of saving a life are found by different routes to valuations of life and are expressed as costs upon the economy. The evaluation focuses on the treatment costs of the health effect and the impact on a national economy of the temporary, or permanent removal, of an individual from the workforce. This method is called the human capital approach and expresses the benefit as the value of saving a life.

Various subjective valuation methods have also been established to include non-monetary impacts of ill-health and death more explicitly. These effects include anxiety, pain, distress and discomfort. The value of each health effect can be determined by two approaches. Firstly, the amount that an individual would be willing to pay to decrease the risk of a particular health effect. Alternatively, the valuation can be based on the amount the individual would accept in compensation for an increased risk. These valuations are known as preference valuations and are based on surveys. Observing what people actually spend, or accept as compensation, is termed the revealed preference. What people state they accept for an increase, or a reduction, in health effect risk is usually called the expressed preference.

From these surveys, monetary values are inferred for incremental increases in risk, enabling a statistical life value to be derived.

Economic Costs

Besides health effect detriments, the adverse consequences of industrial accidents also incur a range of economic detriments. These may be broadly summarised as resulting from:

- damage and loss of plant and private property

- application of countermeasures to reduce harmful exposure to hazardous substances

- impact on the activity with which the installation is associated

- long-term social and political impact

- ecological impact

These categories of economic risk have been quantified in various studies of accident consequences. Further details are provided later in the paper.

Non-Monetary Criteria

Besides quantifiable health and economic detriments, other factors have been considered when judging the acceptability of various decision options that involve risk. A case in point is the operation of nuclear power reactors in Eastern Europe. Continued operation of some of these power stations has been justified by including expressions of risk other than the potential health effects. A significant detriment being the loss of the benefit of electrical services, and the social impact of closing these stations.

Approaches have been proposed that move the emphasis away from probabilistic figures, focusing on qualitative risk characteristics. These characteristics, such as voluntariness, dread, timescale of consequences, familiarity and controllability, would be used to weight the aversion of risk (Reference 3).

NUCLEAR INDUSTRY APPROACHES TO COST BENEFIT ANALYSIS

The operation of nuclear facilities is controlled to ensure that the workforce and public are not exposed to harmful levels of ionising radiation. Permitted levels of radiological exposure being proscribed by national regulatory authorities. Having established that a proposed facility can been operated to meet regulatory criteria, an assessment study would be carried out to demonstrate that radiological risk was ALARP

CBA is one of several appraisal tools for aiding management decisions for radiological protection and safety.

Management of Occupational Safety.

At the design stage of a proposed development, CBA would be undertaken as an input to an ALARP assessment. The CBA study would typically comprise an assessment of the occupational radiation dose to a member of the workforce and its corresponding health detriment. Mortality risk factors are used to estimate the health detriment.

A monetary valuation of the health detriment is used in a design study to optimise the radiation dose. For example, radiation dose exposure could be reduced by providing extra biological shielding. Alternatively reduced exposure could be achieved by providing additional facilities to aid remote operations and maintenance. An optimum position would be achieved when the capital cost of introducing the next increment of protection exceeded the dose detriment cost that the measure would save.

CBA is undertaken for decisions on the optiminisation of waste management practices, by evaluating the health detriment of each alternative design. Usually, the detriment is expressed as the resultant radiation dose to the workforce or public, arising from the waste management option under consideration.

Baseline costs for radiation-induced health effects have been published (Reference 4) for use in occupational cost benefit studies. The costs were calculated using the human capital approach for a broad range of health effects, which were:

- fatal cancer
- non-fatal cancer
- hereditary effects

Sex and age specific annual earnings of the exposed population were used to calculate the loss of earnings resulting from each type of effect. This calculation took account of the probability of death from other causes, in determining the number of years of life lost because of radiological exposure. The cost of providing medical treatment was also included. This model provided monetary detriments ranging from $0.2 million for very low individual doses, up to $2.2 millions for the highest doses received under normal operational conditions (1986 money values).

The base line costs are multiplied by an additional factor, either equal to or greater than unity, related to the individual dose distribution encountered in specific instances. This factor is used to reflect the increased aversion to higher levels of occupational risk.

Application of CBA to Potential Accident Situations

The health and economic detriments, together with a probabilistic estimate of adverse consequences, are examined in situations where an assessment is undertaken to consider potential accidents at nuclear plants.

Models are employed to predict the probabilistic consequences of postulated events involving uncontrolled releases of radioactive materials. The models estimate the health effects of these releases and allow a range of countermeasures to be specified.

Various subjective valuation methods have been established to include the non-monetary values of the health effects of accidents at nuclear facilities. These effects include anxiety, pain, grief and discomfort. Each health effect was assigned a monetary value that represented an individual's aversion to that type of effect. These values represented either the amount of money an individual would accept to compensate for an increase in the risk. Alternatively, the amount they would pay to reduce this risk by a given amount. Monetary values estimated by a subjective approach (Reference 5) are summarised in Table 1.

Health costs estimated in the USA have employed human capital methods to evaluate costs due to loss of income and medical treatment (Reference 6).

Table 1. Values (1992) for Radiation-Induced Health Effects.

Health Effect	Value ($ million/effect)
Early mortality	1.4
Early morbidity	0.1
Fatal cancer	0.7
Non-fatal cancers	1.4
Hereditary effects	1.4

Models have also been developed that estimate the economic impact of radiological accidents, both in the USA and Europe. These models estimate the costs arising from the following consequences:

- forced movement of populations
- food restrictions
- decontamination
- radiation induced health effects
- off-shore consequences

Economic impact models have been used in the general areas of emergency planning, siting policy and plant safety design features. For a hypothetical accidental release from a pressurised water reactor located in south-west England, a total cost has been estimated (Reference 5) to be $2445 millions (1990 values). The scale of the postulated radiological release was sufficiently large to cause significant off-site consequences. The break down of the total cost of the accident is presented in Table 2.

A cost benefit study for risk management would involve combining the cost of reducing a given detriment with a numerical estimate of the likelihood of an accident giving rise to the detriment in question. The health and economic costs would be evaluated to measure the quantity of the benefit.

Table 2. Economic Costs (1992) Arising from an Accidental Release from a Nuclear Power Station.

Health Effect Costs (subjective) $m	Countermeasure Costs $m	Total Costs $m
443	2002	2445

TRANSPORT INDUSTRY APPROACHES TO COST BENEFIT ANALYSIS

Many transport investments are made not to secure journey time savings but to reduce accidents. Since the 1950s the industry has developed CBA methodologies to aid safety management decisions.

The uses of CBA for transport safety investments have historically centred on economic benefit valuations. Early applications of CBA were based on valuations derived in terms of Gross National Product. For example, in 1972 the White House Office of Science and Technology estimated that the average cost of a traffic death to be $0.14 million. This figure was based on an average per capita income, multiplied by the average years forgone (Reference 7).

In the 1980s the basis of the valuation of economic benefits was widened to include treatment costs of the health effects. Latterly, studies have been based on subjective assessments of statistical life values. In the UK, the Department of Transport commissioned an extensive review of subjective valuation methods. For appraisal of proposed highway schemes, the review suggested a value of $0.95 million (1992 money values) for a statistical life, (Reference 8).

A typical application of CBA was a study carried out in 1972 by the National Highway Safety Administration on the quantifiable costs of motor vehicle accidents. This study was used as an input to a decision to reject underride guards for trucks. Similar CBA studies have been used to support projects as diverse as highway crash barriers and new rail investment options.

Recently, British Rail have used CBA as a ranking tool, when appraising safety scheme investments, to aid selection of the most effective allocation of resource. The benefits were measured in terms of fatalities and the injuries that might be avoided by each investment option (Reference 9).

Perhaps the most widely publicised use of CBA in the transport industry was the part it played in a safety related decision by the Ford Motor Company. This decision concerned a retrospective safety modification to the petrol tank on one of their car models.

Besides accident assessment studies, CBA has also been used for pollution control. In the US, CBA was employed to compare two methods of controlling asbestos emissions from motor vehicles. The analysis indicated that there would be a national cost of $10 millions per life saved (1977 money value) if asbestos filters were installed. For an alternative scheme, it was estimated that substitute brake materials would have a cost of $100 millions, per life saved. From this work, it was concluded that neither alternative came close to the $0.3 million valuation that workers in hazardous occupations have, implicitly, given their own lives (Reference 10).

CHEMICAL PROCESS INDUSTRY APPROACH TO COST BENEFIT ANALYSIS

The use of a safety performance index, the Fatal Accident Frequency Rate (FAFR), is advocated in the UK chemical industry. This accident rate expresses the number of deaths occurring in a group of people, during their working lives. The average UK value for the chemical industry is 3.5. Kletz (1977, Reference 11) suggested that for risk management purposes, the aim should be to achieve levels of safety above the FAFR average. However, when the risk reduction measures cost more than $1.4 millions per life saved, it is necessary to look for a lower cost solution.

There is no claimed justification for the chosen monetary value of $1.4 millions. Nevertheless, this figure became a quoted norm for many process industries when arriving at practical improvements to safety.

LIMITATIONS TO COST BENEFIT ANALYSIS

There are several problem areas in the use of CBA for safety management. These areas include:

- the uncertainties involved in estimating the likelihood and impact of the consequences of accidents

- the dependency of valuation of consequences on the scale of the accident

- the comparison of present expenditure with future health detriments

- public perception on the acceptability of valuations of "dread" events

- assignment of monetary values to all of the consequences

It is perhaps the last issue in which the main limitation of CBA lies. Assigning monetary values involves the weighing of various factors and there are many disagreements as to the factors that should be included in the evaluations. These disagreements are reflected in the wide range of health effect valuations that are quoted in this paper. A further disadvantage of monetary expressions, is that they quickly become obsolete as purchasing values change.

With regard to the health effect valuations, there are several criticisms of methodologies for deriving these values. The human capital method ignores public aversion to large scale accidents. It is also recognised that it does not adequately address the non-monetary impacts of accidents such as anxiety, pain and distress. Another objection that is raised, is that the human capital approach does not consider the valuations of those directly at risk.

There are also several flaws in subjective valuation methods. Preference valuation assumes, among other things, that the individual is fully aware of the actual level of risk and the monetary value of the consequences. These valuations also rely on individual decisions on expenditure, which are governed more by income and necessity than the conceived worth of the good concerned.

Each valuation method provides different results since a range of cost elements are used in the derivation of the estimates. Care is therefore required to ensure that the assigned values are consistent with the use to which they are put.

CONCLUSIONS

The term cost benefit analysis describes many different techniques encompassing consideration of the monetary advantages of decision choices. It is used across a wide range of industries, and in public administration, to provide solutions to the problems of resource allocation.

This paper has sought to place CBA within a safety management framework by reviewing applications that aid industrial risk reduction studies.

The CBA approach has limitations, which are discussed, restricting its safety management role to that of making broad comparisons. Despite its limitations, CBA offers a technique that provides a systematic input to management decisions on safety and risk.

It is appreciated that, for judging the acceptability of safety proposals, it would be appropriate to extend CBA methodologies by establishing alternative criteria that evaluate risk in qualitative terms.

REFERENCES

1   British Standard 4778: 1991, IEC 50(191): 1990. Quality Vocabulary, Section 3.2 Glossary Of International Terms.

2   The Tolerability Of Risk From Nuclear Power Stations, UK Health and Safety Executive, 1992.

3   Slovic S, Fisch B. How Safe Is Safe Enough ?: Risk and Chance, selected readings (edited by Dowie J and Lefrere P).

4   NRPB-ASP 9, Cost Benefit Analysis In The Optimisation Of Radiological Protection, (UK) National Radiological Protection Board, 1986.

5   Haywood S M, Robinson C A, Hardy C, NRPB-R242, COCO-1: Model for Assessing the Cost of Off-site Consequences of Accidental Releases of Radioactivity. (UK) National Radiological Protection Board, 1992.

6   Burke R P, Aldrich D C, Rasmusssen N C. Economic Risks of Nuclear Power Reactor Accidents. Sandia National Laboratories, NUREG CR-3673.

7   Linnerooth J, "The Evaluation of Life Saving".

8   Department of Transport, UK, Valuation of Road Accident Fatalities, 1988.

9   HSE Conference, International Conference On Risk Assessment, London, October 1992.

10 Saunders R, Wheeler T, Handbook of Safety Management, Pitman.

11 Kletz T. Evaluate Risk In Plant Design, Hydrocarbon Processing, 56 (5), 297, 1977.

Integration of Quality, Process Safety, and Environmental Management Systems

B. D. Berkey, W. M. Dowd, and P. I. Jones
Hercules, Incorporated

INTRODUCTION

Objective

The objective of the effort described in this paper is to identify means by which various plant level activities affected by quality, safety, and environmental initiatives can be managed efficiently. The ultimate goal is to avoid duplication of effort, make maximum use of in-place resources, and meet the goals of quality, safety, and environmental protection in the most efficient manner.

Background

The last several years have seen increasing efforts and focus within the chemical industry on the three critical issues of quality, process safety, and environmental protection. In most cases, these issues have been dealt with by different persons using different tools and systems. Different guidance documents have been established for each area.

Efforts in the quality area revolve largely around the International Standards Organization (ISO) 9000 series of standards. Many European companies have been certified by outside auditors as having met the requirements set by the ISO 9001 or 9002 standards. Many companies in the United States have also been certified or are in the process of obtaining certification.

Process safety efforts have been guided by various documents, regulations, and standards including: the Codes of Management Practices contained in the Responsible Care initiatives developed by chemical industry organizations in North America and Europe, the Guidelines for the Technical Management of Process Safety published by the Center for Chemical Process Safety of the American Institute of Chemical Engineers, and various national regulations and standards such as the United States Occupational Health and Safety Administration's rule for Process Safety Management for Highly Hazardous Materials (29 CFR 1910.119) and European regulations developed in response to the Seveso Directive.

Environmental matters are governed by a wide variety of government regulations and guidance documents.   For example, the Responsible Care intiatives mentioned earlier deal with all aspects of environmental protection. National regulations such as those by the Environmental Protection Agency in the United States also contain detailed requirements for environmental protection.

There are fundamental differences between the nature and objectives of quality initiatives as defined by the ISO 9000 standards and process safety and environmental efforts. The quality efforts are largely product-oriented. They focus on having a documented system for producing a product which is suitable for use by the consumer on a first-pass basis, without the need for extensive test, inspection, and rework of the product. The ISO 9000 standards are also generic in that they are designed for a wide variety of industries and products. Process safety efforts, as well as many environmental initiatives are more specific to the chemical industry. They are also more concerned with the process which produces the product as opposed to the product itself.

Various efforts have already been made to integrate process safety, quality, and environmental management using quality standards as a model. For example, the Canadian Chemical Producers Association (CCPA) has produced a document which defines the management of their Responsible Care Codes of Management Practice using the framework established by ISO 9001. The British Standards Institute has produced BS 7750 which defines an environmental management system using the framework provided by their quality standard, BS 5750. The Chemical Manufacturers Association in the United States is considering an effort to further integrate the three areas. Most of these efforts involve establishing policy and procedures to control activities in the three disciplines.

The Paper Technology Group within Hercules Incorporated has been actively engaged in obtaining ISO 9000 certification for its plants and the overall business. All of the European plants have been certified to ISO 9001 or 9002. The plants in the United States and Canada are in the process of obtaining certification. Most of the PTG plants around the world are small plants characterized by batch or semi-batch operations. Several of the processes are explicitly covered by process safety regulations in the United States and several other countries. The plants are also heavily impacted by environmental regulations as is most of the chemical industry. The PTG plants do not have extensive professional staffs. The responsibilities for dealing with the documentation, reporting and other requirements of quality, process safety, and environmental protection often lie with one or two persons. Therefore, efficient use of the available resources is critical to achieving the goals in these areas.

APPROACH

The efforts and regulatory initiatives in the areas of quality, safety, and the environment have caused many plant personnel to feel like the person shown in Figure 1. This is particularly true for small plants with limited staff. Everyone in the plant must share in the practices of safety, quality, and environmental protection. However, the considerable efforts required for program development, documentation, and reporting to comply with the various initiatives often weigh heavily on the few persons charged with carrying them out. As stated previously, the Paper Technology Group of the Hercules is characterized by small plants with limited staff, and efficiency in the use of those personnel is most important. Therefore, the focus in this effort has been on identification of specific, mostly plant-level, activities which are involved in two or more of the three areas. The level of overlap between the areas as well as the opportunities for increased efficiency become greater as one goes deeper

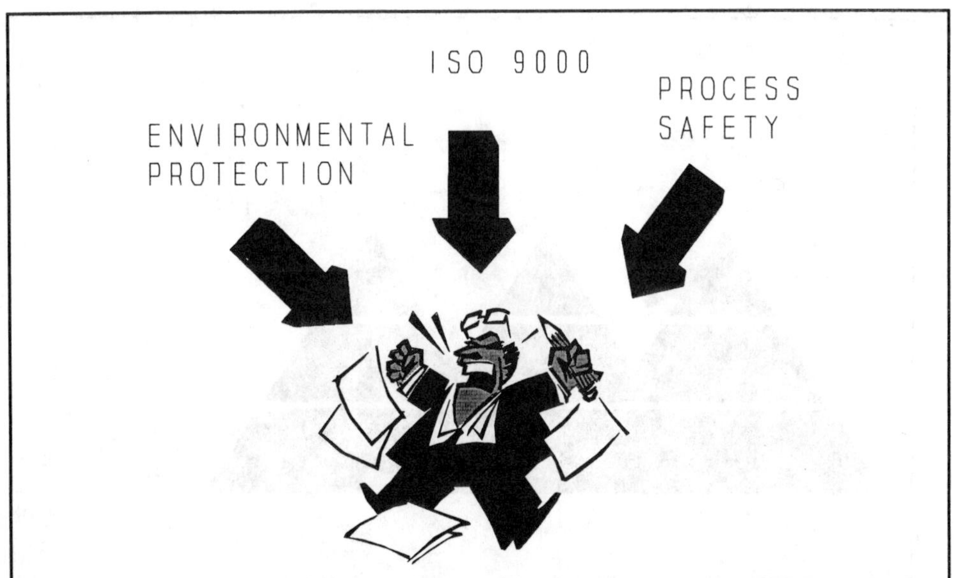

Figure 1. Pressure at the Plant Level

into the specific activities at a plant or process level. Therefore, the approach
has been to work from the plant or process level up to (1) identify the specific
activities which have significant overlap between quality, safety, and
environmental protection, (2) identify the effect of the various requirements on
the activity, and (3) determine the best means to integrate those requirements
into the activity. This type of "bottom-up" approach is intended to focus
attention on those areas where integration will have the most benefit in terms of
reducing duplication of effort and increasing efficiency.

AREAS OF OVERLAP

One means which has been used to illustrate the overlap between the three
areas is the use of the triangles shown in Figure 2. The tip of each triangle
represents the general policy statements governing each area along with the
policy and procedure manuals which give additional guidance. The policy
statements give the "why" behind the program while the procedures state "what"
the program will do. The bottom portion of each triangle represents the actual
activities associated with quality, process safety, and environmental
considerations. These activities represent the "how" of the "why", "what",
"how" sequence. Many of the activities in the bottom portion of the triangles
are characterized by one or more of three traits: (1) they are plant- or process-
level activities, (2) they are generic activities related to operating a chemical

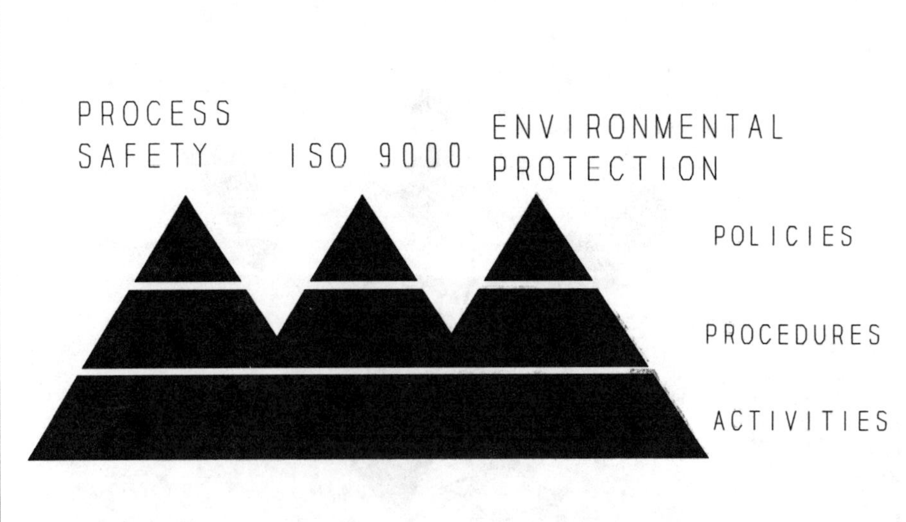

Figure 2. Overlaps Between ISO, Process Safety, and Environmental

plant or process and are not necessarily specific to any one of the three areas of quality, process safety, and environmental protection, and (3) they are activities which in most cases were already in place in some form at many plants prior to the development of the increased emphasis on process safety, quality, and environmental protection. The areas where the triangles overlap are those activities which are common to two or more of the disciplines of quality, process safety, and environment.

A review of the guidance documents mentioned previously has identified six areas of plant activities where integration of quality, safety, and environmental concerns is likely to provide the most benefit. These areas are (1) new process design, (2) process documentation, (3) control of contractors (purchased services), (4) management of change, (5) training, and (6) auditing. These areas are not necessarily the only ones which involve some overlap. However, they are considered to be areas which provide the greatest opportunity for improving efficiency and reducing duplication of effort. Each of these activities is discussed in more detail below. For each activity, key considerations for compliance with process safety, quality, and environmental initiatives are reviewed and the primary opportunities for increased efficiency from integration are identified.

Design of a New Process or Product

During the development of a new product, process safety concerns are usually focused more on the process required to produce the product as opposed to the product itself. If the new product can be made in existing process equipment, the process safety concerns are likely to be handled through the plant's management of change system. If design of a new process is required, the design effort requires close attention from a process safety perspective. Key considerations in this area include: (1) incorporation of best engineering practice in the form of compliance with applicable codes and standards, (2) application of formal hazard review techniques such as Hazard and Operability Study or Failure Modes and Effects Analysis, and (3) application of a hazard minimization strategy which includes selection of low hazard chemicals,

Environmental concerns in new process or product development are similar in scope to the process safety concern. The key considerations include (1) selection of chemicals and process options to minimize waste, (2) definition of waste treatment methods, and (3) compliance with regulations and permitting requirements.

ISO 9000 requirements in this area are focused on the product itself. These requirements include developing a system to control and verify the design of new products. For example, ISO 9001 requires that the system identify the person(s) responsible for each aspect of design and development. It also requires definition of interfaces between different disciplines which may be involved in the design effort. Other requirements of the system include provision of design inputs in the form of performance specifications and design outputs in the form of drawings, operating procedures, and other process specifications. ISO 9001 also requires the establishment of a design verification plan to assure that the design output considers all of the design inputs.

The system established under ISO 9000 provides a framework for the design control effort. Integration of the safety and environmental concerns into the system established under ISO 9000 will greatly improve the efficiency of the process as well as assure that all critical aspects are covered. This can be accomplished by including safety and environmental professionals in the persons responsible for design control and by specifying the interfaces between the disciplines. These actions should assure proper consideration of process safety and environmental concerns in the design effort.

Process Documentation

Process documentation in this context includes process flow diagrams, standard operating procedures, and piping and instrumentation diagrams. Also included is design basis information such as material and energy balances, process chemistry information, and equipment specifications. Most of this documentation comes under the ISO 9000 definition of "work instructions". The OSHA process safety regulation, similar regulations in other countries as well as Responsible Care initiatives all place great emphasis on process documentation. Key considerations for process documentation in the process safety area include: (1) the documentation must be up-to-date, (2) it must incorporate all phases of operations, including abnormal and emergency events, and (3) it must include critical safety related information such as operating limits for a process, the consequences of deviations from the limits, other appropriate cautions and warning statements. Environmental considerations for process documentation are similar with emphasis on waste minimization during normal operations and appropriate handling of upset conditions and emergency events such as spills or other releases.

In the quality area, process documentation is one of the most critical aspects of a quality system. Document control is covered under Section 5 of ISO 9001 while the actual work instructions are included under Section 9, Process Control. Key considerations for process documentation in the quality area is that the documentation must reflect the way in which the process is actually operated. The phrase most often used to express this is "What we say is what we do." Therefore provision of accurate documentation and upkeep of the documentation is one of the most basic concepts of the ISO 9000 system.

Process documentation is probably the area where adherence to ISO 9000 will provide the most benefit to process safety and environmental management. The framework established to manage documentation and keep it updated can also be used to manage process safety and environmental documentation. However, mere compliance with ISO 9000 does not automatically mean the objectives of process safety and environmental documentation will be met. Documentation or work instructions which are perfectly acceptable to an ISO 9000 audit may completely fail to cover subject matter critical to process safety or environmental protection. Therefore, the documentation management system established under ISO 9000 must include provisions for assuring that the critical factors mentioned earlier are incorporated into the work instructions and other process documentation.

Control of Contractors

Contractor activities have been implicated in many of the serious process safety incidents of the last several years. Therefore, any process safety management system must include some means of controlling the selection and the actions of contractors in the plant. Some key considerations in this area include: (1) review of contractor safety programs and performance in the selection process, (2) communication of critical process safety information to the contractors prior to starting the job, (3) proper training of contractor employees, and (4) liaison between the contractors and plant personnel during the job.

ISO 9000 is focused primarily on products as opposed to processes and does not specifically refer to contractors. However, Section 6 on purchasing requires that purchased products or services must conform to specified requirements. Therefore, a system to assess the ability of candidate suppliers and contractors to meet the requirements must be implemented. Expansion of this system to cover the process safety concerns of communication, training, and liaison issues is another opportunity to use an ISO established system to the advantage of process safety management.

Management of Change within Existing Processes

From the process safety perspective, management of change is one of the most critical activities in preventing major accidents. Many of the more serious accidents which have occurred in the industry have been related to some type of

temporary or permanent change within the process. Key considerations related
to managing change in the process safety and environmental areas include: (1)
identifying the fact that a change has taken place, (2) reviewing the change for
its impact on the process safety or environmental issues associated with the
facility, (3) determining the acceptability of the change, and (4) assuring that the
change is reflected in the process documentation.

Managing change is also a critical step in quality management. While
management of change is not specifically referenced as a separate section of the
ISO 9000 standards, it is inherent in many of the sections. The emphasis is on
identifying change (item 1 from above under safety and environmental) and on
assuring that the process documentation continues to reflect the way the process
is actually operated (item 4). The effect of the change on the process capability
must also be considered.

The ISO 9000 emphasis on good process documentation in the form of
work instructions and adherence to the concept of "what we say is what we do"
works to the advantage of managing change in process safety and environmental
management. The documentation updating in itself is an important part of
managing change in all three areas. In addition, strict adherence to ISO 9000
should help the plant identify changes as they are planned. This fact should
help in dealing with a major problem in process safety management of change
which is recognition of the fact that a change has been made or is planned.
This is particularly critical for changes in procedures or changes which are
made under maintenance which do not involve any capital expense. Therefore,
the major advantage of ISO 9000 in managing change is in using ISO 9000 to
identify changes and to initiate the hazard reviews and other requirements of the
plant's management of change system.

Training

Proper training of plant operators, maintenance personnel and staff is
critical to all three areas of quality, process safety, and environmental
protection. Key considerations for training in process safety and environmental
areas include: (1) the content of the training, (2) the schedule and frequency for
the training, and (3) the documentation of the training along with the associated
recordkeeping. Key items for the training content include the type of hazards
encountered in the process, critical control parameters, proper response to upset
or unusual conditions, and emergency procedures. The schedule and frequency
for the training is often established by regulatory guidelines or by company
policies. Training documentation must include the employees trained, some
description of the content of the training, and the means used to verify employee
understanding of the training.

The training program required under the ISO 9000 system includes: (1) determination of training needs, (2) arrangement of the required programs, (3) qualification of personnel, (4) and maintenance of records to demonstrate that employees have met training goals commensurate with their job responsibility. Compliance with these requirements establishes a framework which can be used to the advantage of process safety and environmental programs. The key to integration of quality, process safety, and environmental is in determining the training needs and training course content. As in the case of process documentation, recognition in the ISO 9000 system of the needs of process safety and environmental training will assure that the plant's training program will cover appropriate subjects for all three areas. This allows process safety and environmental programs to take advantage of the features of the ISO 9000 training, including documentation and recordkeeping, scheduling, and testing of competency. Therefore, it is critical that all employees involved in areas or activities covered by process safety or environmental programs be included in the training matrix or program developed under ISO 9000. In addition, the training content areas listed in the discussion of process safety and environmental training be covered in the content of the ISO 9000 training.

Auditing

Quality, process safety, and environmental management all rely on audits to assure compliance with their various requirements. ISO audits focus largely on documentation and on conformance of the process operation to the documentation. Process safety and environmental audits focus more on conformance of plant design and maintenance to various standards and on implementation of the recommendations of hazards analyses and other types of process reviews. The various audits require different types of expertise in the auditors. Due to this divergence in purpose of the various audits, integration of the actual auditing process is not likely to be of great benefit. Efforts to do so are likely to result in difficulty in managing the audit as well as in an increased likelihood that some critical area may be missed due to the increased scope of the effort.

The areas of commonality in the auditing area include the fact that both types of audits must collect evidence, focusing on written procedures, state of the facility and equipment, and the actual practices of the employees and performance of the process. In addition, all of the audits will result in action items which must be identified and addressed.

There are two areas which would benefit from the integration of the auditing component of process safety, quality, and environmental programs. The first benefit could be achieved in the design of the audit schedule, status report, and recordkeeping. Although the content of the audit would vary based on the required expertise and scope, a central schedule and standard formats would provide some efficiency in the administration and tracking of the overall auditing process. The second area which can benefit from integration is the follow-up to the action items from the audits. The formal report from each type of audit will include audit dates and locations, audit objectives, observations and conclusions, and any deficiencies found along with corrective actions. Responsibility for follow-up must be designated as well as a schedule for implementation of the action items to correct the deficiencies. The completion of the action items must also be documented to complete the ISO 9000 audit process. Integration of process safety and environmental audits into this framework will allow a common system to be used for recordkeeping and tracking of action items from all three types of audits.

SUMMARY

This paper has identified specific areas where integration of process safety, quality, and environmental programs can provide benefits at the plant level in terms of increased efficiency. These programs should not be considered as separate and distinct efforts. The primary benefit of the ISO 9000 system of standards is in the provision of a structure or framework for many of the activities. Proper consideration of process safety and environmental protection concerns will have a significant impact on the content of the documentation and other aspects of plant level activities. Some effort in the initial design of the program to comply with ISO 9000 can provide considerable savings in time, effort, and resources in the programs for process safety and environmental protection. The key to success is to consider all three areas while implementing the ISO 9000 system rather than focusing simply on achieving certification.

Integration of Risk Management into Process Safety Standards and Application of Those Standards

M. Reid McPhail

Senior Loss Prevention Engineer, Novacor Risk Management. P.O. Box 2535, Station M, Calgary, Alberta, Canada T2P 2N6

Jan C. A. Windhorst

Associate Engineer, Novacor Chemicals, Ltd., Projects Management Group, P.O. Box 5006. Red Deer, Alberta, Canada T4N 6A1

ABSTRACT

Novacor project management, risk management and manufacturing sites have developed a process safety program based on managing risks. The program is based on numerical corporate risk criteria contained in a loss prevention standard. Derived from this master standard are other standards which allow designers and engineers to achieve desired results, minimization of the frequency and consequences of a chemical incident, in a timely manner. This paper discusses the integration of numerical risk criteria in standards and how the standards are organized and maintained. Results of the application of a quantitative risk analysis at a Novacor site and recommendations to minimize risk at that site are given.

1 NOVACOR PROCESS SAFETY PROGRAM

Novacor's process safety program is part of an overall risk management program. This risk management program takes into consideration all aspects of Responsible Care[1] such as process safety, occupational safety, environmental impairments; and business risk (property loss and the cost of business interruption).

---

[1] Responsible Care, a Registered Trade Mark, is an initiative of the Canadian Chemical Producer's Association and the Chemical Manufacturer's Association.

The process safety program was established by a senior management directive and deals almost exclusively with process safety although it overlaps with other aspects of the overall risk management program. The process safety program is detailed in the Novacor Loss Prevention (Engineering) Standards, or LPS, and incorporates the following CCPS publications:

o   "Guidelines for Technical Management of Chemical Process Safety";

o   "Plant Guidelines for Technical Management of Chemical Process Safety";

as well as:

o   API Recommended Practice 750, "Management of Process Hazards"; and

o   OSHA 29 CFR 1910 "Process Safety Management of Highly Hazardous Chemicals".

2 LOSS PREVENTION (ENGINEERING) STANDARDS (LPS)

2.1   History

Novacor LPS were developed to avoid (or to reduce the effects of) industrial accidents, to satisfy CCPA and CMA Responsible Care initiatives and to address concerns arising from the public. LPS, which form the core of the Novacor process safety program, are comprised of compulsory and quantifiable criteria, i.e. descriptive policy statements, and ways of implementing these criteria through either mandatory standards or recommended guidelines, see **Table 1**. LPS apply to new plants and changes

KEY SECTIONS OF LPS:	DESCRIPTION:	JARGON:
CRITERIA	What has to be achieved.	MUST
STANDARDS	How to achieve it.	SHALL
GUIDELINES	Recommended practices.	SHOULD

Table 1. LPS Keywords

to existing facilities, however, risk minimization is a principle that has to be applied across the board. The purpose of LPS is to have Novacor staff focus on objectives while the selection of methods for implementing these objectives is of secondary importance. For example, when considering the potential of release from a pipe between a tank and a pump the objective is to diminish potential spills not to install extra isolation valves.

Novacor has a large number of LPS in place that cover subjects such as fire protection, alarm management, plant design, explosion protection, etc.

LPS were developed through the representation of Novacor manufacturing sites on a LPS committee. This committee has two representatives from olefin plants, two from plastics plants, one representative from the risk management department and one from the projects management group.

The committee, which reports directly to senior corporate management, is responsible for site participation in the review processes, editing, obtaining executive approvals, distribution, maintenance, quality, handling deviations and the organization of LPS related training courses. The flowchart for LPS development is given in **Figure 1**.

2.2   Effectiveness

In order for LPSs to be effective they have to be pro-active. This has been achieved on the organizational side by making the LPS committee the intersection of senior corporate management, plant operations and engineering/project groups and the risk management department, see **Figure 2**. On the technical side LPS endeavour to promote inherent process safety into a design at the earliest stage. This allows problems to be designed-out and avoids or otherwise minimizes the need for mitigation equipment.

Another aspect of effectiveness is user-friendliness. A gauge for the user-friendliness has been the increased demand for LPS documents and their use as training tools in courses. User-friendliness is enhanced by the LPS process which focuses on an early response to the needs of manufacturing sites. LPS committee members are viewed as a process safety resource rather than "impractical theorists".

The effectiveness of LPS is measured through:

(i)   an auditing LPS (LPS 12.1 "Audits") which is based on ISO 10,011-1, -2 and -3 standards "Guidelines for Auditing Quality Systems";

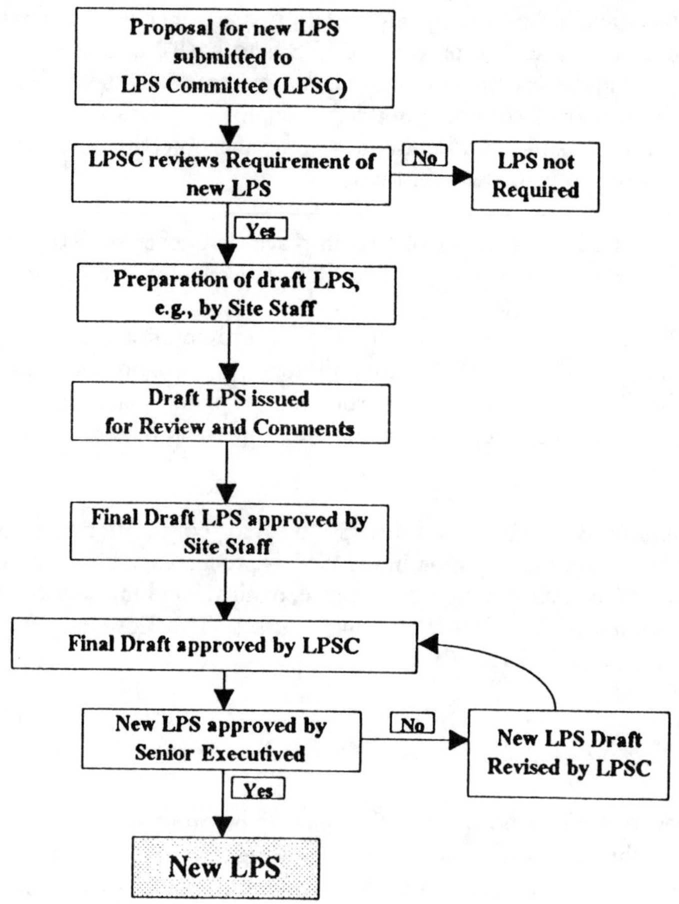

Figure 1. LPS Development chart.

(ii)   the LPSs for hazard reviews (see **Table 2**) which are the foundation of Novacor's Management of Change and Capital Project Reviews; and

(iii)   deviation control. Deviations from LPS criteria are not permitted, deviations from LPS standards are strictly controlled through an approval procedure. Documentation of deviations serves as feedback for future LPS revisions.

All LPS are revised or reconfirmed on a three year cycle, depending on feedback from users and technological developments.

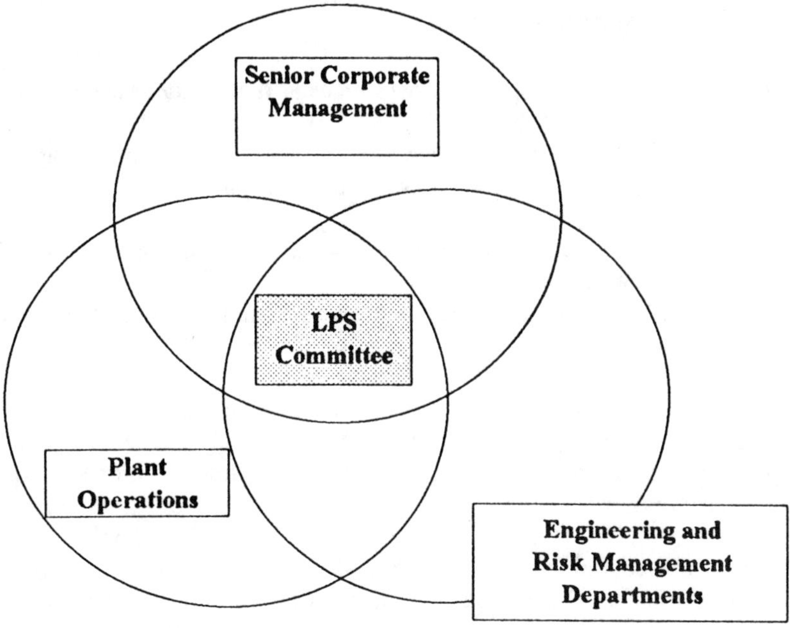

Figure 2. LPS Committee Organization

LPS GROUP 9	TITLE
LPS 9.1	Preliminary Hazard Review
LPS 9.2	Process Hazard Review
LPS 9.3	Detailed Design Review
LPS 9.4	Construction Reviews
LPS 9.5	Facility Turnover Inspection
LPS 9.6	Pre-Startup Commissioning Checks
LPS 9.7	Pre-Startup Audits
LPS 9.8	Decommissioning and Restoration Reviews

Table 2. Process Hazard Reviews.

2.3 Risk Management LPS

Early on in LPS development it became clear that a numerical risk driven "master"-criterion would be needed to link different LPS criteria and standards. Risk driven decisions are often made during the design and operation of a plant. Examples of risk driven decisions are:

(i) selection of equipment spacing based on fire exposure prevention or on explosion protection; and
(ii) the decision to make a building explosion resistant based on the number of occupants.

Making a poor risk driven decision means either the creation of risk which should be deemed "unacceptable" or alternately the squandering of financial and human resources that could have gone toward risk reduction in other plant areas.

The LPS committee decided that design engineers, designers and managers should not be placed in a position of un-informed risk decision making. A special risk management LPS was written for this purpose which enables Novacor staff to compare their hazard reduction activities with defined numerical risk criteria and standards. Values for risk criteria and standards were not arbitrarily chosen but based on an analysis of available data, some of which have been published recently. The risk management LPS includes aspects of informed decision making, based on best available technologies, and quality management. In its simplest form, LPS risk management means that corporate resources should be applied to unacceptable risks so that potentially serious process related accidents can be dealt with, but as risks become less serious or likely, they deserve fewer resources, down to a point where diminishing returns make it unreasonable to do more.

3 PROBABILISTIC RISK MANAGEMENT

3.1 Introduction

Novacor's process safety risk program utilizes common risk assessment methodologies[2] within the realm of classical risk management. More

---

[2] Guidelines for "Chemical Process Quantitative Risk Analysis, by Center For Chemical Process Safety of the American Institute of Chemical Engineers (1989).

specifically, the management of process related accidents involves decisions as to what is considered an "unacceptable" risk and what must be done to reduce unacceptable risks to a tolerable level. A basis to evaluate "unacceptability" of process related accidents has been established in LPS 12.2 "Probabilistic Risk Management".

Key elements LPS 12.2 include:

o   Use and interpretation of risk management criteria, standards and guidelines as well as the interpretation of results obtained by applying them is restricted to trained and qualified Novacor personnel and consultants retained by Novacor.

o   Risk management takes the assumptions and limitations which are inherent to this management technique into consideration. On the management principles side, there are elements such as decision making, communication, etc., with their inherent assumptions and limitations.

o   On the quantitative risk assessment side there are limitations which include:

(i)   absence of any guarantee that all possible process related accident scenarios have been considered;

(ii)   uncertainty in the consequence analysis which is almost exclusively based on simplified mathematical models;

(iii)   uncertainty in the failure characteristics of equipment/components. Actual plant data will depend on management systems, quality control and training of individuals who look after the equipment and components.

o   A risk assessment methodology which is given in **Figure 3.**

o   Process related accident quantitative risk analysis (QRA) results must be reproducible.

o   Quantitative Individual Risk Criteria[3] which depend upon personnel relationship with operating plant or as public are given in **Figure 4.**

---

3   Individual (Geographical) Risk - a risk measure that gives the frequency of fatality to an individual due to the impact of ALL specific Process related Accident if that person is at a specified location relative to the source of those impacts.

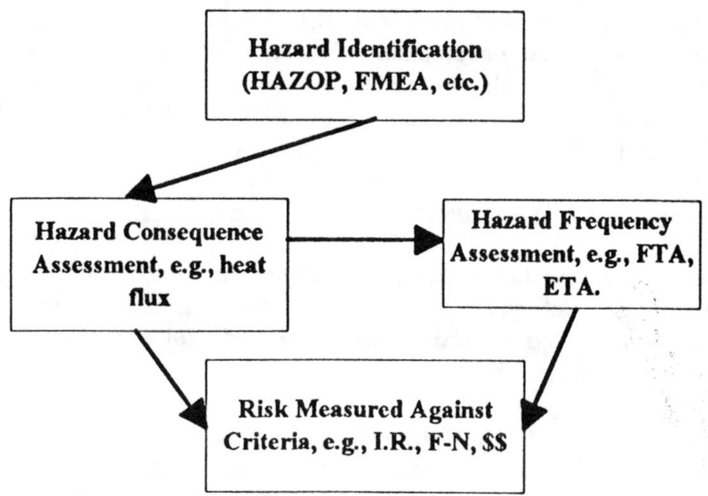

Figure 3. Quantitative Risk Assessment Methodology

o   Societal Risk Criteria[4] are given in **Figure 5**.

o   Unacceptable criteria are considered as the ultimate maximum threshold risk exposure level which can be tolerated by the corporation for the process related accidents being evaluated. The unacceptable criteria apply to existing Novacor facilities with no "grandfathering" allowance. In case criteria are not satisfied for an existing facility the "As Low As Reasonably Practicable" (ALARP) principle has to be applied to achieve a tolerable risk level.

o   ALARP involves management techniques including optimization of the inspection and predictive/preventative maintenance intervals, quality management, human engineering, and design measures to lower risks.

o   New facilities or plant modifications are designed with an individual risk and societal risk measures one-tenth of that corresponding to the upper-limit of the unacceptable risk criteria value. The factor of 10 is a confidence factor intended to make up for potential uncertainties and unknown events in the risk assessment.

---

4   Societal Risk - a risk measure that gives the possible impacts to a large off-site exposed population who may be affected by one or more process related accidents. Societal risk in LPS 12.2 is defined as a plot with the cumulative frequencies (F) of events which can cause N OR MORE fatalities versus the number of fatalities (N).

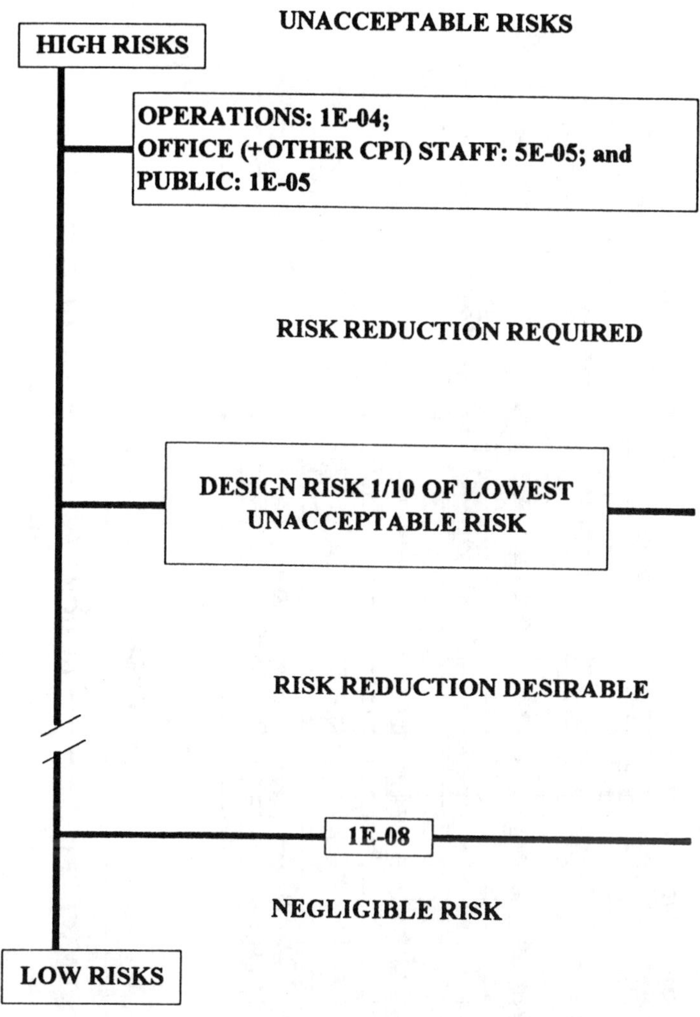

Individual Risk Criteria (Fatalities/Year)

Figure 4. Novacor Individual Risk Criteria

o   If it is not possible to meet the Design Risk Standards, i.e. the calculated risk is less than the unacceptable risk level criterion but greater than the design risk standard, then ALARP risk principle has to be applied to lower the risk as much as possible. If the risk calculated, after ALARP, is still greater; then the person responsible for plant risk management (in consultation with LPS committee) shall decide if the actions taken are adequate or if further mitigation is required.

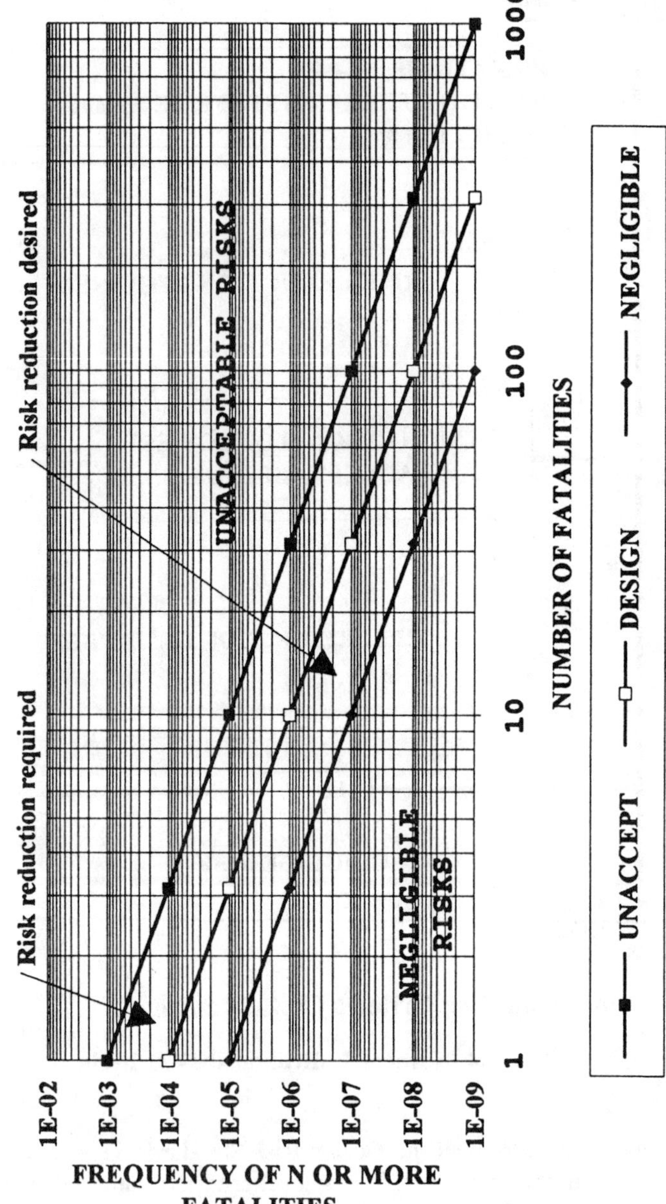

Figure 5. Novacor F-N curve for Societal Risk

110

o   It should be noted that detailed QRA is only performed on severe hazards, e.g. possibility of fatality. Alternately, the QRA is performed for those hazards whereby previous experience or knowledge cannot adequately establish a relative risk magnitude of the process related accident hazard.

We believe LPS 12.2 will be satisfied, automatically, when all other LPS criteria and standards are satisfied. This assumption is based on the fact that Novacor risk criteria have been integrated into all other LPS for the purpose of corporate consistency, simplification and streamlining of design and associated activities. One example is in the dealing with "Buffer Zones" as stipulated in Responsible Care. To simplify the risk evaluation process, maximum hazard consequence exposure levels to off-site areas have been established in LPS 2.4 "Buffer Zones". Meeting these exposure threshold levels will result in satisfying the societal risk, as established in LPS 12.2, regardless of the frequency of process related accidents.

Auditing of plant changes includes process related accident risk assessments as per LPS 12.2. This ensures that Novacor's due diligence meets the intent of Responsible Care with respect to hazard analysis and risk assessment and that elements of OSHA's 29 CFR 1910 PSM are also being addressed.

The following example in section 4 of this paper further displays Novacor's approach to integration of probabilistic risk management into process safety design standards.

4   EXAMPLE

4.1   Introduction

Probabilistic risk management has been applied to various facilities within Novacor, such as toxic gas storage containers, a research/pilot plant facility and a manufacturing site with three process plants.

4.2   Manufacturing Site Quantitative Risk Analysis

The QRA of the manufacturing site was initiated as a means of identifying and evaluating hazards and to establish risks in order to advise on appropriate actions by management. The scope of the study covered risks posed by the manufacturing site to its neighbours and staff as well as the evaluation of buffer zones in compliance with the Responsible Care initiative.

Deliverables included a hazard analysis[5], a risk assessment[6], risk mitigation recommendations; we were specifically interested in Individual Risk contours, F-N curve, on-site fatality rate and the 10 highest contributors to the latter.

PMG commissioned an outside consultant to quantify the risk from this manufacturing site and to evaluate potential risk reducing measures as necessary and appropriate. Benefits received from this study are:

(i)   establishment of a baseline risk which for future benchmarking;

(ii)   communicatable risk information for the public, e.g. as part of the Responsible Care right to know policy;

(iii)   elimination of local and organizational biases and the independency of the analysis which is important when releasing information to neighbours of the manufacturing facility; and

(iv)   prioritized risk information applicable for informed decision making by senior corporate and facility management.

4.3   Discussion of Results

o   The Individual Risk for in plant personnel was determined to have a value which exceeded the Novacor unacceptability criterion of 10^{-4} fatalities per year.

o   Individual Risk for the public marginally exceeded Individual Risk criteria.

o   The Societal Risk criteria (F-N curve) to the public are satisfied.

o   Buffer zones criteria are presently met and will be met as long as development of public facilities, such as housing, within the 1×10^{-6} fatalities per year Individual Risk contour are avoided.

o   A list of the 10 main contributors to the on-site fatality rate is given in **Table 3**.

---

5   Hazard analysis: the identification of undesired events that lead to the materialisation of a hazard, the analysis of the mechanisms by which these undesired events could occur and the estimation of the extent, magnitude and likelihood of any harmful effects.

6   Risk assessment: the quantitative evaluation of the likelihood of undesired events and the likelihood of harm or damage being caused together with the value judgments made concerning the significance of the results.

RANK	INCIDENT	% OF TOTAL FAT. RATE
1	Chlorine in Plant 3	15
2	Chlorine for Plants 1 and 2	8
3	Rupture of natural gas feed to convection section in Plant 1	8
4	Combined risk from the three fired heaters in Plants 1,2 and 3	6
5	Rupture of fuel gas line to fired heater in Plant 1	3
6	Large leak in natural gas line to convection section of fired heater in Plant 3	3
7	Rupture of feedline to Plant 1	3
8	Rupture of natural gas feed to convection section of fired heater in Plant 2	2
9	Rupture of feedline to Plant 3	2
10	Rupture of tank car at loading dock	2

Table 3. Ten Highest Contributors to Site Risk.

PMG's course of action for the public Individual Risk was to apply the ALARP principle. PMG's course of action for in plant personnel was, firstly to consider that the Individual Risk for in plant personnel was within the range of accuracy of the criterion, secondly that human exposure within the high individual risk areas was negligible and thirdly to apply the ALARP principle. PMG has therefore judged that LPS 12.2 "Probabilistic Risk Management" can be satisfied, provided the facility adopts actions that reduce individual risk and updates the risk assessment study in 3 years, using more accurate data. PMG and Novacor Risk Management deems these actions necessary in order to substantiate satisfaction of the individual risk criteria.

4.4   Design and Operational Recommendations

o   For toxins: eliminate chlorine. This allows Individual Risk criteria for the public to be satisfied and reduces the on-site Fatality Rate by 21%.

o   For explosions:

(i)   audit startup and shutdown procedures of fired heaters; and
(ii)   put transparent film on windows and review buildings for better fastening of cladding.

o   For pool fires:

(i)   install remotely operable isolation valves;
(ii)   install curbs to guide spills towards sewers, where applicable;
(iii)   keep existing sewers and drains in good condition;
(iv)   audit firewater system performance;
(v)   audit performance and closing times for remotely operated isolation valves; and
(vi)   consider an equipment integrity program.

On the calculation side: improve accuracy of consequence, probability and risk numbers (to be completed prior to next hazard analysis and risk assessment) and reevaluate simulation parameters.

5   CONCLUSIONS

o   Results from QRAs heighten risk awareness.

o   Results of QRAs have resulted in a distinction between geographical and personal individual risks (within Novacor).

o   Quantified risk criteria are needed for responsible and consistent process management decisions.

o   QRAs can easily be managed.

o   Risk criteria should be established before performing QRAs.

6   ACKNOWLEDGEMENTS

The authors would like to thank G.L.W. Clark and D.J. Boomer, senior corporate executives of Novacor for their sponsorship of this paper and all of those individuals throughout Novacor who have contributed to this paper. Further thanks also go to Phil Myers of DNV Technica and DNV Technica for their contributions to this paper.

7   REFERENCES

1   Frank M. Renshaw, "Major Accident Prevention Program of the Rohm And Haas Company", CCPS conference in Toronto (1990).

2   B.J.M. Ale, "Risk Analysis and Risk Policy in the Netherlands and the EEC", J. Loss Prev. Process Ind., 1991, Vol 4, January pp 58-64.

3   Henry Ozog, R. Peter Stickle, "Process Hazard Management Documents, Practices compared", Oil and Gas Journal, Jan. 28, 1991.

4   Arendt, J.S., Lorenzo, D.K., Lusby, A.F., "Evaluating Process Safety in the Chemical Industry", published by the Chemical Manufacturers Association, December 1989.

5   Concawe (The Oil Companies' European Organization for Environmental and Health Protection), "Quantified Risk Assessment", Report no. 88/56, July 1988.

6   Kolluru, R.V., "Understand the Basics of Risk Assessment," Chemical Engineering Progress, March 1991, pp. 61 - 67.

7   Wilson, R. and Crouch, E.A., "Risk Assessment and Comparisons: An Introduction," Science, 236, April 17, 1987, pp. 267 - 270.

8   Ballard, G., "Guest editorial: Societal risk-progress since Farmer", Reliability Engineering and System Safety 39 (1993), pp.123-127.

Early Safety, Health, and Environmental Process Guidelines

Clifford J. Walk
Rohm & Haas Company

ABSTRACT

The **Early Safety, Health and Environmental Process Guidelines** was a project of the Process Hazard Analysis and Environmental Engineering Department (PHA/EE) in the Corporate Engineering Division (CED) of the Rohm and Haas Company. The idea for the **Guidelines** grew out of our perception of a need for standardization of the critical work process of early SHE review. Management support for the project was and is critical to its success. This support took the form of time for us to produce the **Guidelines**, management review of the **Guidelines**, and support for and encouragement of outreach efforts to the entire staff of the Engineering Division.

INTRODUCTION

The **Early SHE Process Guidelines** was a project of the Process Hazard Analysis and Environmental Engineering Department (PHA/EE) in the Corporate Engineering Division (CED) of the Rohm and Haas Company. The quality effort in the Engineering Division takes the form of Natural Quality Teams. An NQT is composed of people who naturally work together. The team's goal is to serve its customers better. Our team was made up of all the department members and several representatives of both suppliers to and customers of our department. A CED Quality Consultant was also an integral member of the team. Altogether about 14 people were on the team of which seven did the actual writing.

The idea for the **Guidelines** grew out of our perceptions of a need for standardization of the critical work process of early SHE review. Checks with our customers (mostly within CED) before starting the project and an outside review of a draft copy of the **Guidelines** (both by people within and outside CED) showed that this need was real, and that we were on the right track.

Mission
Provide a guideline document on
how to identify and resolve SHE issues
Knowledge
• Experience as SHE Professionals
• Customer survey
• Publish Guidelines to CED
• Give presentation to CED
• Follow-up with survey users and SHE process participants
Act
• Prepare document outline
• Make writing assignments
Plan
Study
• Collect review responses
• Compare comments to our vision and re-write where necessary
Do
• Write sections
• Assemble document
• Internal review
• Revise
• Send out for customer review

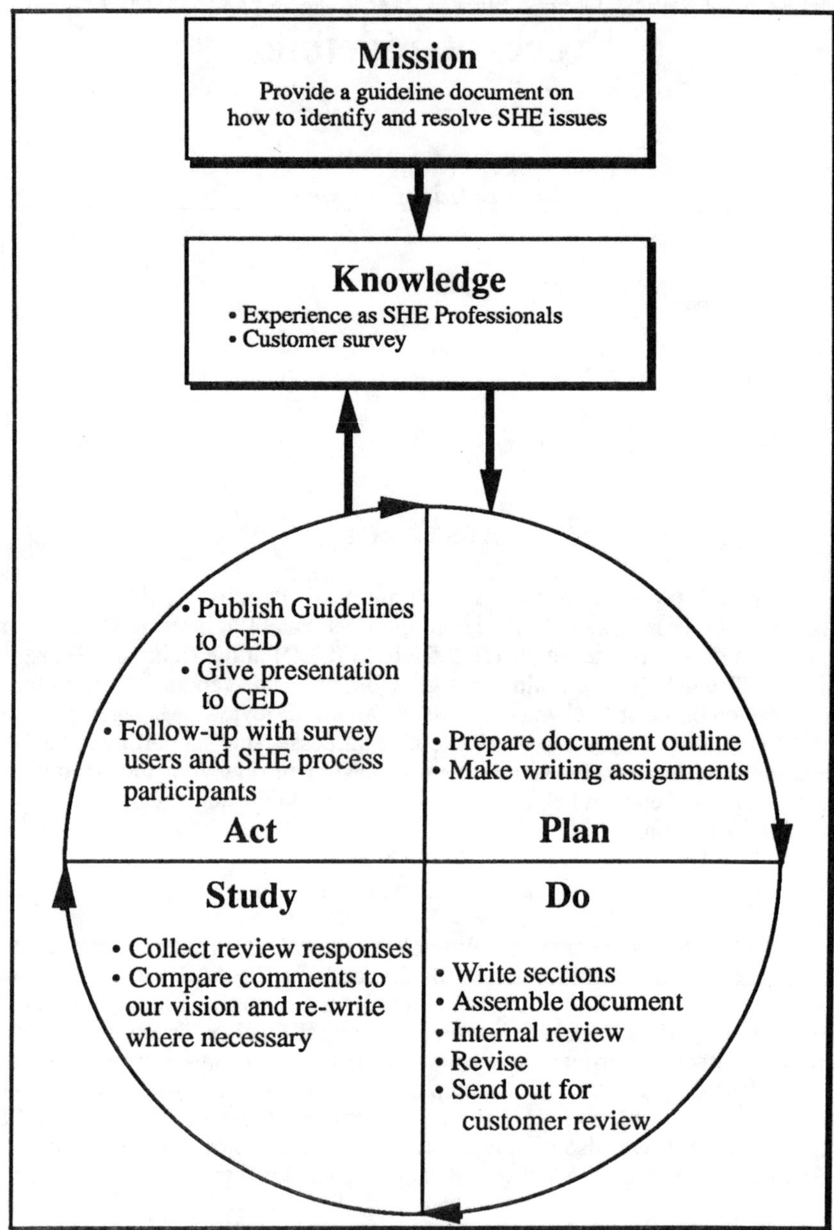

Figure 1

The **Guidelines** were a finalist for the 1992 Rohm and Haas company wide Total Quality Leadership award.

MANAGEMENT MODEL

Our NQT used a **Quality Management Model** (Figure 1) which emphasized the cycle of Plan-Do-Study-Act. The cycle is powered by knowledge of the process gained through our own experience and by interaction with our customers.

RANKING OF PROJECTS

When our team was formed our first goal was to come up with a project which would have impact on our customers (the process and project engineers in CED) and which would fulfill a critical need. We **brainstormed** a list of customer needs. Then we brainstormed a list of projects which we felt needed to be done.

We agreed that some formalization of the Safety, Health, and Environmental (SHE) Review process was the most important project on our list. Our experience is that SHE reviews done late in a project often uncover problems which could have been more easily dealt with earlier in the project. This increase in non-value added cost clearly does not meet the needs of the CED customer. A SHE review process would also meet a number of our internal customer's needs such as consultation on safety and environmental issues, education, training and providing leadership of or participation in SHE Reviews.

We questioned ourselves on why early SHE reviews aren't being done. We developed a list of reasons and picked those we thought most important. We put the same question to some of our customers. The reasons (see box) fall into two broad categories: manpower limitations and uncertainty about the process. While we felt that we couldn't do much to address the manpower issue, we could clear up the uncertainties around the SHE process.

At this point we developed a **Cause and Effect Diagram** (or fish bone) of the SHE Process around the issues of "not understanding" or "not owning" the process

---

Why aren't early SHE reviews being done?

Our perception

- Manpower/time constraints
- Reluctance to invest resources at an early stage in projects that may never progress beyond the early stage
- SHE not integrated into CED thinking
- Engineers don't know what/how to do SHE reviews
- Don't anticipate future projects well enough to have sufficient time to do early review

Customer survey

- Unclear who "owns" the SHE process
- Manpower/time constraints
- Engineers don't know what/how to do SHE reviews
- Don't anticipate future projects well enough to have sufficient time to do early
- SHE not integrated into CED thinking
- Engineers not aware of the corporations expectations in the SHE area

(Figure 2). From this we agreed that a key problem was the lack of a clear, written definition of the SHE Process. We envisioned a set of SHE Process guidelines which would address issues of how to do a SHE review, who's responsible (who 'owns' the process) and the timing of the SHE Process.

TEAM FORMATION

Up until this point most of our effort had been in our own department. We felt that as SHE consultants to CED and Rohm and Haas plants we were in a unique position to see how the use of Early SHE Reviews in CED projects could be effective. Having been attendees at and organizers of many SHE reviews made us qualified to describe the ideal process. We also recognized the value of customer input and asked representatives of our CED customers (process and project engineers) and corporate SHE professionals to join the team early in the project. Another valuable member of the team was a CED Quality Consultant. The Engineering Division employs several full-time quality consultants who are trained in all aspects of TQL. The consultant was invaluable in keeping us focused on producing a quality product. The consultant's sharing of experience with other NQTs in CED also helped us.

Training is important to any quality effort. Early in the life of our NQT, the group underwent the one day **Process Improvement Team Startup and Orientation Program** (PITSTOP). This is a standard training program for Quality efforts at CED. It teaches mission statements, quality improvement strategy, and a wide variety of "quality tools". In addition, all team members have had the CED three half-day **Management Skills for TQL** seminars which cover team building, conflict resolution and effective meeting skills. We have also all received **Performance Management** training. Performance Management stresses the importance of positive reinforcement of behaviors to get desired results.

The project was a **team** effort all the way. We decided that the **Guidelines** should be written by PHA/EE department members, each member concentrating on one or two sections. The rest of the team was used as a sounding board and to help review successive drafts of the **Guidelines**. One member of the writing team also served as a general editor, providing a consistent style and look to the document.

Tools
Measurement
Environment
Methods
Person
Don't own or understand the process
Resources not easily identified
No checklists or not updated
Hazards not easily identifiable
Mechanical hazards
Reactivity hazards
Accessibility
Resources
Don't know what is available
Short range objectives
Goals
Not clear
Prioritize
Short range perception
Cost
Pressure perception
SHE items are add-on
Not a priority
SHE not integrated into CED thinking
Program schedule pressures
Lack of
Rewards & recognition
Reviews don't take SHE issues into account
Viewed as 'them' (not part of team)
Split responsibility among S. H. & E.
No perceived insurance benefits
No measurement of cost justification
Little between groups
Communications
Consistency of representation
Roles
Plant people
Right people at right time (scheduling)
No assignment of responsibilities
No decision maker
No followup
Hidden agendas
Training
Process too formal, too long and too large
No Corp. policy
No clear purpose to review
No clear definition (nothing written)
Lack of consistency
Legal and cultural differences
Consistency of technique
lack of
Turf issues
Fear
Lack of SHE training
Lack of any kind of training

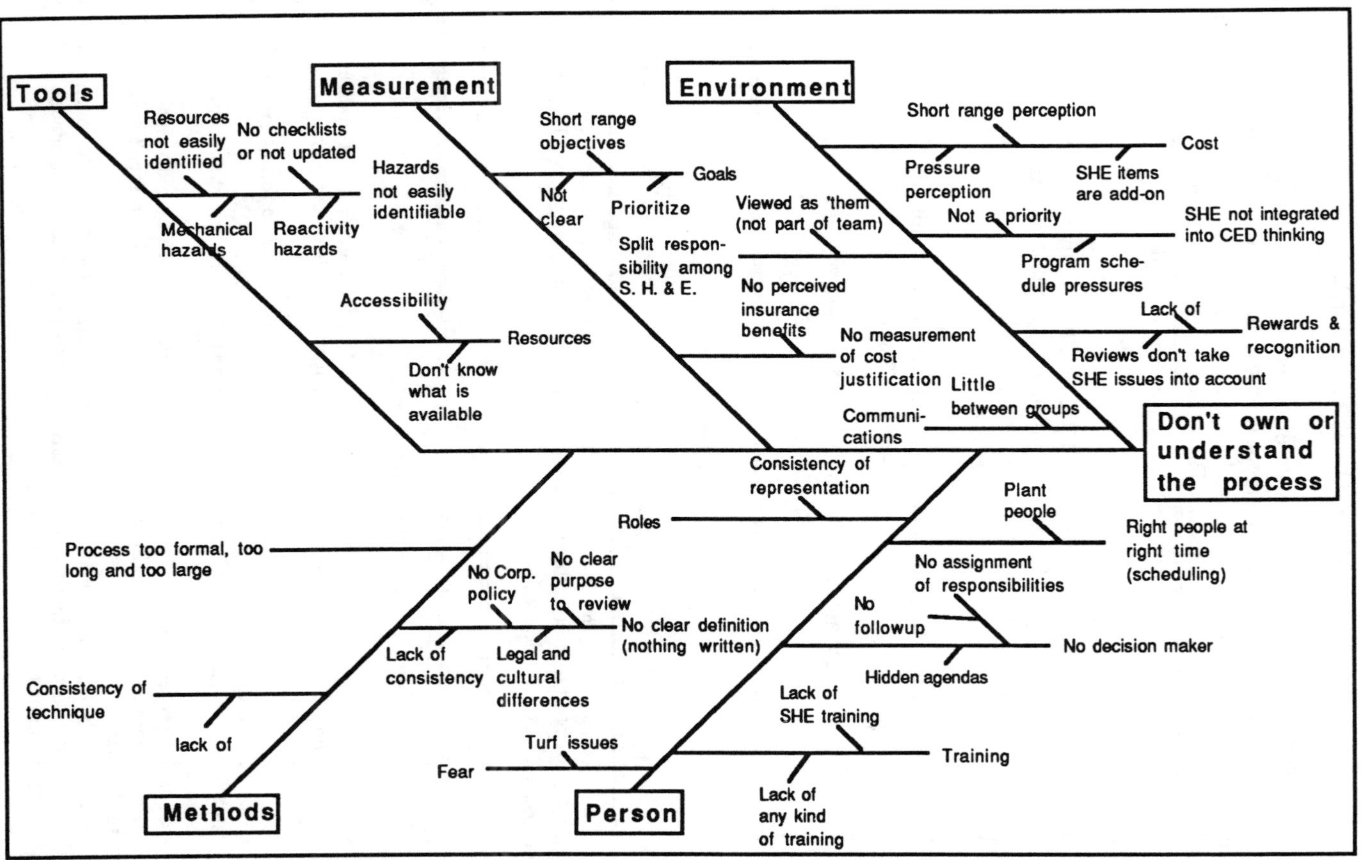

Figure 2

MISSION STATEMENT

One of the keys to a successful team effort is knowing what it is you're trying to accomplish. A clear mission statement is essential. Our mission statement evolved through the life of the project as we learned more about the needs of our customers but it did not change substantially. The final version was as follows:

Early SHE Guidelines NQT
Mission Statement

To identify and resolve Safety, Health and Environmental issues <u>early</u> in the development/design phase at CED to ensure plants are safe, environmentally sound, and cost effective.

Provide a guideline document on how to identify and resolve SHE issues;

Increase awareness within CED of the advantages of early consideration of SHE issues.

CUSTOMER NEEDS

In addition to a Mission Statement a quality team needs a list of objectives. Our objectives took the form of a list of criteria on which to judge the final product (see below). The table shows the characteristics we believed to be desirable in the final product. Although these criteria were developed internally in our NQT, our customer representatives provided valuable input.

Desired Outcomes

• Written in sections.	• Table of contents
• Brevity, conciseness - target 20 pages or less.	• Index.
• Clarity.	• System for distribution of updates.
• Use flowcharts and graphics in place of text where possible.	• Put revision dates and numbers on every page.
	• Consider electronic form - put on-line
• Use key words, highlighting, underlining for emphasis of important points	• Should be 81/2" x 11"
	• Don't make the print too small
• Consider professional publication.	• Sections written by individuals
• Obtain user feedback on draft before final issue.	• Ease to update - consider 3 ring binder.
• Provide a "road map" to assist the various audiences in finding the sections of the document most relevant to their needs.	• General editor to provide consistent "look and feel" to the document

The first step in writing the Guideline was an attempt to understand how we thought the SHE Process should work. We used a **flowchart** at this stage. As with our mission statement the flow chart evolved over the course of the project as we learned more about the early SHE process. The final version is shown in Figure 3. It eventually became the outline for the document, the framework which holds it together and a 'road map' for the users.

We made the document easy for the customer to use by creating a detailed Table of Contents, and keying each section of the document to our flow diagram of the SHE Process.

Although the document is longer than our original expectations, it goes into greater depth. This makes the **Guidelines** a better tool for the engineer planning a SHE review. Part of the goal of keeping to 20 pages was a desire for brevity and conciseness. We tried to do this as much as possible without sacrificing readability.

CUSTOMER REVIEW

The essence of the quality effort is understanding the needs of your customers. Although we had representatives of customers on our team we felt we needed a wider viewpoint. In addition to our direct customers (the project and process engineers in CED) we wanted to get the opinions of other SHE professionals in the company (some of which were potential users of the manual). After several rounds of individual writing and review within the NQT, we pulled the sections into a complete draft. This was polished up and sent out to 24 people for review.

We asked for general comments as well as comments on each section of the document. The individual comments were combined (along with the text being referred to) in one document for review by the NQT. This format enabled us to compare comments from different reviewers and to relate the comments to the document as a whole. We received many comments, all which were well thought out and helpful.

Along with the draft we also sent a questionnaire to elicit further comments about the usefulness of the document. The questionnaire was composed of questions to be answered on a 1-5 scale with a space for suggested improvements to the topic covered by the question. The box shows one section of the questionnaire where we asked our reviewers to rate the completeness, correctness and detail of each section. This helped us to pinpoint which sections needed improvement.

In addition to asking people about the contents of the **Guidelines** themselves, we also asked for suggestions on how to implement the process outlined in the **Guidelines**. We recognized early that just writing these **Guidelines** and putting them on everyone desk was not going to significantly change the way people work.

Early SHE Process

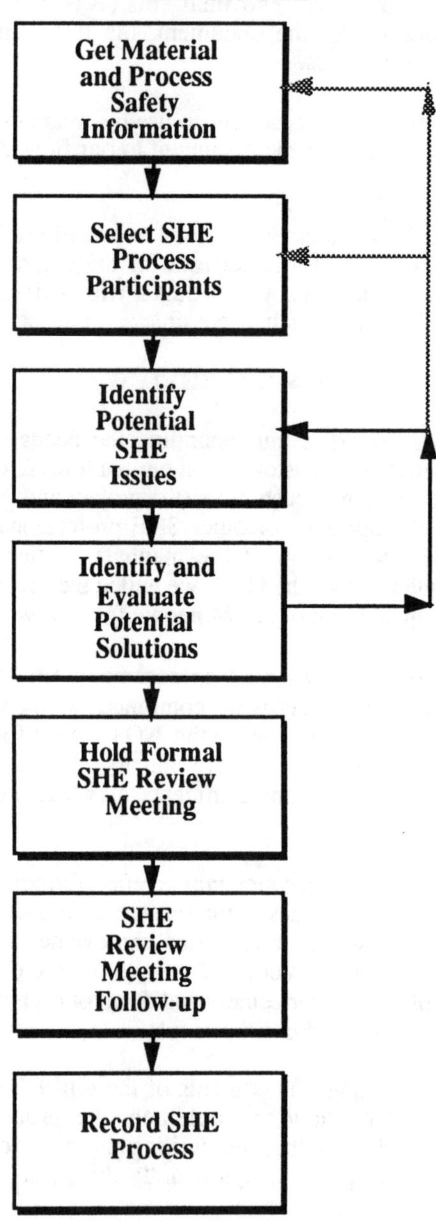

Figure 3

Questionnaire Results

Section	Complete	Correct	Detail
Introduction/Summary	3.75	4.75	3.75
Material and Process Safety Information	3.8	3.8	3.6
SHE Process Participants & Roles	3.8	4.6	3.8
Identification of SHE Issues	4	4.75	4
Potential Solutions	3.5	4.75	3.75
Consensus on SHE Issue Strategy	4.25	4.75	4.5
SHE Review Meeting Follow-up	4.5	4.75	4.75
Record Keeping	4.3	4.7	4.7
Acronym List	4.5	4.7	4.7
Glossary	3.75	4.5	4.25
Resource Guide	3.7	4.7	3.7
Process Checklist*	4.7	4.7	4.7
Materials Checklist*	4.7	4.7	4.7
Identifying Solutions Checklist*	4.7	4.7	4.7
Environmental Checklist*	4.7	4.7	4.7
Meeting Participants Checklist*	4.3	4.7	4.3
SHE Review Package Checklist*	4	4.7	4
Bibliography	4.7	4.7	4.7

We received 18 returns (of the 24 review copies issued). This gave us a broad range of opinion which led to many new ideas. Although this affected our time table for publication we believe that the final product is much better than it would have been otherwise. We also found that our reviewers had their own ideas about what the purpose of the Guideline should be (based on their own particular needs and circumstances). The next box highlights the major suggestions on which we took action. We made some changes in direction due to this (as witness our evolving mission statement) but were able to resolve conflicting ideas by remembering that our primary customers are the project and process engineers in the Engineering Division. In addition to soliciting opinions from around the company we also asked our Division Management Team to review the document. Their support would be critical to the success of our effort.

DATA AND FACTS

When we started our project there was no formal system for doing early SHE reviews in the Engineering Division. SHE reviews were being done but perhaps not as early as they should have been. Our early customer survey work confirmed this. Most of the data collection portion of our work was not so much on how the current (ill-defined) system works but on how an "ideal" system would work. The major data

collection effort was in getting a review of a draft of the **Guidelines**. We asked
both customers and other SHE experts to review the document.

<u>Changes Prompted by Reviewers Comments</u>

- Made more mention of the Product Integrity Department
- Added a Preface and an Introduction
- Made section on Applicable Guidelines, Laws and Practices (1.6) more concise (i.e., shorter)
- Put more emphasis on the H of SHE
- Added industrial hygiene checklist to Appendix
- Put more emphasis on SHE Process flowchart
- Put less emphasis on process engineer (contrasted to project engineer)
- Dropped sample SHE Review meeting agenda
- Made section titles the same as SHE Process flowchart blocks
- Made more detailed Resource Guide (Appendix D)
- Added Rohm and Haas 20 Safety Principles (Appendix C) and SHE Policy Guidelines (Appendix F)

PROJECT TIMELINE

Early in the project we developed a **Project Timeline**. (Figure 4). The timeline
was constructed at about the time the outline of the document was started. The
Guidelines took about 9 months longer to issue than we originally planned. Most
of the extra time was spent in the revision stage. Our outside reviewers provided us
with a wealth of material to think about. We received much more and better
comments on the draft than we expected. Based on the returns we made several major
changes in the text and changed the focus in some areas.

IMPLEMENTATION AND FOLLOW-UP

The introduction of the **Guidelines** to CED was through several **formal
presentations** to the division by the PHA/EE Department. The presentations
focused on why the Early SHE Process was important and introduced the
Guidelines as an aid to the process. Three presentations were held in order to reach
the approximately 150 engineers in the Engineering Division.

Copies of the **Guidelines** were also distributed to plant SHE personnel by the
Safety Health and Environmental Affairs Department. The head of this department
was a member of our NQT and saw value for the plants as well as CED. The
Guidelines have been well received, especially in our smaller plants which do not
have large SHE staffs. We have given several presentations to our Research Division
and several plant presentations as well.

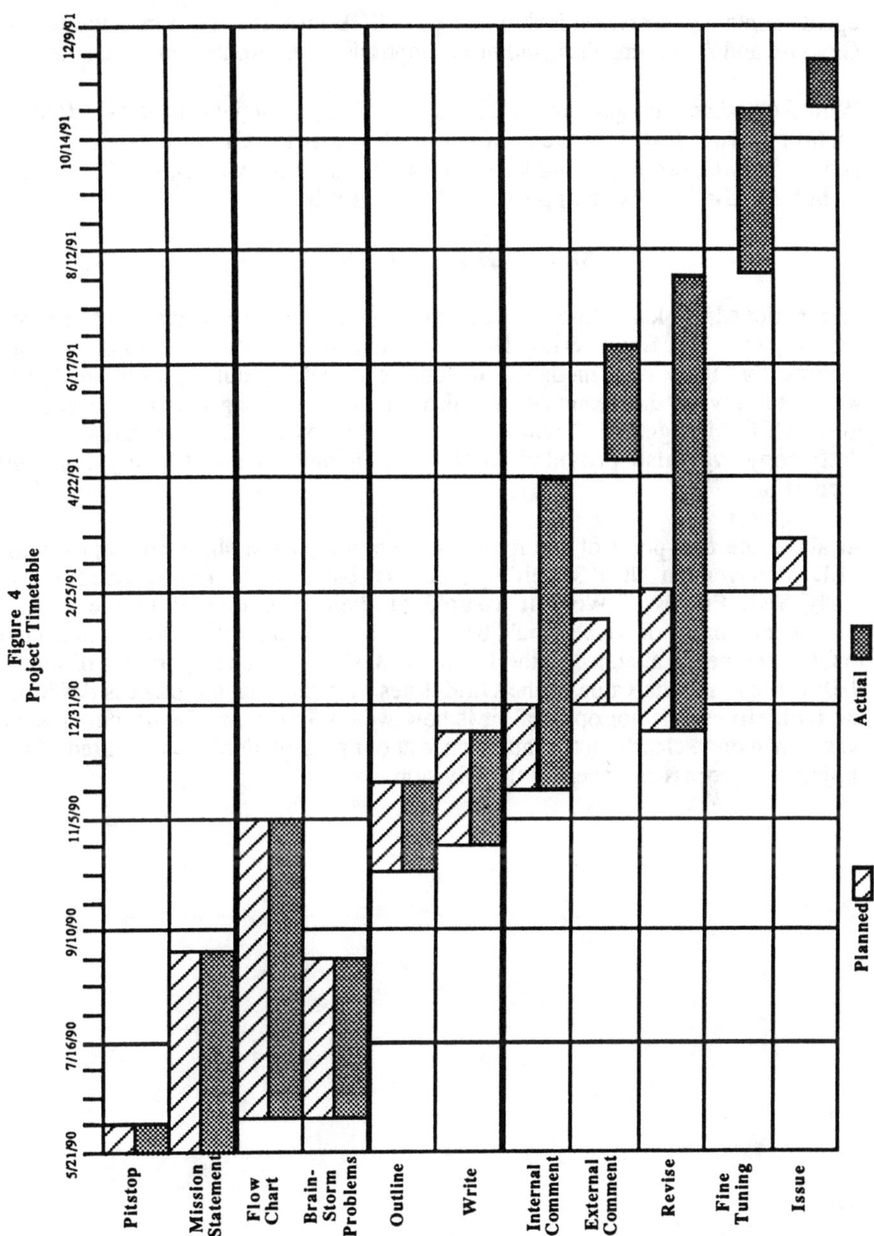

Three months after the release of the document we issued a questionnaire to all CED recipients asking for comments on their use of the **Guidelines** and suggestions for improvement. Based on those replies and other factors, such as the OSHA Process Safety Management regulation, we issued an update to the **Guidelines** about eight months after the original. Along with information on the OSHA PSM standard, the update contained a revised Rohm and Haas SHE resources section, additions to the Glossary and Acronyms lists, and more emphasis on environmental concerns.

With the update we again asked the owners of the document if the **Guidelines** are useful to them. Based on the returns it is clear that the **Guidelines** are being used and are helpful, not just to our CED customers but in the plants as well. We plan to update the **Guidelines** on approximately a yearly basis.

MANAGEMENT SUPPORT

As mentioned, we knew that management support would be crucial to the success of the project. The Engineering Division management was kept informed of our progress through several methods. In addition to sending out copies to the 24 people selected to review the document, we also sent a copy (along with the questionnaire) to the CED Management Team. We received many helpful comments. The final draft copy was also provided to the Management Team for any last minute suggestions.

At about the mid-point of the revision stage, we gave a short presentation to the CED Management NQT soliciting their assistance in the implementation of the Early SHE Process. We felt assured of their commitment to the follow-up implementation work which would be required. The Director of CED issued a memo to CED engineers, attached to the **Guidelines**, stressing the importance of the Early SHE Process and the value of the **Guidelines**. He said that the process described in the **Guidelines** was not optional, it is how we do SHE now. Management support was shown quite clearly in that attendance at our presentations was required of all the project and process engineers in the Division.

Team Make-Up—An Essential Element of a Successful Process Hazard Analysis

W. L. Frank

DuPont Engineering, P.O. Box 6090, Newark, DE 19714-6090

J. E. Giffin

Union Carbide Corporation, Safety Engineering, P.O. Box 8361, South Charleston, VW 25303

D. C. Hendershot

Rohm & Haas Company, Corporate Engineering Division, P.O. Box 584, Bristol, PA 19007

ABSTRACT

The use of a team approach is generally recognized to enhance the productivity and improve the resultant quality of process hazard analysis. Indeed, many experienced hazard analysts regard team member selection to be at least as important as technique selection in the success of the analysis. In the second edition of "Guidelines for Hazard Evaluation Procedures," CCPS notes that "...most high quality hazard evaluation studies require the combined efforts of a multidisciplinary team," and provides guidance on how such a team should be constituted. The first part of this paper elaborates upon these concepts, discussing the importance of a varied mix of educational background, work experience, area of expertise, etc., among the various team members.

However, in addition to the "technical" aspects of team selection, there is a second aspect that is less well recognized (or, at least, less frequently addressed). This is the recognition that all team members bring a unique set of personality traits, behavioral patterns, and biases along with them to the process hazard analysis meetings, all of which affect the quality and efficiency of the safety study. This presents a real challenge to the review Leader, who must take this group of individuals and meld them into a cohesive team. The second part of this paper describes some of the more challenging types of individuals that may be encountered in process hazard analysis meetings and suggests ways of responding to them.

INTRODUCTION

The American colloquialism "Two heads are better than one" can be appropriately generalized as "Multiple heads are better than one" when discussing process hazard analysis. It has been widely recognized that a team approach brings many advantages to the process hazard analysis task, and most safety study methodologies have been specifically designed to use such an approach.

The intent behind a team-based approach is to bring to the study an appropriate mix of knowledge, backgrounds, and expertise to ensure that the study is conducted in a comprehensive, yet cost-effective manner. While the size and constitution of the team will vary depending upon a number of factors, e.g., the analysis technique used and the nature and objectives of the study, the underlying team concept remains the same.

The importance of team makeup was discussed in "Guidelines for Hazard Evaluation Procedures, 2nd Edition" ("HEP Guidelines"), a CCPS publication to which the authors of this paper were fortunate to have had the opportunity to contribute (Ref. 1). In the first part of this paper, we will elaborate on some of the concepts discussed in "HEP Guidelines" with regard to team makeup and the roles of the various participants. For example:

- What type of expertise and backgrounds should be represented?
- What are the training requirements for the participants?
- What is the role of the meeting leader?

Like any committee, the process hazard analysis team is a group of individuals that must be melded into a cohesive working unit to ensure optimum effectiveness. Occasionally, this task of team building can be almost as daunting as the task of completing the analysis itself. In the second part of this paper, we describe some of the personalities and behavioral traits that may be encountered in the course of a process hazard analysis and attempt to provide guidance on how the interpersonal aspects of the hazard evaluation task can be addressed.

TEAM MAKEUP

Typically, there are four roles that should be represented on a process hazard analysis team, the:

- Responsible Party,
- Meeting Leader (or Facilitator),
- Scribe (or Secretary), and
- Expert(s).

The Responsible Party is just what the title implies, i.e., the individual who has been chartered by management to ensure that a thorough, accurate analysis of the process is conducted. Often, this will be a member of the line organization of the facility under study. In other circumstances, this might be a senior individual from outside the facility who has been brought in to ensure objectivity or, perhaps, to bring a fresh perspective to the analysis. In any event, this individual will typically provide input as to the makeup of the team, will likely be responsible for coordination (including scheduling of the meetings and, ultimately, issuing of the study report) and will have consulted with management on the scope of the analysis (i.e., what portion of the facility is to be studied and what are the purpose and goals of the study).

As expected, the Leader has the responsibility for the actual leading of the meeting. As such, the Leader must be both knowledgeable and experienced in the application of the technique(s) to be used during the study. Herein lies a potential pitfall that can seriously limit the value of a process hazard analysis; i.e., the lack of an <u>experienced</u> Leader. Recent Federal regulatory initiatives have created a new industry—the training of process hazard analysis meeting Leaders. While such training can be a valuable first step in the development of a Leader, it is the "on-the-job" training, where leaders become <u>practiced</u> in applying the various techniques, that ultimately determines their effectiveness. For this reason, many organizations, if of sufficient size, will designate individuals to train and regularly perform as process hazard analysis Leaders, sometimes full-time and sometimes as an adjunct role to their day-to-day responsibilities.

Finally, but just as importantly, the Leader must be a good "people-person." The reasons for this will be made clear in the second part of this paper.

The discussions involved in a process hazard analysis can be lengthy, detailed, and often somewhat circuitous. To achieve the goals of the analysis, it is imperative that a clear, concise record of the results of the study be documented at its completion. This report must often be distilled from the content of many days of deliberations. While the responsibility for the preparation of the final report might be assigned to any of a number of different parties, its preparation will be made more difficult and its accuracy and utility will likely be diminished unless concise "real-time" records are maintained throughout the course of the analysis.

This is the role of the Scribe. As such, it should be apparent that the Scribe would require good communications skills, both oral and written. For example, the Scribe must be able to focus on the current discussion, distilling from it the key points that need to be recorded. This requires good concentration and the ability to paraphrase and condense. The Scribe must also have the confidence to speak up and seek clarification when required.

A variety of recording media is available for use by the Scribe. If handwritten notes are taken, e.g., on an easel, a printing white board, or a chart pad, the Scribe should be capable of writing rapidly yet legibly. Electronic media (such as specialized programs run on lap-top computers) are sometimes used for recording the minutes of hazard analysis meetings. If such are used, the Scribe should be a capable typist who is computer literate. Some of the issues associated with hazard analysis documentation and the use of hazard analysis computer software are discussed in References 2 and 3.

The bulk of the membership of the team will typically be comprised of the Experts. These are the persons that the Responsible Party counts upon to provide the process knowledge necessary to ensure an adequately detailed, accurate analysis. Generally, the Experts will represent the following functions:

- Operations,
- Technical and R&D,
- Engineering and Maintenance,
- Health, Safety, and Environmental, and
- Specialists (e.g., Analytical Lab or Computer Systems).

It has been suggested that the appropriately sized process hazard analysis team is the "smallest group of people that knows everything about the process" (Ref. 4). Unfortunately, most organizations are graced neither with omniscience nor with unlimited personnel resources. Consequently, the staffing of the team results from careful deliberations intended to provide the best mix of process information within the constraints imposed by the availability of personnel. While there may be some overlap in the contributions made by the various participants, the input expected from the functional groups listed above would be as follows.

Operations

The Operations representatives should be intimately familiar with the process equipment and would be the authorities on operating procedures, past operating incidents, etc. Operations is typically the "owner" of the physical plant and represents a custodial interest in the process hazard analysis proceedings.

Technical and R&D

Those representing the technical side of the organization would be familiar with the bases for operating limits, would be familiar with the process flow sheet and would be in the best position to research the answers to "What If?" questions that extend beyond the experience of the analysis committee.

Engineering and Maintenance

Representatives of the engineering organization should be familiar with the design bases for the process equipment and should be conversant on applicable industry and national consensus standards and codes. Maintenance (both mechanical and instrument/electrical) personnel are normally more familiar with maintenance histories, test and inspection programs, specifications and procedures.

Health, Safety, and Environmental (HSE)

The input provided within this category could be as broad in scope as the name implies. At the minimum, it would include familiarity with site and corporate policies, practices and procedures in the HSE area, as well as any applicable state or federal regulations. Also included under this category could be specialized process hazard analysis support such as consequence analysis.

The majority of the team members would be expected to represent the above organizational groups. Typically, a core group of meeting participants would be formed, representing a cross section of these functions. These team members would be expected to attend <u>all</u> process hazard analysis meetings. In the event of an unavoidable absence, a team member should be represented by a replacement. However, absences should be discouraged, because it is important that team members know what has already been covered as well as what has not been covered.

Specialists

Specialists provide more detailed knowledge of areas of technology relevant to all or part of a hazard analysis. Specialists might participate in a complete analysis, or may be called in to serve as *ad hoc* members of the team on an as-needed basis. Drawing on such resources in this manner allows the most efficient use of the time and talents of valuable human resources, keeps the meeting attendance to a more manageable size, and keeps the interest level high among the Experts that might otherwise become bored and feel that their time is being wasted by attendance in the analysis meetings when their contribution is not required.

Table 1 shows some typical backgrounds and functional responsibilities that would be represented in either the team core group or *ad hoc* members. As alluded to previously, the underlying intent in team formulation is to provide the proper mix of expertise and experience to ensure a thorough, adequate analysis of the safety of the process. The nature of this mix of expertise and experience will be dependent upon the nature of the process and the goal of the analysis. This can be illustrated with a few examples:

- A hazard analysis addressing the addition of a distributed control system (DCS) would suggest the involvement of the site DCS Expert and, perhaps, a DCS vendor representative if this technology is new to the site. If vendors or other "outside" personnel are brought to the team, protection of trade secret or proprietary information must be ensured.

Table 1

Candidates for Hazard Evaluation Team Members and Resources
(Prepared from Table 2.3 of Ref. 1)

Chemical engineer	Mechanical engineer
Chemist/R&D engineer	Medical doctor/nurse
Civil engineer	Metallurgist
Electrical engineer	Operations supervisor
Environmental engineer	Operator/technician
Expert from another plant	Outside consultant
Fire protection engineer	Process engineer
Hazard evaluation expert/leader	Process control programmer
Human factors specialist	Project engineer
Industrial hygienist	Recorder/secretary/scribe
Inspection engineer/technician	Safety engineer
Instrument engineer	Shift first-line supervisor
Interpreter	Toxicologist
Maintenance supervisor	Transportation specialist
Maintenance planner	Vendor representative
Mechanic/pipefitter/electrician	

- A hazard analysis chartered in conjunction with an incident investigation might require the contribution of a Human Factors Expert if human error appears to have played an important role in the incident.
- A preliminary analysis to evaluate the feasibility and hazards of a new reaction process and to ensure maximum application of inherent safety concepts might require the input of the R&D chemist responsible for its development as well as some occasional consultation from the fluid flow/emergency relief consultant who would ultimately be expected to design the relief valve/vent system for the reactor.
- A hazard analysis addressing a process that is entirely new to one plant site would surely profit from the involvement of someone familiar with that process at a "sister" site within the corporation.

The team, as assembled, will typically represent a mixture of both technically degreed and nondegreed personnel. While recent federal regulatory initiatives have made the issue of "employee involvement" a *cause célèbre*, many organizations have long recognized the substantive role that can, and should, be provided through the participation of operators and mechanics in the process hazard analysis meetings. In the final analysis, these employees, and their first-line supervisors, are the most familiar with the way that the process *is actually operated and maintained* at 2:00 AM on a Sunday morning.

Another criterion that should be considered in addition to type of work experience, is the length and quality of work experience. Certainly, the team must have a sound foundation of experienced personnel intimately familiar with the process. It is they who bring the detailed process knowledge necessary to ensure a sufficiently detailed and productive analysis. However, a judicious addition of some relatively less experienced personnel can serve several purposes. First, some of these will become the "experienced" hazard analysis team members (perhaps, even Leaders) of the future, and their participation serves as a form of apprenticeship training. Additionally, and more to the purpose at hand, it is often advantageous to have a relatively uninitiated, unbiased viewpoint represented at the meeting. Relatively new personnel may bring fresh insights into the hazards of the process to which more experienced personnel may have become inured. It is vital that the Leader encourage these people to contribute and ensure that they are not intimidated by the presence of their more experienced colleagues.

One mistake that is often made and should be avoided, is putting the recently hired college graduate in charge of the hazard analysis. In an organization that does not appreciate the value of process hazard analysis, this task can be perceived as "busy work" to occupy someone that "doesn't know enough yet to be of much value in the day-to-day operations."

The size of teams can vary significantly, depending upon factors such as the complexity of the study, the goals of the study, and the available personnel resources, as influenced by the size of the organization. At one extreme, a two-member team may assemble to review the advisability of some equipment modifications. Alternatively, a team of seven or eight may be assembled for a "front-door-to-back-door" review of an entire facility. It is important to keep teams to a manageable size to facilitate the Leader keeping the meetings on track.

It should be noted that we have described the four principal roles as if they were staffed by different individuals. This will not always be true, particularly in the case of small teams where, for example, the Leader may have to assume the role of the Scribe, or the Responsible Party may be one of the most Knowledgeable Experts.

Finally, we would be remiss if we did not cite the rather specific hazard analysis team requirements now contained in OSHA's Process Safety Management regulations:

"The process hazard analysis shall be performed by a team with expertise in engineering and process operations, and the team shall include at least one employee who has experience and knowledge specific to the process being evaluated. Also, one member of the team must be knowledgeable in the specific process hazard analysis methodology being used." (Ref. 5)

The various considerations discussed in this section of the paper define a hazard analysis team that would meet all of the requirements of the OSHA rule.

THE "PERSONALITIES"

It must be remembered that any team is, at least initially, just a group of individuals. One of the challenges (and this can often be a significant challenge) posed for the Leader is that of taking these individuals and melding them into a smoothly functioning unit.

This goal should not be misconstrued to mean that all sources of dissent must be ferreted out and removed from the proceedings. Quite the contrary; well-intentioned differences of opinion are essential to the critical review required during a process hazard analysis. What we do wish to achieve, however, is a forum in which all participants feel comfortable to, and actually do, contribute freely and on an equal basis to the deliberations.

It is up to the Leader to foster such an environment. As such, the leader must be capable of recognizing and responding to the various personalities, behavioral traits, and biases that the individuals bring to the meetings. The balance of this section contains some "tongue-in-cheek" characterizations of some of the more notable types of personalities who may be encountered on a hazard analysis team and suggests some ways in which the Leader can respond to ensure that these individuals contribute positively to the team effort.

"The Dictator"

The Dictator is a higher level manager responsible for the facility, for whom a number of the participants in the hazard analysis work. You can immediately tell that all of the subordinates are afraid to say anything until they know what the Dictator thinks, so they won't find themselves on the wrong side of an issue. The hazard analysis deteriorates into a one or two participant meeting, with a large number of people in the meeting unwilling to make any original contributions.

Suggestions. The Dictator is best dealt with before the hazard analysis starts, if the potential problem is recognized. It may be possible to convince the Dictator to spend valuable time on other projects and leave the hazard analysis to subordinates, who will then be more willing to voice their own opinions. If the Dictator's participation is considered essential to a good analysis because of a unique knowledge and understanding of the process, consider bringing in some outsiders who are knowledgeable in the plant and/or process—people who don't work for the Dictator and would not be afraid to speak their mind. The Leader can also attempt to allow others to express an opinion by establishing a specific order for contributions. Also, the Dictator may not even realize that the subordinates feel too inhibited to participate openly. Discuss the problem—the Dictator may be more than willing to try to reassure the others that their own contributions are essential to a good analysis, and that they can disagree without fear of reprisal.

"Detail Lovers (DLs)"

DLs want to do the hazard analysis to a much greater level of resolution than the rest of the team. They can spend hours reviewing a single feedback control loop, so that it would take years to review a complete process. This slows down the entire analysis, and also results in the rest of the team losing interest in the hazard analysis, getting into side conversations about other subjects or going to sleep.

Suggestions. DLs can actually be useful in an analysis if kept under control. <u>Details can be important.</u> Abraham Lincoln once observed that one cannot equate a horse chestnut with a chestnut horse (Ref. 6). But DLs cannot be allowed to drown the entire review in a morass of detail and turn off the entire team. It is important to agree to ground rules at the beginning of the meeting as to the scope and level of resolution of the analysis. Agree to record concerns that will require a great deal of time to resolve fully and arrange to have them reviewed in greater detail outside the meeting, and make sure that the DLs are involved in these meetings. It is important for the Leader to maintain control, realize when further discussions are not productive, and stop the discussion by recording any concerns or questions that will need further investigation off-line.

"Run It Right and It's OK (No Problem or NP)"

NP is often a process design engineer who knows that the process is safe if properly operated, so all you have to do is hire people who will always run it the right way. If asked "What would be the consequence of leaving manual valve 123-456 open during start-up?", NP's response is "Why would anybody ever do that? It says right in the instructions to make sure the valve is closed, and everybody knows that it would be dangerous to leave it open."

Suggestions. NP needs a better understanding of human error. If we all did everything the right way every time without fail, we would never need any kind of safety devices, alarms, interlocks, etc. NP has undoubtedly included a number of these safety features in the design—if everybody would do everything correctly every time, why are they there? It is up to the Leader and the rest of the team to insist that human error be addressed and followed up, and to remind NP of the many examples of things that people have done that don't make sense. The penalty for making a mistake must not be a serious injury, because people *will* eventually make mistakes.

"Monday Morning Quarterback (QB)"

QB is not involved in the hazard analysis, but reads the report in great detail after the meeting looking for things that were left out. Of course, there will always be something to find because no hazard analysis is ever complete or perfect, and the hazards identified are dependent on the background and experience of the participants in the analysis. If it is a good analysis with a good team with an appropriate level of knowledge of the process and dedication to the hazard evaluation process, QB's additional findings will probably be minor. If the team is missing some important area of knowledge, QB may identify important hazards that were not identified.

Suggestions. QB may be annoying, but actually should be regarded as useful. In effect, QB is a quality-control inspector for the hazard analysis. All team members must always remember that the purpose of a hazard analysis is to make the plant safer and identify and evaluate all hazards, regardless of who identifies them. If QB identifies something that the original team missed, the team should be contacted and the findings added to the report. It may even be necessary to reconvene the team to discuss the new issues. If QB identifies an important issue that the team missed, this may be an indication that an important area of expertise was missing from the original team. QB or somebody with similar knowledge and background should have been a part of the team! It is essential in forming teams to make sure that all relevant expertise and experience is represented. If the QBs are repeatedly finding issues that have been missed in hazard studies, the team selection process needs to be reviewed and improved.

"Tom and Jerry"

Tom and Jerry just don't get along; they argue with each other automatically. If Tom says it's red, Jerry says it's green, perhaps without even thinking about it. If Tom had said it is green, Jerry would have been just as happy to say it is red. These two can totally disrupt any meeting, causing it to degenerate into arguments and shouting matches. They can be particularly damaging to the team effort if they differ on a design or operations issue and seek to steer the results of the hazard analysis to support their own points of view. Other team members may either take sides and get into the fight, or just drop out, sit back, and watch the action.

Suggestions. Avoid putting people with serious incompatibilities on the same team, if at all possible. It is usually possible to identify more than one person with the required knowledge and expertise, and eliminate serious personal conflicts. It is up to the person organizing the team to be aware of any such potential problems and select a team that avoids them. If Tom and Jerry wind up on the same team, the Leader must be strong enough to maintain control of the meeting and handle disagreements and arguments. Meeting ground rules must be established, and the Leader must remind participants of them when necessary.

"The Neophyte (Neo)"

Neo is a young engineer or scientist, right out of school, who is assigned to a hazard analysis because the responsible manager has more important jobs for the experienced engineers. Neo may be a brilliant engineer, eager and interested in the hazard evaluation, but just does not have the experience and background to be an effective contributor to the hazard evaluation process. If Neo is supposed to represent an important aspect of the facility operation, the team will not be adequately represented in that area, and important hazards may be missed.

Suggestions. It is essential that a hazard study team include a high level of experience and expertise in all relevant areas. If a commitment is made to do a hazard evaluation, that commitment must include an agreement to provide the appropriate level of personnel resources to do the job right. A hazard study represents a significant allocation of time and effort. If the team membership is not appropriate, the product will be inadequate, and the analysis may have to be repeated. It is more efficient to devote appropriate resources to the analysis to begin with and do it right the first time. If it becomes apparent during the analysis that someone's experience or background is inadequate, the Leader should be willing to ask management to add another member to the team to fill in the gap. Neos are often aware of their lack of adequate background and would welcome help. A hazard analysis can be an excellent training tool for new personnel; they will examine the process in great detail and come away with a much better understanding of the plant and process. However, such a team member should be considered to be a secondary contributor and cannot replace an experienced person in a key role. Often a Neo can be very effective in the role of recorder or secretary, removing the documentation responsibility from other team members and allowing them to concentrate completely on the hazard evaluation study.

"Beeper"

Beeper is "essential to the operation of the plant" and cannot be out of touch for even a few minutes, much less for a series of several hour-long hazard evaluation meetings. Beeper is frequently paged, and disrupts the meeting by making long phone calls from the meeting room, leaving the room to make phone calls, or visiting the plant to preside over the latest crisis. Beeper is often the last to arrive for the start of a meeting, extends lunch breaks, seeks early adjournment, and is not available to provide input when needed. Sometimes Beeper will send a substitute to the meeting because "something came up." Of course, the substitute has not participated in the analysis to date, and walks in cold without knowing what is going on or what has been discussed previously.

Suggestions. Before the start of the hazard analysis, all participants must make a commitment that the analysis meetings will be their top priority during the meetings. Schedules should be worked out ahead of time to allow key personnel to participate fully without interruption. Meetings could be scheduled for part days to allow participants to have the remainder of the day free for other necessary activities. Coverage by others to relieve key personnel of their normal responsibilities should be set up well ahead of time. Consider the hazard evaluation participants to be "on vacation," and make the same arrangements you would make for those circumstances. In many cases it is best to arrange to have the hazard evaluation meetings off-site to reduce interruptions. The off-site location should be near enough to the plant so that it is still easy to visit the plant for tours, to confirm information about the facility, to obtain information from plant records, and to have suitable access to process experts who are not team members but may be needed for consultation on certain aspects of the process operations.

"The Judge"

The Judge has a very high self-regard when it comes to opinions and has relatively less respect for those of others. The standard reply is "*I'm* not convinced yet," as if the Judge is the final arbiter for the team. The impact on the team may be somewhat similar to that of the Dictator even though the Judge may not have the same level of managerial authority.

Suggestions. Some of the approaches used for the Dictator may also work for the Judge. The task becomes somewhat easier if the Judge lacks the high-level managerial authority that the Dictator has, in which case there is no need for other team members to fear reprisals for disagreeing. Emphasize early in the meetings the equality of each participant's opinions and that the final report of the team must represent consensus. Remember that a strong Leader can "take the floor" and steer the nature of the discussion away from dominant participants and towards less open participants, thus allowing a mixture of opinions to be heard.

"The Old Codger (OC)"

This individual has been with the facility since Day One (or before) and knows "every detail" of its equipment and operation. This is the individual that everyone comes to on a daily basis for any arcane bit of process information, so it is natural that most will defer to OC's opinion within the process hazard analysis meeting. As far as most are concerned, the day OC retires is the day the company will have to sell the business.

Suggestions. By all means, tap into and take full advantage of the years of experience and breadth of knowledge that such senior employees bring to the meetings. At the same time, it is important to emphasize that each hazard analysis is a fresh opportunity to reevaluate the process for new hazards or for old hazards that may have been tolerated previously but can no longer be accepted in the current regulatory and social climate. In other words, not all of the old answers may still apply. Here, again, is where the fresh insight of relatively recently hired employees may be of particular value. In the final analysis, many senior employees, such as OC, are quite willing to accept a mentoring role in helping younger personnel "find" the right answers rather than providing them *a priori*.

"The Wallflower"

Wallflowers are just too timid to open up and offer an opinion. However, if they have important contributions to make, they must not be left off the team. Their reticence may just be a manifestation of personality, or they may be intimidated by longer service or more senior team members, as discussed previously. They may be afraid of embarrassing themselves by suggesting something that they fear would sound ludicrous to the rest of the team.

Suggestions. Process hazard analysis meetings are, regardless of the technique used, "brainstorming" sessions, and one of the basic rules of brainstorming is that "There are no dumb ideas." The Leader, when laying out ground rules at the start of the study, should emphasize to all that any concern should be brought forth for discussion. The skilled Leader should be able to deal with trivialities (and there won't be that many) graciously so as not to embarrass or "turn off" the suggester. The Leader can also draw out the truly shy individuals by directly asking for their opinions when the discussion touches on their area of expertise. Eventually, most Wallflowers will gain sufficient self-assurance to contribute more freely.

"The Defensive Design Engineer (DDE)"

DDE has been working evenings and weekends for months to ensure that the project comes in on schedule and under budget. Now, an already too-full schedule is made even busier because DDE is forced to sit in meetings with a "bunch of non-contributors who still recognize their spouses and children and are nit-picking the design and running up the cost with their silly criticisms."

Suggestion. This situation represents one of the true tests of a Leader's skill. Tempers can get short when people are under pressure and feeling under-appreciated. Extreme levels of tact are called for in dealing with DDE. Emphasize that the process hazard analysis is meant to be a constructive exercise to maximize the success and safety of the project. Steer the discussions away from any judgmental tones. Spend some time acknowledging the success stories inherent in the project effort thus far, but without detracting from the emphasis on the importance of a comprehensive, rigorous review.

"Ivory Tower"

Tower is a brilliant but rebellious Research and Development scientist who thinks safety studies should be staffed by persons of lesser importance. Tower tries to convince the rest of the team that they are wasting their time and often falls asleep during the study.

Suggestion. If at all possible, find a replacement and make sure Tower doesn't get assigned to any future safety study teams (after all, that's what Tower wanted, wasn't it?). If a replacement cannot be made, keep Tower awake during periods when R&D input is needed.

"Don't-Bother-Me (DBM)"

DBM is careful not to participate in the study except when an R&D question comes up, and then answers with a complicated response that leaves the team wondering what the answer really is. DBM spends a lot of time reading the Wall Street Journal during the meetings.

Suggestion. During the pre-study team-building, it's important to stress that all members of the team are encouraged to participate fully in the study, not just in their areas of expertise, and that breaks will be set aside for reading newspapers, making telephone calls, etc. An effective Leader will keep all members of the team challenged to participate at all times during work sessions.

"Talkathon"

Talkathons love to talk more than anything, even if only to hear the sound of their own voices. Nothing is too inconsequential to argue about. In spite of the Leader's repeated admonitions, this incessant talking results in safety studies extending to the wee hours of the morning, the extended sessions becoming necessary to complete the task in time to meet other schedule constraints such as pre-arranged return flights from foreign countries.

Suggestion. The Leader needs to ensure that all team members understand "up front" that adherence to the agreed-upon schedule depends upon everybody sticking to the subject, and that excessive rabbit-chasing, war stories, or filibusters will be done "indirectly" on unpaid overtime after normal working hours. Additionally, a skilled Leader can steer the discussion away from unduly loquacious participants and towards other team members by repeatedly using "cut-off" techniques such as "Yes, I hear what you're saying, but let's see what opinions others have on this matter..." Sooner or later, most people will start to get the message. Peer pressure also helps, especially after midnight.

"Sotto Voce (SV)"

You can't hear what SVs say most of the time, because they speak at such a low volume. But after straining to hear a few times, you learn to quiet the rest of the team in order to hear them. Value of contribution does not correlate directly with volume of expression.

Suggestion. Accommodate soft-spoken and hard-of-hearing people, as well as those with vision problems, by requesting seating to minimize difficulties. For example, tactfully arrange for those who are soft-spoken to sit where their voices will project better, e.g., at the end of a rectangular table, and for those with hearing problems to sit where they can be near the center of the group yet see most of the team members (people with hearing problems usually read lips, either consciously or unconsciously). Sit close to SVs and repeat, paraphrase, or play back their gems so the rest of the team can benefit.

"The Know-It-All"

Know-It-Alls are simultaneously the most intelligent and obnoxious persons you ever met. There is no subject on which they are not instant experts, including safety study methodologies never encountered before. Just ask them how to do it, because they divine their knowledge intuitively. The infuriating truth is that they are often right. But then, the Know-It-Alls are always right by definition, irregardless. They'll even inform you that irregardless is not a word. Their biggest problem is that they alienate the entire team.

Suggestion. A Leader needs the highest degree of skill, patience, tact, and forbearance to deal with the Know-It-All. But it can be done. Show them the respect their intelligence and knowledge deserve; paraphrase their less-than-tactful statements with more tactful language; resist the temptation to shut them up, put them down, or tell them off. Call on your best interpersonal skills to deal with the Know-It-Alls, and most of your team members will take your cue and look for ways to facilitate their contributions rather than figure out ways to give them their comeuppance.

"The Turn-around (Turn)"

Because of a past reputation as an agitator and consensus splitter, Turn was initially avoided as a potential team member. Turn had arguments and difficulty getting along with almost everyone in the organization. But because of a very narrow and specialized field of expertise, no suitable replacement could be found. However, Turn had recently retired from the company and came back as a "contractor." Imagine the team's surprise to discover what a mellowing influence the "retired status" had upon Turn. The entire team had difficulty believing the cooperation and actual consensus-building Turn facilitated. Finally, someone had found a way to tap this genius—retire, then rehire.

Moral. Don't always judge team members' potential contributions based totally on their past performances. There is room for improvement, especially if their motivation has changed, as Turn-around's did.

"The Whiz"

Whiz's one year as a Process Engineer didn't come even close to the 5-year experience requirement the Leader had stipulated for team membership. However, the unit manager was confident that Whiz was very accomplished and a quick study. As it turned out, Whiz proved to be the best contributor on the team and knew the operating unit better than the Production Engineer knew it. The new Plant Manager was incredulous that the team had found so many Major Concerns with the unit, about five times the average. Whiz made the whole team look good.

Moral. Don't get stuck on your guidelines; remember they are guidelines, not immutable commandments. All safety studies should have detailed and specific guidelines on how to execute the studies, but there should be a door for good judgment to enter through.

"The Addict"

The study occurred in a foreign country that is one of the cigarette-smoking capitals of the world, and The Addict was an inveterate smoker, determined to smoke in the safety-study conference room, poorly ventilated though it was. The Leader was equally determined no one was going to smoke in there. The Leader won the battle, but lost the war. The Addict later convinced the Director of the Eastern Region to put on the Leader's evaluation that the Leader needed to cultivate a "sense of humor."

Moral. The Leader's inability to handle all situations with supreme tact and finesse will eventually follow the adage "What goes around, comes around."

"The Bulldozer"

Although the Leader had heard rumors about this takeover artist, in short time the Bulldozer was, in practice, the actual Leader, usurping all the decisions on scope, protocol, and technical content. Everything the rest of the team said was disputed. Bulldozer was always right, snickering every time the Leader or anyone made a mistake. This sort typically serves on a team just once and is never invited back.

Suggestion. Find other, more challenging work for the Bulldozer. The importance of team-building is just too great to tolerate a disruptive influence such as this.

"The Imbiber"

The study took place in a country where the culture dictated that alcoholic beverages would be served at lunch. After the fourth round, the worried Leader started hinting broadly that it was time to resume the safety study. But the Imbiber could no longer follow the fault tree branches after lunch. Since several on the team were having trouble, though not asleep on the table like the Imbiber, the team adjourned until the following day. After all, what's another day in a foreign capital?

Answer. About $1,000 U.S. (An expensive lesson that justifies a no-alcohol-during-meeting-time rule!)

"Romeo and Juliet"

Two chemical engineers in Production in the same unit on the same safety study team—why were they getting along so well? Why did he always agree so completely with her, and she with him? The answers soon became obvious. They were married before the final report was issued.

Moral. Never let safety studies get in the way of true love!

REFERENCES

1. <u>Guidelines for Hazard Evaluation Procedures, Second Edition with Worked Examples</u>, American Institute of Chemical Engineers, 1992.

2. Hendershot, D.C., "Documentation and Utilization of the Results of Hazard Evaluation Studies," <u>Plant/Operations Progress</u>, Vol. 11, No. 4, pp. 256-263, October 1992.

3. Freeman, R. A., "Documentation of Hazard and Operability Studies," <u>Plant/Operations Progress</u>, Vol. 10, No. 3, pp. 155-158, July 1991.

4. Mr. Thomas H. Pratt, Personal Communication.

5. 29 CFR Part 1910.119, "Process Safety Management of Highly Hazardous Chemicals; Explosives and Blasting Agents; Final Rule," <u>Federal Register</u>, February 24, 1992.

6. Wills, Garry, <u>Lincoln at Gettysburg</u>, Simon and Schuster, New York, 1992, p. 163.

Additional Reading

1. Arendt, J.S., Frank, W.L., Giffin, J.E., "A Survey of Hazard Evaluation Techniques (with Guidance on the Application)," Chemical Process Safety Report, Tab 400, Thompson Publishing Co., March 1992.

A Practical Approach for Integrating Process Safety and Total Quality Management Programs

J. J. Rooney
Process Safety Institute

M. F. Smith and J. S. Arendt
JBF Associates

Most experts believe that there is an inextricable link between process safety management (PSM) systems and total quality management (TQM) systems. Both represent management systems that should contain the essential features of planning, organizing, implementing, and controlling. However, many PSM systems fall short when compared to analogous TQM systems. This paper will discuss the relationships between advanced PSM and TQM programs.

First, the essential features of several PSM systems (i.e., management of change, training, and process risk management) will be compared to the structures and organization imbedded within several TQM-related systems (i.e., the Malcolm Baldridge National Quality Award criteria, the International Organization for Standards [ISO] 9000 series standards, and the American Society of Mechanical Engineers' Management Control System [ASME MCS] standard). Traditional links to these TQM structures will be discussed as well as notable shortcomings in existing PSM programs (e.g., the issue of PSM "customer satisfaction"). Examples from actual company experience will be used to illustrate these similarities and differences.

Second, one of the most fundamental precepts of TQM is that what is measured can be controlled. TQM control systems currently in use by industry will be described. This paper will present the concept of the PSM control system as a way of linking PSM to TQM. Examples of PSM feedback loops and continuous improvement actions will be presented. This paper will lay out a blueprint for companies to use when integrating total quality concepts with PSM systems, resulting in a more cost-effective way of managing process safety in industrial facilities.

PROCESS SAFETY MANAGEMENT

A review of worldwide chemical and petroleum industry performance (i.e., losses) between 1957 and 1986 suggests the need for improved approaches to the handling of hazardous materials. A majority of the 100 largest property losses of these industries (on an adjusted, constant-dollar basis) occurred during the last 10 years. Today, the public, customers, in-plant personnel, and government regulatory agencies all demand that companies take necessary actions to reduce the possibility of episodic hazardous material incidents. In the U.S., two government agencies (i.e., Occupational Safety and Health Administration [OSHA] and Environmental Protection Agency [EPA]) have attempted to help reverse this trend toward increasing numbers of larger losses. In 1992, OSHA enacted a regulation entitled *Process Safety Management of Highly Hazardous Chemicals* (29 CFR 1910.119); this year the EPA, through the Clean Air Act, will enact a risk management program regulation that reaches well beyond OSHA's PSM regulation in the areas of offsite consequence assessment, listing of equipment and controls to prevent loss of containment, and establishment of a formal management system for risk decision-making. The chemical process industries also have developed more sophisticated ways to improve process safety by introducing safety management systems to (1) augment process safety engineering activities and (2) reduce the risk of chemical accidents. This paper will outline and compare two PSM systems that the chemical industry organizations/societies (American Institute of Chemical Engineers' Center for Chemical Process Safety [AIChE CCPS] and Chemical Manufacturers Association [CMA] Responsible Care Process Safety Code) developed with the two federal regulations and with four example TQM systems.

Process safety is the operation of facilities that handle, use, process, or store hazardous materials in a manner free from episodic or catastrophic incidents. However, the handling, use, processing, and storage of materials with inherent hazardous properties can never be performed in the total absence of risk. In other words, process safety is an ideal condition toward which one strives and also a dynamic condition involving the technology, materials, people, and equipment that constitute a facility. PSM ensures that a properly designed facility is kept that way, and is operated according to the safe manner intended.

PSM is the application of management principles and systems to the identification, understanding, and control of process hazards to prevent process-related injuries and incidents.[1,2] PSM systems — which use the four essential features of management systems: planning, organizing, implementing, and controlling — are comprehensive sets of policies, procedures, and practices designed to ensure that barriers to major incidents are in place, in use, and effective. These systems serve to integrate process safety concepts into the ongoing activities of everyone involved in operations — from the process operator to the chief executive officer.

PSM has a different focus than personnel safety management activities that have been in place in industry for years. Process safety excludes: (1) personnel safety issues (e.g., slips, trips, and falls) or security issues (e.g., theft) that may occur at process operations, and (2) chronic releases to the environment (although those issues are generally present whenever process safety is a concern). Although PSM does not explicitly address traditional personnel safety or environmental pollution controls, these two areas overlap the area of process safety. PSM concepts are applicable to all industries where hazardous materials are handled, used, processed, or stored.

Prevention of process accidents requires the implementation of effective, comprehensive PSM systems. Because of the infinite variety of plant-specific conditions that exist, effective PSM systems can, and do, vary a great deal in how they are implemented. However, PSM systems all address the need for managing the process safety-related aspects of technology, facilities, personnel, hazardous materials, and emergency responses.

The structure and titles used within a company's PSM system are unimportant, so long as the appropriate content is present. Clearly, there is no single "correct" PSM system applicable universally; the creativity of managers should be encouraged (rather than limited) in developing management systems particularly suitable for their own organizations and operations. Each company should adopt a management organization and structure that are appropriate for its culture, its mission, and its business; the PSM system should then be tailored to that organization and structure. While a comprehensive PSM system will cover each of the elements described in the four example PSM systems, the management system for each facility of the organization must be tailored to that specific facility.

While a comprehensive, integrated system is the objective, it need not be reached in one step. PSM systems are usually implemented in stages. New PSM systems should be linked to and integrated with the systems already in place. For example, operating management should be involved in planning, organizing, implementing, and controlling the PSM system to incorporate the structure and culture of operating management into the PSM system. PSM should be an integrated activity, with appropriate linkages among the various PSM elements and their components, rather than a series of discrete activities. Through this "holistic" approach, process safety can be effectively managed. Each step toward a comprehensive, integrated PSM system contributes to enhanced safety performance. Strategic links to successful TQM systems within a company can increase the performance and effectiveness of PSM programs. Moreover, applying lessons learned in "growing" an effective TQM system can pay great dividends when looking for ways to continually improve PSM programs.

TOTAL QUALITY MANAGEMENT

Total Quality Management (TQM) is a term initially coined in 1985 by the Naval Air Systems Command to describe its Japanese-style (i.e., participative) management approach to quality improvement. Since then, TQM has taken on many meanings. Simply put, TQM, a management approach to long-term success through customer satisfaction, is based on the participation of all members of an organization in *improving* processes, products, services, and the culture in which they work. The methods for implementing this approach are found in the teachings of such quality leaders as Philip B. Crosby, W. Edwards Deming, Armand V. Feigenbaum, Kaoru Ishikawa, and J. M. Juran.

The roots of TQM can be traced to the work of Walter A. Shewhart, who is known as the father of statistical quality control. He brought together the disciplines of statistics, engineering, and economics in his book, *Economic Control of Quality of Manufactured Product*.[3] His approach to quality improvement is the Shewhart Cycle, better known as the Plan-Do-Check-Act (PDCA) cycle. PDCA is a four-step process (Figure 1). In the first step (Plan), a plan to effect improvement is developed. In the second step (Do), the plan is carried out, preferably on a small scale. In the third step (Check), the effects of the plan are observed. In the last step (Act), the results are studied to determine what was learned and what can be predicted. The cycle returns to the plan step for the next quality improvement project. PDCA focuses on a single process rather than an organization which is a system of processes, and it is the basis of all subsequent quality improvement processes. The PDCA cycle can be used by companies starting to implement their PSM programs.

In the 1990s, four schools of thought on quality improvement dominate the current literature in the United States:

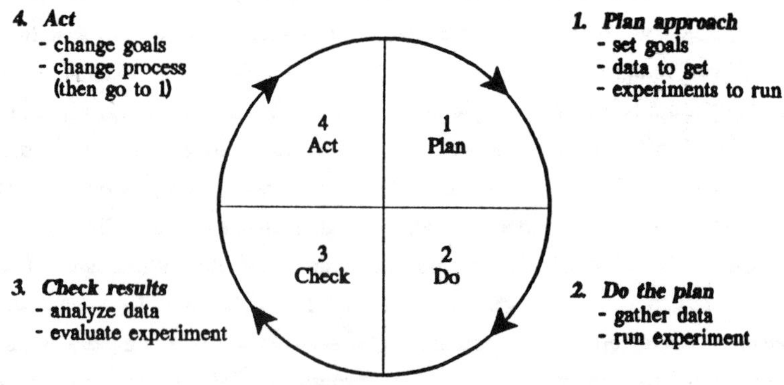

Figure 1 Shewhart Cycle

- TQM,
- The ISO 9000 Series Standards,[4-8]
- The Malcolm Baldridge National Quality Award (MBNQA), and
- ASME MCS Standard.[9]

As noted previously, TQM is a management approach to long-term success through customer satisfaction. The ISO 9000 is a set of five individual but related international series standards on quality management and quality assurance. These standards help companies effectively document the quality system elements to be implemented to maintain an efficient quality system. Congress established MBNQA in 1987 to raise awareness of quality management and to recognize U.S. companies that have implemented successful quality management systems. Well-structured organizations can receive ASME accreditation by implementing a system of management control conforming to the ASME MCS standard. These four quality improvement approaches/standards can be used by companies implementing their PSM systems.

TQM, a set of guiding principles that represent the foundation of a continuously improving organization, is the application of quantitative methods and human resources to improve the materials and services supplied to an organization, all the processes within the organization, and the degree to which the needs of the customer are met, now and in the future. The principles of TQM are:

- *Customer satisfaction* — Since customer satisfaction is paramount in TQM, identifying and measuring customer satisfaction is crucial to TQM success.

- *Management leadership* — In implementing TQM, senior management must lead the way. This responsibility cannot be delegated, or TQM will not reach fruition.

- *Process improvement* — The focus is on the process rather than the product. This means defect prevention rather than inspection of the final product.

- *Education and training* — An aggressive training program is necessary to deploy management's vision and create a quality culture. Training in job skills and TQM tools is also required.

- *Statistical methods* — Statistical methods allow measurement of the process. Measurement is necessary to determine where a company is in the improvement process.

- *Teams* — Employee participation in a team format eliminates sectionalism in a company and promotes cooperation in achieving the corporate vision.

- *Continuous improvement* — A process is never perfect. Continuous improvement is necessary as competitors improve their processes and customers' expectations increase.

Many of these concepts are emerging in the implementation of PSM in industry. For example, it is widely recognized that management leadership is the cornerstone for effective PSM. Moreover, employee participation and the use of teams has empowered workers to look for ways of preventing process accidents. Finally, workers are receiving extensive education and training to perform their assigned PSM-related tasks.

On the other hand, industry is lagging behind in looking for ways to apply the remaining TQM principles. For example, mention the need for a PSM coordinator to satisfy their "customer's" needs for PSM services and products, and you'll likely receive a blank stare. Probably the most thought provoking and promising area for extrapolating TQM principles to PSM systems is in the area of process improvement and statistical methods.

The measurement of PSM performance on which to base a continuous improvement ethic is an essential gap that needs to be filled before PSM systems can approach the level of success achieved by good TQM programs.

Interest in the five ISO 9000 standards and the technically equivalent U.S. standards, the ANSI/ASQC Q90 series, is at a fever pitch. ISO 9000 (Q90) explains the philosophy behind the other four 9000-series standards, giving an organization a road map for applying the standards. ISO 9001 (Q91) and ISO 9002 (Q92), which are contractual standards to be used in customer-supplier relationships, are very similar. Companies that both design and manufacture a product use ISO 9001; companies that only manufacture a product or service use ISO 9002. A company that is 'ISO registered' is usually registered to ISO 9001 or ISO 9002, which does not mean that the product or service is registered, but rather the processes used to produce the product or services are registered. ISO 9003 (Q93) covers final inspection and testing, and the final standard, ISO 9004 (Q94), is a beginner's textbook in total quality. ISO 9004 presents the philosophy behind the quality program elements and provides guidance on how to implement a quality system within a company.

The ISO 9000 series represents agreed-upon methods which ensure that customers get what they ordered. The standards do not represent a TQM system, but they can form the basis on which to build such a system. The systems of auditing, procedures, and documentation requirements of ISO 9000 (summarized in Table 1) can be directly used by companies starting to develop their PSM systems. Compliance with ISO 9000 ensures that management systems are in place and that they are continually monitored.

The MBNQA criteria are built on the framework shown in Figure 2. The framework has four basic elements:

- *Driver* — Senior executive leadership creates the values, goals, and systems and then guides the sustained pursuit of quality and performance objectives.

- *System* — The system comprises the set of well-defined and well-designed processes for meeting the company's quality and performance requirements.

- *Measures of progress* — Measurements provide a results-oriented basis for channeling actions to deliver ever-improving customer value and company performance.

- *Goal* — The basic aim of the quality process is the delivery of ever-improving value to the customer.

The framework sends a clear message: only through personal visible leadership can world-class customer satisfaction be achieved. The criteria provide a basis for building a world-class company. Organizations can use the criteria for self-assessment, and an organization that performs a sincere self-assessment cannot help but know better, at all levels, why it is in business and how each part of the organization contributes to staying in business. The organization will know where it needs to improve its processes but not how to improve its processes. The MBNQA framework can be used by a company in managing the PSM process.

In 1990, ASME issued a Management Control System (MCS) standard that sets forth guidelines for implementing a system of management control. The ASME standard addresses every phase and activity of an organization, including marketing, project planning, data collection, design engineering, product development, procurement, and

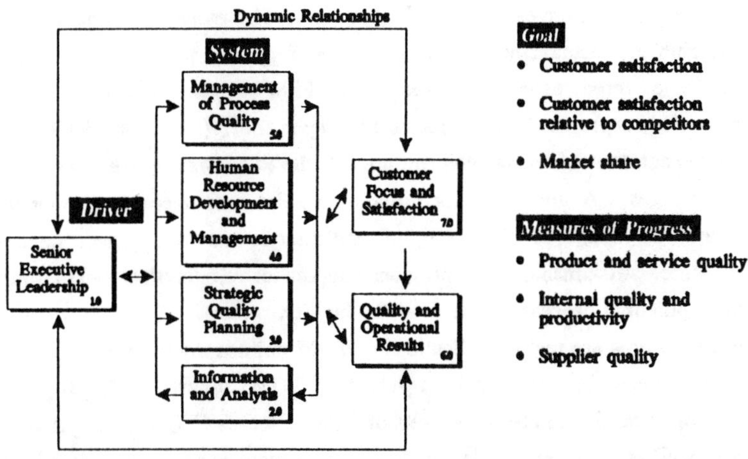

Figure 2 Malcolm Baldridge National Quality Award Criteria Framework

sales and distribution. ASME will accredit well-structured organizations which meet ASME-defined criteria in their adherence to quality, cost, and production/performance objectives. To receive accreditation, a company must have all activities that affect these variables under adequate and continuous control. According to the ASME MCS standard, a management control system is a network of planned and systematic actions necessary to provide confidence that a product is produced or a service is performed in accordance with specified requirements. The basis of management control is a company's commitment to the effective utilization of its resources, as well as to customer satisfaction through TQM. *How many product specifications exist for PSM-related products?*

The ASME MCS accreditation program serves both the customer and the supplier. MCS accreditation provides the customer with a reliable means to evaluate suppliers based on their compliance to rigid requirements for management control set by an outside organization. An accredited supplier can use ASME's certificate of accreditation as an effective customer relations tool: to communicate that it has successfully established and executed management control in all phases of operation, and as a result, has the capability to produce quality products or perform quality services. The ASME MCS standard can be used as a guideline in establishing control of PSM operations.

COMPARISON AMONG TQM AND PSM SYSTEMS

Table 1 presents a comparison of elements among four example PSM systems and four example TQM systems. A comprehensive PSM system will cover all of the elements in the four example PSM systems. Additionally, a PSM system can incorporate the attributes of the four example TQM systems. The essential elements of effective PSM systems are management of change, training, and process risk management. Without these three essential elements, a PSM system could be a reactive, unsophisticated, segregated, unimplemented "paper trail" (i.e., a smoke screen). An effective management of change element updates the PSM system to account for specific types of changes. A good training program enables employees to participate in the management of change program and informs all relevant facility personnel of these changes. Process risk management provides a sophisticated analytical tool to troubleshoot existing or potential process problems at the facility.

PSM systems are really applications of TQM philosophy to process safety. Thus, PSM is considered by many to be a subset of TQM; other facility programs (e.g., configuration control) can also be subsets of TQM. Because TQM concepts and systems have been well received and understood by management over the past decade, showing the parallel between PSM and TQM could hasten acceptance of PSM concepts by management familiar and supportive of TQM concepts. Both PSM and TQM systems

Table 1 Comparison of PSM and TQM Systems

PSM SYSTEMS				TQM SYSTEMS			
AIChE CCPS	CMA RESPONSIBLE CARE	OSHA 29 CFR 1910.119	EPA PREVENTION PROGRAM	ISO 9000	MALCOLM BALDRIDGE	ASME MCS	ISR
Accountability Process Knowledge and Documentation Project Reviews and Design Procedures Risk Management Management of Change	MANAGEMENT LEADERSHIP: Commitment Accountability Performance Measurement Incident Investigation Information Sharing CAER Integration	Employee Participation Process Safety Information Process Hazard Analysis Operating Procedures Training Contractor	Management System Process Hazard Analysis Process Safety Information Standard Operating Procedures Training Maintenance	Management Responsibility Inspection, Measuring, and Test Equipment Inspection and Test Status Contract Review Procedures Control of Nonconforming Product	Leadership Information and Analysis Strategic Quality Planning Human Resource Development and Management Management of Process Quality	Evaluating the Management Control System Control of Measuring and Test Equipment Nonconformance Corrective and Preventive Action Procurement Post-Production Functions	Leadership and Administration Organizational Rules Accident/Incident Investigation Accident/Incident Analysis Purchasing and Engineering Controls Records and Reports

Table 1 Comparison of PSM and TQM Systems (cont'd)

PSM SYSTEMS				TQM SYSTEMS			
AIChE CCPS	CMA RESPONSIBLE CARE	OSHA 29 CFR 1910.119	EPA PREVENTION PROGRAM	ISO 9000	MALCOLM BALDRIDGE	ASME MCS	ISR
Process Equipment Integrity Incident Investigation Training and Performance Human Factors Standards, Codes, and Laws	TECHNOLOGY: Design Documentation Process Hazards Information Process Hazard Analysis Management of Change	Pre-startup Safety Review Mechanical Integrity Hot Work Permit Management of Change Incident Investigation	Pre-startup Review Management of Change Accident Investigation Safety Audits	Document Control Procedures Corrective Action Purchasing Handling, Storage, Packaging, and Delivery Purchaser Supplied Product Quality Records Product Identification Traceability Internal Quality Audits	Quality and Operational Results (Measures of Progress: Product and Service Quality, Internal Quality and Productivity, Supplier Quality) Customer Focus and Satisfaction (relative to competitors, market share)	Production or Service Process Control (Needs Engineering) Human Resources Training Design Scheduling Marketing Procedures	Planned Inspections Program Evaluation System Management Training Job/Task Analysis and Procedures Job/Task Observation Employee Training

Table 1 Comparison of PSM and TQM Systems (cont'd)

PSM SYSTEMS				TQM SYSTEMS			
AIChE CCPS	CMA RESPONSIBLE CARE	OSHA 29 CFR 1910.119	EPA PREVENTION PROGRAM	ISO 9000	MALCOLM BALDRIDGE	ASME MCS	ISR
Audits and Corrective Actions Enhancement of Process Safety Knowledge	FACILITIES: Siting Codes and Standards Safety Reviews Maintenance and Inspection Multiple Safeguards Emergency Management	Emergency Planning and Response Compliance Audits Trade Secrets		Process Control Training Inspection and Testing Statistical Techniques Design Control and Servicing Human Resource		Data	Hiring and Placement Emergency Preparedness Personal Protective Equipment Health Control and Services Off-the-job Safety Personal Communications

Table 1 Comparison of PSM and TQM Systems (cont'd)

	PSM SYSTEMS			TQM SYSTEMS			
AIChE CCPS	CMA RESPONSIBLE CARE	OSHA 29 CFR 1910.119	EPA PREVENTION PROGRAM	ISO 9000	MALCOLM BALDRIDGE	ASME MCS	ISR
	PERSONNEL: Job Skills Safe Work Practices Initial Training Employee Proficiency Fitness for Duty Contractors						Group Meetings General Promotion Human Resources

have the essential management system features of planning, organizing, implementing, and controlling.

TQM's main thrust is to implement continuous improvement activities involving everyone in an organization (managers and workers) in a totally integrated effort toward improving performance at every level. Two problems that could keep PSM systems from being integrated are: (1) continuous improvement feedback might be curtailed once the PSM system meets its goals and objectives, and (2) PSM system elements might be organized and implemented in a segregated fashion by facility personnel establishing clear lines of authority, strong sponsorship, explicit roles and responsibility, detailed work plans, specific guidelines, and initiating mechanisms. For example, if the goals and objectives of a PSM system are just to meet the OSHA and/or EPA PSM system regulations, then no effort is made to improve the PSM system beyond the criteria dictated by the government agency regulations. In other words, the customer (i.e., OSHA or EPA) is satisfied, but no effort is made to improve the PSM system beyond the customer's expectations. Solely focusing on satisfying federal regulations, state laws, and industrial standards (by implementing quick-fix, myopic, bare-minimum, internal PSM system element policies) stifles continuous improvement; a proactive long-term strategy is more cost-effective. Besides, the customer might even be off-track by emphasizing the wrong objectives and noncritical path activities. At the facility, creating clear lines of authority; promoting strong sponsorship and explicit roles and responsibilities; and defining detailed work plans, specific guidelines, and initiating mechanisms tend to produce a segregated (incomplete integration) PSM system organization and implementation effort because some, but not all, personnel at the facility will be involved in the PSM system organization and implementation processes.

CONTROL SYSTEMS

All chemical engineers learn the fundamentals of process control as part of their undergraduate education. A conceptual control system for a plant is shown in Figure 3. Some phenomenon of the plant, called the controlled variable, produces an initial effect upon a measuring means. This may be an equipment sensor or an operator. The measured value of the controlled variable is stored in the technology data base. The data collection and storage system provides the measured value and the desired value of the controlled variable to the decision maker. Additional data may be requested from the data storage system and possibly other additional information. The decision maker determines appropriate corrective action based on the difference between the measured and desired values and the direction of the difference. The equipment controls institute the appropriate corrective action, and the corrective action restores the value of the controlled variable to the desired value.

A fundamental precept of TQM is that you can only control a process which you can measure. ISO 9000, MBNQA, and ASME MCS require an accurate and detailed description of a process and measurement of that process. The detailed description of the process allows the development of meaningful measures. A well-defined system of measures improves performance by providing a framework for making decisions. Otherwise, managers may base some decisions on false assumptions and personal objectives. Performance measures enable managers to see relationships between variables, providing insight into how processes work. A lack of proper performance measures is a big factor in the failure of many TQM programs. Therefore, if you want to manage and improve performance, you must first measure performance. CCPS, in collaboration with the U.K. Health and Safety Executive, is conducting research to fill this critical gap.

The PDCA cycle is a control system. The Plan step begins by identifying the data needed to analyze a process (i.e., define a measure of the process). The Do step gathers the data from the process, while the Check step evaluates the data and any other information. The Act step implements the changes to the process.

Applying TQM methods such as PDCA to PSM enables management to develop a PSM control system (Figure 4). Developing a measure system is the first step. Measures that describe the performance of important PSM process variables must be developed and must be truly representative of changes in process conditions. The

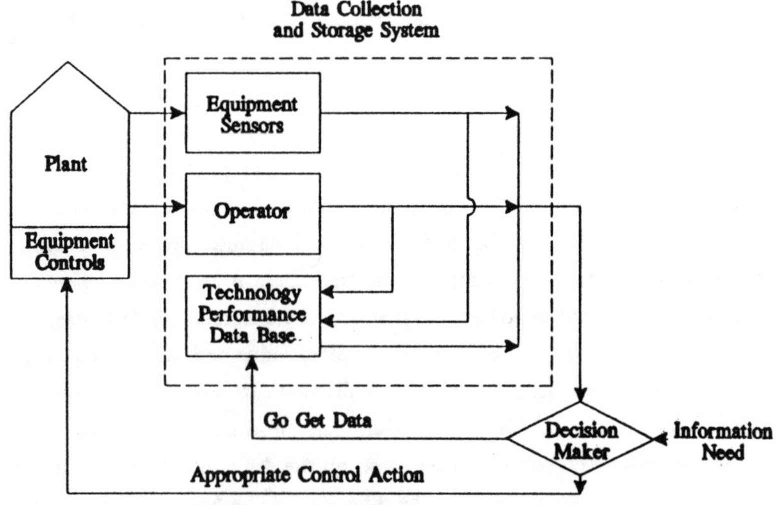

Figure 3 Technology Control System

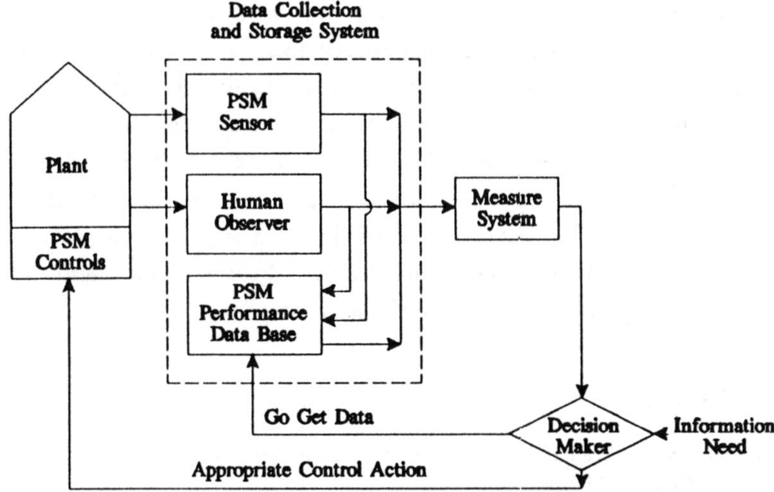

Figure 4 PSM Control System

measures must not become constant, erratic, or nonquantitative under some operating conditions of the PSM system, nor should they be easily contaminated by measurement disturbances, signal processing, and transmission. With these measures in hand, the facility can design an appropriate data collection and storage system. The facility can install PSM sensors to monitor its controlled variables. A PSM sensor is a device that responds to a stimulus from the plant and then transmits a signal to the data base and measure system. Human observers can also respond to a stimulus from the plant (e.g., significant events or near miss events) and also transmit the necessary signals. The decision maker can use the measure system and the PSM performance data base to recognize problems and opportunities, identify trends and patterns, allocate resources, and develop and motivate the organization. These control actions are implemented at the plant by PSM controls.

CONCLUSIONS

A blueprint companies can use for integrating/linking TQM concepts with PSM systems was laid out, resulting in a more cost-effective way of managing process safety in industrial facilities. Four example PSM systems (AIChE CCPS, CMA Responsible Care, OSHA 29 CFR 1910.119, EPA Prevention Program) were compared with four example TQM structures (ISO 9000, Malcolm Baldridge, ASME MCS, ISR). Using the TQM concept of continuous improvement using a totally integrated effort in implementing

PSM systems should produce a very effective PSM system. However, measuring PSM performance to establish continuous improvement (perhaps by creating certain PSM product requirements) is a deficiency in current PSM systems that needs to be overcome before PSM systems can approach the level of success achieved by good TQM systems. For example, creating certain PSM product requirements (as in ASME MCS) that PSM performance must meet might instill improvements (perhaps statistically) to existing PSM performance methods. TQM and PSM were also linked through two examples of control feedback systems; PSM can model itself after any of the example control feedback systems in establishing its own control feedback system.

REFERENCES

1. *Guidelines for Technical Management of Chemical Process Safety.* New York: AIChE/CCPS, 1989.

2. *Plant Guidelines for Technical Management of Chemical Process Safety.* New York: AIChE/CCPS, 1992.

3. Walter A. Shewhart, *Economic Control of Quality of Manufactured Product*, D. Van Nostrand Co., Inc., New York, 1931.

4. ISO 9000, *Quality Management and Quality Assurance Standards — Guidelines for Selection and Use.*

5. ISO 9001, *Quality Systems — Model for Quality Assurance in Design, Development, Production, Installation and Servicing.*

6. ISO 9002, *Quality Systems — Model for Quality Assurance in Production and Installations.*

7. ISO 9003, *Quality Systems — Model for Quality Assurance in Final Inspection and Test.*

8. ISO 9004, *Quality Management and Quality System Elements — Guidelines.*

9. ASME MCS-1-1990, *Management Control System*, May 15, 1990.

Evolution of Process Safety Management in a Mexican Company

Alejandro Lorea-Hernandez
Grupo IRSA, S.A. de C.V. Bosque de ciruelos 99, 11700 Mexico, D.F., Mexico

INTRODUCTION

At present, our Company, the Industrias Resistol Group (GIRSA) is one of the major chemical companies in Mexico, is dedicaded to the manufacture of both organic and inorganic chemicals like synthetic rubber, polystyrene, sodium phosphates and sulphuric acid. GIRSA was formed in 1971 with the participation of Resistol, a Mexican Chemical Company founded in 1941, Monsanto Co. and DESC, a private Mexican Investment group.

Two of the corporate policies established at that time were on Pollution Prevention and Safety matters; the first one oriented toward compliance of the regulations from the Government and the second to provide a safe workplace for the employees.

With that scope, the environmental area worked based on projects to treat or control emissions trying to satisfy the increasing requirements and, on the other hand, the Safety area was mainly devoted to fields such as personnel safety and property protection. Nevertheless, during the last twenty one years, GIRSA has been recognized by both Government and Industry sectors as a responsible company with consistent achievements in these areas.

In 1985 after the Mexico City LPG explosion and the toxic material release at Bhopal we were requested by the Government to conduct and submit a risk assessment study in one of our facilities that handles hydrogen cyanide to produce acetonecyanohidrine.

The study was a complete success not only in terms of risk management, but in terms of industrial an governmental cooperation leading to the very first official decree on land use restrictions around industrial areas in our country.

We consider that study as a landmark in the evolution of the Environmental Safety and Health (ES&H) issues at GIRSA: first it showed the close relationship between the three areas and second, it was the starting point to get involved in the worldwide Highly Hazardous Materials Management Program that Monsanto had launched following the Bhopal incident.

TRYING TO IMPLEMENT A MANAGEMENT SYSTEM

After three more risk assessment studies at different facilities, a corporate policy was established regarding acute risks on the community by the operation of our facilities.

This policy specifies that GIRSA must do any reasonable effort to avoid unacceptable risks that could affect our neighbors. Because the lack of both a well defined structure and follow up, not to mention strong committment, the results in our Process Risk Management were scarce and the Program time consuming and without relevant decisions.

In April 1992 ocurred the gasoline explosion in the sewer system of the Mexican City of Guadalajara; the first action from our Goverment was the request to each critical facility to submit a risk analysis within a period of six months.

To GIRSA, this request covered fifteen facilities and, as it can be understood, to us and in general for the Mexican Chemical Industry this represented a huge effort and a tremendous pressure to comply.

In accordance with the requirement, we finished our studies; but one thing was clear: if we want to progress on ES&H matters we must change our current strategy.

AN INTEGRATED ES&H MANAGEMENT SYSTEM

At the end of 1990 the ES&H Corporate Group (in Spanish CASH that stands for Control Ambiental, Seguridad e Higiene), proposed to the Board of Directors ten major actions in orden to establish an strategic plan covering the three areas, beeing the first and most urgent a decentralized body to promote both self-regulatory and implementing responsibilities at the operating Divisions. We would like to see (as in other Companies, Monsanto for instance), a Manufacturing Council to evaluate and take continued action in pollution prevention, safety and industrial hygiene.

The GIRSA-Manufacturing Council was established in 1991 in order to support the manufacturing facilities to obtain and mantain a competitive level based on: Management skills and Systems thinking. The following key areas were identified: Quality, Costs, Human Resources and ES&H. The next step was the constitution of the correspondant working Committees. For the ES&H or CASH Committee the initial objective was to increase the effectiviness at our facilities or plants through:

1. Propose an ES&H Direction (Vision, Mision, Policies,..),

2. Define Management Systems for the ES&H function, and

3. Propose the structure needed.

Through a higly participative scheme whith operation people and after the revision of different management systems, we have conformed a model for our managemet system (see Figure 1). The foundations of the system are:

Mision: To protect the environment and to care safety and health along the life-cycle of our products.

Basis: Information Management, Hazard Identification, Risk Assessment and Risk Management.

Process: Continous Improvement.

The model takes into account the basis of a management system; ie,. first there is a set of elements that gives DIRECTION to the system, being a relevant one our Vision on ES&H whose main points are:

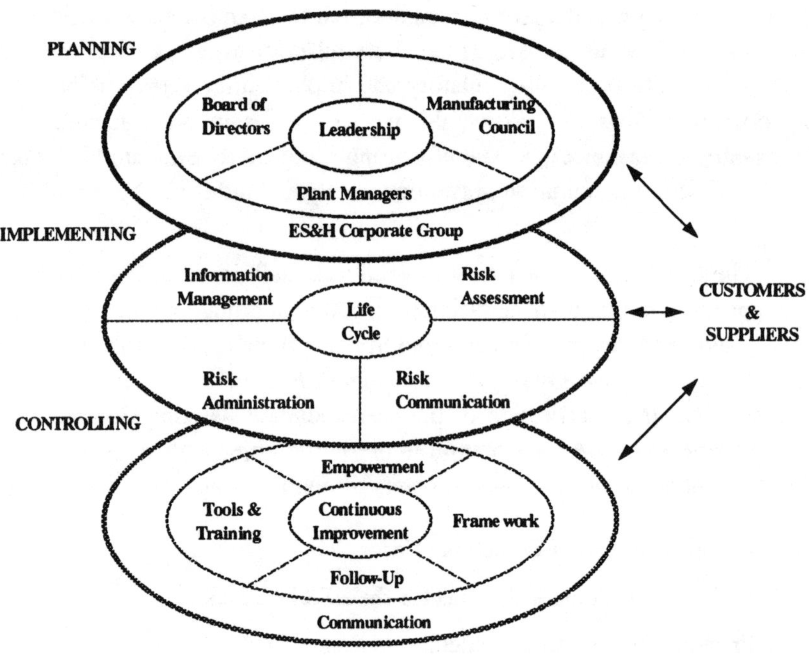

Figure 1. Model of the Integrated Environmental, Safety and Health Management System of GIRSA

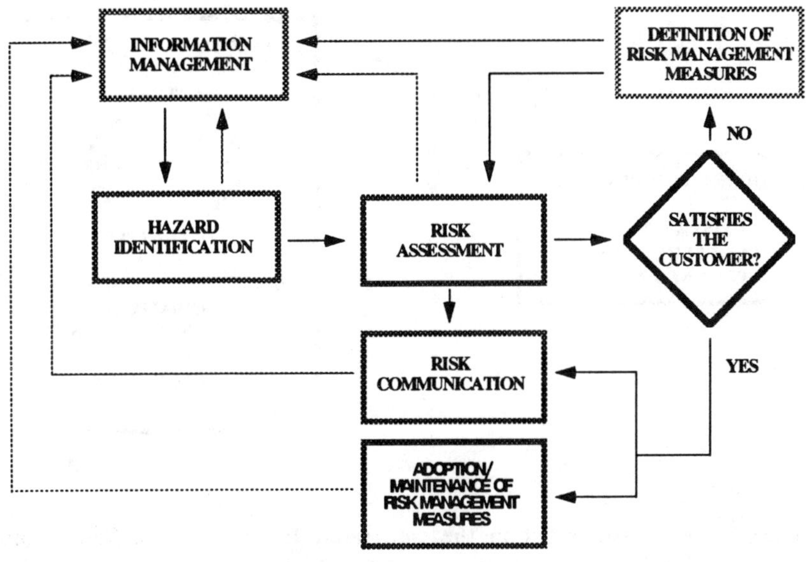

Figure 2. Risk Analysis Process in the Integrated Environmental, Safety and Health Management System of GIRSA

+ Safe Products.

+ No Effects on Community, Personnel, Environment and Facilities.

+ Openess to Community.

+ Waste Reduction.

Second, the operative core of the system or IMPLEMENTING phase comprises the elements of Risk Analysis (see Figure 2) including Information Management, Hazard Identification, Risk Assessment, Risk Management and Risk Communication. The model depicted in Figure 2 is the same to analize contamination, safety or health risks. The scope is to assess actual and potential risks (see Figure 3) along the life-cycle of our products and to manage them accordingly (see Figure 4).

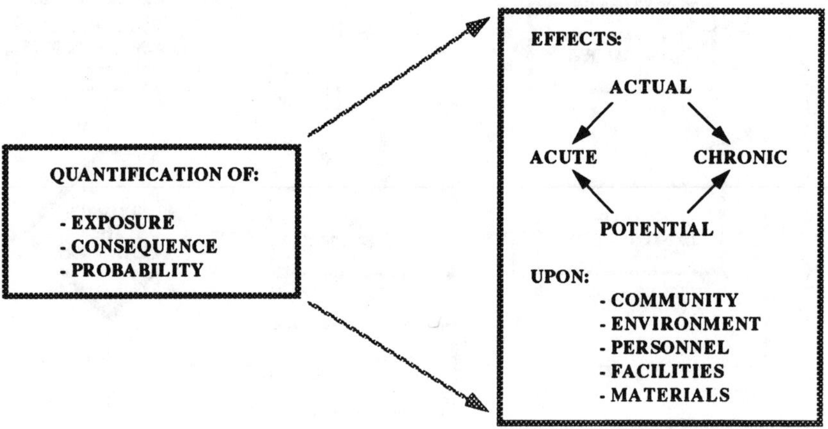

Figure 3. Risk Assessment in the Integrated Environmental, Safety and Health Management System of GIRSA

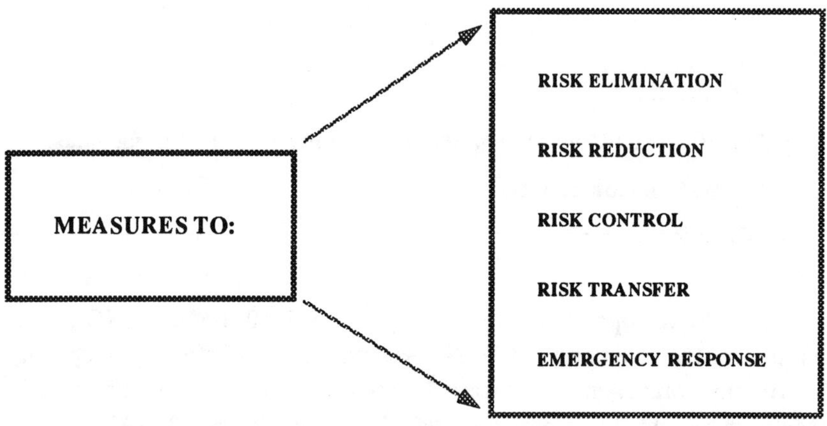

Figure 4. Scope of Risk Management of the Integrated Environmental, Safety and Health Management System of GIRSA

Third, the CONTROLLING phase which includes a follow-up and reporting system of activities, internal coordination and communication, performance measurement of the whole System and the offering of tools and training needed across the Integrated ES&H Management System. A key element is the empowerment of the personnel through recognition of contributions and the involvement in the decision take process.

The focus of the Integrated ES&H Management System is our CLIENTS and SUPPLIERS along the life-cycle of our products; that is why the magnitude of the risk (see Figure 2) must satisfy the requirements of our clients. The role of our suppliers must be an active one since their participation improve the system as a whole and more specifically in the Risk Management and Risk Communication elements.

Considering that focus; ie, CUSTOMER SATISFACTION, and supplier cooperation among others concepts like WORKER INVOLVEMENT and STRATEGIC PLANNING, we can say that our Integrated ES&H Management System resembles one for Total Quality Management. So, to evaluate its performance we consider an adaptation of the method used to define the Mexican Quality Award. This was done again through the ES&H Committee of the Manufacturing Council involving the operating personnel (mainly the Plant Managers since they are the primary responsibles for ES&H performance at their facilities).

The model for the evaluation method comprises eight categories with a maximum total scoring of 1000 points (see Figure 5 and Table 1); to evaluate each category and specific theme it is examined the approach, the scope and the results of the ES&H management at the site, the operating Division and eventually GIRSA as a whole. An example of the criteria used to establish the score in any specific theme is given in Table 2.

At present (February 1993) we are in the documenting phase in order to provide a useful reference as well as a tool to our operating people; preparing the schedule for evaluators training workshops with the goal of starting the adoption process of the Integrated ES&H Management System across GIRSA in April 1993. Our expectation is that in the second

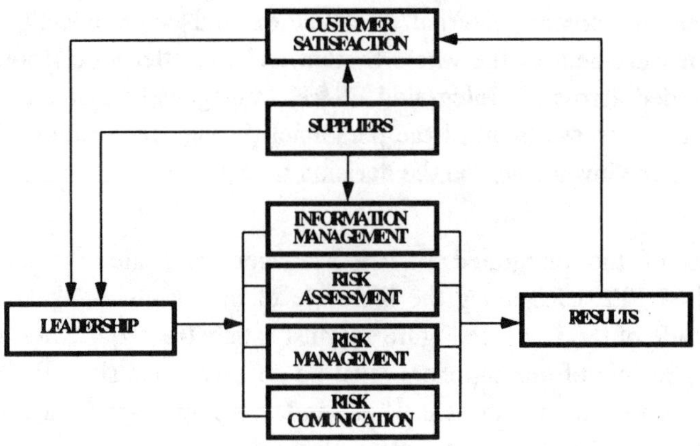

Figure 5. Performance Evaluation Method of the Integrated Environmental, Safety and Health Management System of GIRSA

quarter of 1994, when the first Code of Management Practices of Responsabilidad Integral, the Responsible Care type of program adopted in 1991 by the Asociacion Nacional de la Industria Quimica (Mexican Chemical Trade Association), GIRSA will be in an advantge position to satisfy the requirements of the program.

CONCLUSIONS

In a highly competitive world, it is not sufficient just to comply with the apparent requirements of our clients, it is necessary to satisfy all their expectations continously and efficiently; we are confident that the Integrated ES&H Managemenet System is the way of do it so at GIRSA.

Table 1. Evaluation Method for the Integrated ES&H Management System of GIRSA (Categories and Specific Themes).

	Score
1.0 CUSTOMER SATISFACTION	**150**
1.1 Commitment to Customers	60
1.2 Systems to Improve Customer Service	60
1.3 Determining Customer Requirements and Expectations	30
2.0 SUPPLIERS	**100**
2.1 Thorough Knowledge of Suppliers	100
3.0 LEADERSHIP	**150**
3.1 Senior Executive Leadership	30
3.2 ES&H Values	30
3.3 Strategic Planning	30
3.4 Operative Planning	30
3.5 ES&H Human Resources Development	30
4.0 INFORMATION MANAGEMENT	**120**
4.1 Sources, Data and Information	80
4.2 Analysis of Data and Information	40
5.0 RISK ASSESSMENT	**120**
5.1 Life-Cycle Analysis	60
5.2 Methods	60
6.0 RISK MANAGEMENT	**120**
6.1 Priority Areas	50
6.2 Management Strategy	50
6.3 Emergency Response	20
7.0 RISK COMMUNICATION	**90**
7.1 Priority Areas	30
7.2 Customer Participation and Commitment	30
7.3 Community ES&H Culture Diffusion and Promotion	30
8.0 RESULTS	**150**
8.1 Improvement in ES&H Indexes	90
8.2 Improvement in Customer Satisfaction	60

Table 2. Examples of Criteria to Define the Score for Specific Themes in Evaluation Method for the Integrated ES&H Management System of GIRSA.

	APPROACH	SCOPE	RESULTS
0 %	No or scarce evidence of clear concepts and systems. Isolated or non integrated attention to ES&H areas.	Unfinished in any of the stages of any of the products life-cycle.	Limited and scarce.
25 %	Some evidence of a sistematic and preventive approach. Beginning of an integrated attention to ES&H areas.	Beginning at the life-cycle of the main products.	Some positive tendencies in the main indicators.

We consider the Integrated ES&H Management System as the starting point of managing these areas as a whole process, shifting the paradigms that rule them at our Company with a brand new approach much more oriented to a system thinking way of doing and improving things, having in mind the concept of prevention rather than correction to manage risks. Also, with its scope, the System give us a lot of room to improvement.

Plans are under way to relate the performance evaluation of the ES&H Management System with the evaluation of Total Quality Management at GIRSA and, eventually, to modify the procedure used to define the Mexican Quality Award in order to include in it an evaluation covering ES&H since at present it covers just environment protection as an specific theme for evaluation.

REFERENCES

Guidelines for Technical Management of Chemical Process Safety, Center for Chemical Process Safety, American Institute of Chemical Engineers, 1989.

Specification for Environmental Management Systems, BS 7750:1992, British Standards Institution, 1992.

Total Quality Environmental Management, The Primer, Global Environmental Management Initiative (GEMI), 1992.

Total Safety Management, Arthur D. Little.

ACKNOWLEDGEMENT

The work reported here has been carried out with the effort and enthusiastic participation of many people at Grupo IRSA, S.A. de C.V. and truly represents the product of a team. As coordinator of the task group established for this porpouse by the GIRSA Manufacturing Council the author would like to express his appreciation for all their contributions. Moreover he wishes to thank Jaqueline Estrada, Corporate Coordinator for Quality Management at GIRSA for her cooperation, useful comments and suggestions along this endavour.

Norwegian Experience of Legislation Concerning Quality Assurance of Safety in Industry

Endre Nagell Bjordal
Norsk Hydro a.s., Oslo, Norway

ABSTRACT

More than 15 years ago the Norwegian Petroleum Directorate(NPD) took the first step into a new legislation regime - by claiming Quality Assurance(QA) systems of all Safety-, Health- and Environment-(SHE) activities in the oil companies operating on the Norwegian continental shelf.

During the last 15 years Norway has developed new legislation (for offshore and "onshore") for the SHE-field which can be looked upon as a QA-system on a national level.

The paper is presenting the structure of this legislation, is discussing advantages and problems and presents some results so far.

IMPACT OF LEGISLATION ON THE REAL SAFETY LEVEL IN PLANTS

The safety level of a company is normally a result of an interaction between rules and regulations laid down by society and the company's own goals and corporate culture. Some companies, however, are far ahead of the legislation, others far behind. The ideal situation can perhaps be considered to be one in which the average safety level of the industry is equivalent to the level set by society.

Over the years, the rules and regulations concerning safety in the process industry have become increasingly stringent. (This is not the case for all activities in society. It is more dangerous, for instance, to walk or cycle along streets without a pavement today than it was fifty years ago).

The process industry has to admit that, over the years, society has not been satisfied with the average safety level. We are also aware that the nature of new regulations has changed with the passage of time - from detailed, direct instructions, into regulations requiring the employment of certain management tools, such as analyses and documentation of the management system.

There is no single reason for this change. It is important for the process industry to realize that the change is partly ascribable to deep-rooted changes in society itself, such as higher education, and greater respect for life and the environment. The legislation merely reflects these changes.

But there are also other reasons. Today we are more aware of the real reasons behind all kinds of accidents, and we know more about the importance of managerial responsibility in this respect.

The increasing power of the media is also a factor, especially their tendency to focus on all kinds of negative events. This has led to the saying that in an emergency, it is more important to handle the media properly than to deal properly with the accident.

We must also remember that as major accidents become rarer, the focus placed on them by society and the media becomes more intense.

The combined effect of these factors has been the introduction of new types of requirements into the legislation:

The first is concerned with the "right to know" principle.

The second is requirements relating to safety management, and not merely the real safety level.

Thirdly, there is the tendency to concentrate explicitly on process hazards, such as explosions, fires, release of toxic substances etc. Examples of such regulations are the PSM regulation in the US, and the Seveso Directive in the EC.

In order to win acceptance in society, it is crucial for the process industry to succeed in reducing the frequency of process accidents to a very low level.

It has been questioned whether it is possible to achieve such a level solely by concentrating directly on preventing these accidents. As a result there has been a growing tendency to establish "holistic" systems. Norway has gained quite a lot of experience through taking this approach during the last 15 years.

THE NORWEGIAN PROCESS INDUSTRY

The process industry is very important to Norway. It's strength lies in the natural resources within the country. The hydroelectric power produced in Norway does not only cover the normal consumption of the population. It also covers a substantial part of the European consumption of aluminium, magnesium and various ferro-alloys, all of which require large amounts of energy.

In addition the country has huge reservoirs of oil and gas on the Norwegian continental shelf. These provide the input for big plants which produce petrochemicals, fertilizers and various oil refinery products.

Norway also exports hydroelectric power as such, and oil and gas equivalent to the total energy consumption of a country like France.

DEVELOPMENT OF SAFETY LEVEL AND SAFETY REGULATIONS IN NORWAY

In the seventies, all the Scandinavian countries got new basic legislation pertaining to safety, working environment and product control in industry. This legislation was ambitious. The working environment acts also established the "participation principle", which specifies employee rights. The employees have a safety representative (verneombud) and they are represented on the company safety committees, which report directly to the Directorate of Labour Inspection, the Norwegian counterpart of the American OSHA. (In Norway the employees are also represented on the company's Board.)

The Scandinavian countries expected that these laws would be sufficient to bring about the desired improvements in industry for the rest of the century.

After a few years, this turned out not to be the case. Some improvements were achieved, but the number of industrial fires increased substantially, the number of fatalities remained constant, as did the number of accidents. Despite a lack of reliable statistics, it was generally agreed that the number of occupational diseases (due to solvents, carcinogenic substances, etc.), the number of muscular-and-skeletal problems as well as stress and absenteeism were not decreasing. In addition, the focus on unlawful pollution intensified.

In many companies the effect of the "participation principle" in the legislation was to make management believe that safety was being taken care of by the new safety representatives and the rest of the safety organization. This led to management withdrawing to some extent from SHE activities.

The various government agencies felt that it was difficult to establish control which ensured the realization of the visions in the new legislation. However, it takes time to change well established government strategies. The young Norwegian Petroleum Directorate (NPD) was in a better position.

THE NEW OFFSHORE LEGISLATION: A QA-SYSTEM

The Norwegian offshore oil and gas business is only 20 years old. In the early years, Norway depended heavily on big international companies in both exploration and production.

A big oil blow-out from the Bravo platform in 1976 had about the same effect on Norway as the British Piper Alpha accident of 1989, in which 160 lives were lost, had on Great Britain.

The importance of the offshore industry to the Norwegian economy led to a whole new body of legislation for this industry.

As far as safety is concerned, in the 70's the Norwegian Petroleum Directorate (NPD) found some basic control principles well established in the economics departments of big companies. The economists called their system "internal control". NPD found these principles very suitable for government control of safety within the oil companies.

Norwegian petroleum legislation is now based on several important principles which function together as a QA-system on a national level. This legislation has some basic elements:

1. The normal Norwegian principle, that all the individual companies are subject to control by a government agency, has been dropped offshore. Only those compancis which are formally licensed by the Norwegian Parliament to operate a particular oil or gas field, are responsible towards NPD for compliance with the legislation offshore. It is the duty of the chief executive of the oil companies to implement the legislation. Consequently NPD only has 15 big companies directly under its jurisdiction.

    Thereby responsibilities are clear-cut.

2. Defined responsibility is not enough. It was regarded as crucial that the companies had comprehensive, efficient safety management systems. Already in 1979 NPD put its internal control guidelines into operation. From 1981 these guidelines also state that " the internal control normally will be taken care of by a total quality assurance system that shall ensure conformance with the company's own requirements." Subsequently, all the oil companies working offshore established QA systems which also included SHE activities.

3. Safety is defined in the regulations as safety against:
    - process-based accidents
    - working environment hazards
    - pollution

- damage to equipment and materials
- regularity of production.

4.   NPD requires in a special regulation that operators must perform risk analyses in all stages of a project: planning, design, construction, commissioning, operation, maintenance, modifications and final destruction of the platforms.

The original guidelines on risk analysis of 1981 required that if the risk analysis revealed that the probability of any of nine defined hazards occurring exceeded 10 to the minus four per year, the company should take and document special precautions. Among these nine types of accidents were explosions, blowouts, fires, collisions etc.

A newer regulation in force since 1991 states that the chief executive of each company is responsible for establishing the company's own safety goals and safety acceptance criteria. If the results of a risk analysis show that the criteria are not satisfied, risk reduction action must be taken, documented and reported to NPD.

5.   NPD is established as the overriding authority offshore, and coordinates all other governmental control.

NPD has established a reporting system which includes reports on acceptance criteria, risk analyses performed, accidents and near-misses etc.

6.   All the oil companies offshore have to establish their own auditing function, which reports directly to the chief executive. The companies' internal reports are available to NPD.

7.   The system-auditing principle is used by NPD itself. They carry out system audits on the oil companies as the chief means of NPD's own control.

Thus two QA-loops have been established to control the SHE activities in the companies.. Of chief importance, of course, is the one which the legislation requires be established within the individual company. In addition, however, NPD has its own QA-loop: Goal-setting, planning, performing, follow-up, corrective action, auditing.

DEVELOPMENT OF SAFETY IN PETROLEUM INDUSTRY OFFSHORE NORWAY

However, legislation can never be a goal in itself. It is one of several tools used as means to certain ends. As far as safety is concerned, it is the safety activities within the companies that count - and the resulting safety level.

The effectiveness of legislation as a tool for achieving certain ends lies partly in its ability to motivate the companies, and partly in its ability to impose sanctions. How has Norwegian petroleum internal control regulations succeeded in this respect?

As far as motivation is concerned, the reaction of the companies to the regulation has been largely positive - by comparison with reactions in Norwegian companies to legislation generally.

There are two main reasons for this.

Firstly, the fact that these regulations are based on accepted business management principles.

Secondly, NPD is perceived of by the companies as a powerful organization. They feel that NPD plays an important part when the government appoints operators to new prospective oil and gas blocks offshore. As a result, the companies more or less accept the legislation as part and parcel of their licence.

The other aspect of legislation, the possibility of sanctions from NPD and the police, has been used quite often by comparison with other safety agencies. This has caused problems for the safety activities of the companies. It is more difficult to get people to report near-misses when they know that both individuals and companies may be penalized as a result of such a report. It seems that for lawyers it is more important that details in the regulations have been followed - than the goals set in the legislation: Better safety.

The real development of the level of safety in the offshore industry has at any rate definitely been positive.

The frequency of LTAs (Lost-Time Accidents per one million working hours) has been carefully followed. During the 70s the frequency was very high, ranging from 40 to 60 on the various platforms. Today there are platforms which have reduced the LTA frequency by 90%, below 2, and most of them have a frequency between 5 and 10. The frequency of fatalities has been reduced with 50% over the last 10 years, and the number of serious accidents have also been drastically reduced.

The Norwegian State Pollution Control Authority (which corresponds to the American EPA) claims, however, that the environmental part of the companies' management systems is far from good enough.

In order to strengthen its own safety activities, the biggest Norwegian company Norsk Hydro became the first in Scandinavia to hire Du Pont's Safety Service Center for a period of three years in order to get insight in du Pont's famous safety program. This proved a success for Norsk Hydro both onshore and offshore. Some years later, the offshore sector of Norsk Hydro hired the American International Loss Control Institute(ILCI) to establish their safety audit system (Safety Rating System, SRS) in the company - also a success for Norsk Hydro. This shows that the Norwegian legislative model can be coordinated with other systematic tools.The LTA frequency in the whole company fall 50% over a three year period. On a big construction site in Canada Norsk Hydro's safety program reduced the LTA frequency for the contractors from more than 100 to below 20 over a ca 3 years period.

Working along the same lines. Norsk Hydro's Oil and Gas segment has successfully combined the auditing function of both QA, safety and economy in one staff department.

The greatest weaknesses of the offshore industry in total compared with traditional onshore industry are probably documentation and participation.

Overdocumentation has led to bureaucracy and sometimes to a lack of a sense of responsibility. The documentation of acceptance criteria can also easily lead to unwillingness to take advantage of new technology.

The level of constructive participation from the employees' side, by comparison with the onshore industry, is a major disadvantage, from a Scandinavian point of view. Scandinavia has a long tradition of employee involvement, and is looking at the same principle in TQM (Total Quality Management) with great interest.

GENERAL INTERNAL CONTROL REGULATIONS IN NORWAY

The positive experiences of the offshore legislation led the Norwegian government to investigate the possibility of applying internal control regulations onshore as well, in order to improve the national safety level in industry. A government committee recommended this in two reports of 1987.

In 1990 seven Norwegian acts were amended to make it possible to introduce the principles, and January 1st 1992 saw the implementation of regulations pursuant to the following acts:
- Worker protection and working environment
- flammable goods
- explosives
- fire prevention (in buildings and other constructions)
- pollution and waste

- product control
- electrical installations and equipment and
- civil defence (emergency preparedness).

These regulations for "onshore Norway" are much wider than the offshore regulations, and is therefore made more flexible. They apply to:
-all public and private organizations in Norway which have one or more employees.
-everyone owning a building or similar construction
-all employers owning electrical installations
-anyone producing or importing products which constitute a safety hazard to the man-in-the-street buyer.
-anybody who is, or may be, a polluter on a significant scale.

However, these regulations employ the same fundamental QA-principles to form the internal control system. The top manager is responsible to establish the system and to follow up. In this work the top manager has to organize participation of the employees.

Eight different government agencies under 4 different government departments are supervisory authorities for the regulation. Cooperation on this scale has never before been seen in Norway.

REACTIONS TO THE GENERAL INTERNAL CONTROL
REGULATIONS FOR SHE ACTIVITIES

To start with, both employer and employee organizations were doubtful about a regulation of this kind - but for quite different reasons. After 2-3 years of discussions, however, all parties agreed upon the principles.

Some of the governmental agencies were afraid that the idea of requiring QA systems for the whole range of SHE activities, for all enterprises, big or small, would fail. Thus the idea arose of applying QA principles to the actual implementation of the regulations as a whole.

At the time, the Royal Norwegian Council for Scientific and Industrial Research had already established a research program on internal control with a budget of about one million dollars per year for four years. The program focused on problem areas such as small companies, psycho-sociological working environment, audits, coordination of government agencies etc.

A new research project was launched, with a steering committee on which all the governmental agencies supervising the new regulations are represented. The goal of the project is to monitor the implementation of the internal control regulations in the companies. This activity is just according to basic QA principles.However such a systematic follow-up system has never been used in connection with implementation of legislation in Norway before. About 250 companies have been followed for a period of three years, and 20 or so have been studied in more depth in the same period. This has given valuable information to the companies, and also enabled the governmental agencies to obtain useful information which allows them to correct their tactics. The companies can learn from each other what is effective, and what has gone wrong.

The majority of the agencies involved had had little previous contact. Over the past three years, that contact has increased considerably. The two main government departments and the seven government agencies concerned have established a special committee holding regular meetings. A common strategy is being developed. Subcommittees, one on information to the public, one on internal education within the agencies and one on coordination of audits have been established.

Through one of its funds, the most important employers' organization, the Confederation of Norwegian Business and Industry, has started several projects on implementing effective internal control in the companies in more than 15 of its different trades, from hairdressing to electrometallurgical industry. Some of the projects combine internal control with quality improvement. Total funding over the past 2-3 years has amounted to about 4 million dollars.

In parallel with the activities mentioned above, the government has launched a campaign to reduce the numbers of accidents at home and away from work. This campaign is being run on the municipal level throughout Norway.

HAVE THE NEW REGULATIONS HAD AN EFFECT ON SAFETY?

It is often difficult to evaluate specific "causes and effects,"especially over just a couple of years. So also in this case. What we hope to be trends may be statistical variances The following has happened, at any rate:

-The number of fatalities in industry has fallen by 25% over the past two years.
-Compensation after fires in industry has fallen by more than 50%
-In leading businesses, such as aluminium, pulp and paper, chemicals, the LTA frequency has fallen by more than 25%.
-Pollution due to sulphur dioxide and CFCs has been reduced at a rate greater than that made mandatory by international agreements.
- Absenteeism in the companies has fallen more than 13% over a two years period.

Through the research project mentioned above, among several important results, we have learned something by making a comparison of the companies which have established good internal control systems and those which do not have such systems. The result was that the companies with internal control systems reported stastistical significant greater improvements than those without on the following factors:

- LTA frequency,
- absenteeism over more than 4 days,
- continuous pollution,
- regularity of operation.

A gallup revealed that one year after the internal control regulation entered into force, 60% of all companies in Norway.. small or big, knew about the regulations. This is looked upon as a very encouraging result as most of the companies are very small and because the strategy in the beginning has been to concentrate on companies which are supposed to be positive.

We have also learnt that 75% of the companies are combining their internal control systems with their QA or TQM activities.By doing this, the companies achieve a synergy effect. Many, including Norsk Hydro. are using their systematic SHE activities as a spearhead to get the organization used to working systematically and creatively in general towards common goals.

IS NATIONWIDE QA OF SHE POSSIBLE?

The programme described above can be regarded as a big experiment. Is the time ripe for winning acceptance, and do we have the staying power? Norway has today economic problems and growing unemployment - about 6% at present. In such a situation it can be difficult to get through with the message that it will be economically profitable to introduce new systems in the companies.

But another still more difficult question to answer is the following: Is there any good alternative?

The problems for people and environment are increasing. The answere must be to concentrate more on organizing the activities.

We have. however. also seen several weaknesses and difficulties in the implementation:

- Over the years past the industry has been complaining - again and again - about too detailed regulations. Now quite a lot of the companies are complaining of the fact that the internal control regulations give to much freedom in establishing their own systems. They are not used to take real responsibility themselves.

- in order to succeed, it is crucial to establish good working processes within the company. Many smaller companies are opposed to all regulations. They will do as little as possible. As a result they will incur all the expenses of what they are compelled to do, but will not reap the financial benefits of working effectively, being goal-oriented and geared to improvement, etc. These people are likely to become even more set against reforms.

- an understanding of the importance of real changes in attitudes, management style and organizational thinking is still lacking in many companies.

- the coordination between the agencies is not good enough to realize the visions in the legislation.

- the unions are not interested enough in the changes. The participation of the employees is depending also upon the participation of these organizations.

- the government's own organizations are not being zealous enough in their own implementation to enable them to act as locomotives.

The next couple of years will be important. We have seen in the offshore industry that it takes many years to implement changes like these. In order to succeed, Norway will have to find out what is not effective, and take corrective action or change its goals or its strategy. If we succeed in this way of working, our underlying goal will have been achieved. The exiting question will be how fast we can get real improvements on a national level - for people's health and safety, for our environment and for our econonic development.

Incorporating Human Performance Variability in Process Safety Assessment

Jeremy C. Williams
Electrowatt Engineering Services (UK) Ltd.

ABSTRACT

Analyses of the causes of many industrial accidents show that human behavior, and in particular its unpredictability, can affect judgements that might be made regarding overall process safety. Although human performance is variable, recent research shows that it is actually relatively predictable in a statistical sense, and its effect on process safety can be estimated with reasonable precision, providing that the identification and assessment of human interactions is sufficiently comprehensive.

The identification and assessment of interactions and potentially error-producing conditions within systems can be accomplished via human factors review techniques or certain types of safety management assessment program. Guidelines and criteria necessary for identifying human error potential are available, and these can be used to determine the degree to which generic failure frequencies should be modified for process safety Quantified Risk Assessments (QRA). Identification of the important influences on human behavior can then be used to modify generic failure rates, as well as the crucial system design and operational features that may then require more detailed investigation, re-design or modification.

When managing human reliability factors in process safety assessments, it is important for the assessor to be able to understand and take full and proper account of the impact that human performance can have on engineered systems. This paper, therefore, explores the effects that human reliability can have on safety cases, identifies some techniques for modelling the impact of human behavior on QRA, and considers methods for modifying estimates that might be made regarding the probability of system failure, when QRAs are undertaken. By using this approach, it is then possible to initiate a program of continuous improvement, whereby human factors issues become an integral part of any Process Safety Management (PSM) process, and are kept under constant review.

INTRODUCTION

The overall general impact of human performance on process safety is now well recognised within PSM circles (Wilson, 1990). Although estimates vary, it is apparent that at least half of the plant operation errors that are observed may have a human factors origin (Doyle, 1969; Rasmussen, 1980; Joschek, 1981 and 1986; Garrison, 1983; INPO, 1985; Bodsberg et al., 1987; Instone, 1989 and Suokas, 1989). Because so many studies have indicated that human action is a contributing factor in a large proportion of the incidents and accidents that are observed, many organisations are now assessing the effects of human behavior upon process safety and reliability.

Although the probability of human failure has been established as having an important impact on process safety, a continuing and increasingly prominent problem for analysts and industrial safety assessors is their relative inability to predict human performance with sufficient accuracy and confidence. Not only is there a significant absence of quantifiable human failure data, but until recently there has been little or no worthwhile information available to characterise or predict the variability of the human component in process safety assessments. This shortcoming in our ability to characterise and predict the relative sensitivity of systems to human failure in a given environment has the potential to alter process safety and reliability assessments in a profound and unexpected manner, and therefore merits investigation.

Even though we may have a shrewd idea about the general likelihood of failure, when humans are required to perform tasks, we cannot easily specify the limits on the performance that is likely to be emitted. To make matters worse, most human reliability assessment techniques devote relatively little attention to the question of human variability, preferring instead to concentrate on nominal or generic likelihoods of task failure, and leaving the question of computational uncertainty relatively unresolved. Therefore, not only can the analysis result in considerable mathematical uncertainty (e.g. Kletz, 1992), but this can be heightened by the superimposition of any additional uncertainties regarding the variability of human performance. In order to guard against such uncertainties Layfield (1987), for example, thought it prudent to allow the nominal calculations of human reliability to be incorrect by a factor of ten. Despite papers such as those by Williams (1989) and Kirwan et al. (1990), which have attempted to characterise some facets of the variability of human performance, it is clear that there is still a void regarding the nature of human variability with respect to its overall impact on process safety.

Recognising that our knowledge of human performance variability is limited, this paper nevertheless attempts to explore the positive and negative effects that human reliability can have on safety cases and process safety judgements. It also outlines techniques for modelling the impact of human behavior on QRA, investigates the stability of human performance and considers methods for modifying estimates that might then be made regarding the probability of system failure to reflect individual or group variability, when QRAs are undertaken.

OVERALL PROCESS SAFETY JUDGEMENTS

When making human reliability assessments in relation to process safety it is most important that the analyst should feel some confidence whilst making his judgements. Currently, when required, this is achieved by making ranged assessments, with bounds, and trusting that there is sufficient characterisation of the predicted human variability that the sensitivity of the analysis will not be changed markedly by the variability of the human contribution, or that if it is, it will at least be apparent how sensitive the analysis is to such effects (e.g. Delboy et al., 1991). The accuracy of process safety judgements is limited by the effectiveness of hazard analysis, which in any case can be out by a factor of 3 or 4, and sometimes up to a factor of 10 (Snaith, 1981). This degree of imprecision is not markedly different to the imprecision reported by human factors practitioners when dealing with the assessment of human reliability against known data points (Kirwan, 1988a).

Improvements in process safety have been pioneered by a number of practitioners including Hawksley (1984), whose work has shown that process safety improvements alone can make a substantial contribution to overall site safety performance. For example, reduction of the probability of accidents such as the transposition of an inlet and exhaust port connection, (Health and Safety Executive, 1990) which caused the amputation of an operator's hand on a hydraulically operated guillotine would almost certainly have reduced the probability of a similar error which is thought to have caused the Phillips 66 Houston Complex Accident (US Department of Labor, 1990; Kletz, 1991). Whilst both accidents are dreadful, it should be noted that the context of the latter error produced a major process disaster, whereas that of the former produced an occupational accident.

Research currently being undertaken by the Center for Chemical Process Safety (CCPS) is designed to bring about further overall safety improvements by tackling process safety from the point-of-view of the management systems that are employed to ensure safety. This work, which is described by Sweeney (1992), is investigating relevant aspects of PSM such as process knowledge, risk management and the management of change. The approach is focused on formulating a face-valid model of the PSM process, together with a set of safety performance indicators, audits and performance measurement that are designed to investigate quantitative ways of indicating that process safety performance has indeed been improved.

As accident frequencies and incident severities are being reduced largely as a consequence of such initiatives, it is ironic (but fortunate for safety) that one of the principal means available to assess process safety and ensure that assessments remain robust is being removed. Therefore the PSM challenge is to ensure that safety claims can be sustained despite a growing paucity of incident-based data, and at the same time ensure that such claims are not extravagant or compromised by "point-chasing" or "paper improvement" to the detriment of real process safety improvement.

There is evidence in the literature that there is a movement away from hard technical approaches to process safety judgement towards a more management-centered approach. For example, Reber et al. (1990) have demonstrated that overall safety performance can be improved by means of exhortation and leadership with goal setting, and feedback. Much of the output of Juran (1979), Deming (1982) and Taguchi and Clausing (1990), is motivated by these insights, and it is interesting to note that there are now many organisations that have employed such methods and attitudes who report improved overall performance, as a consequence. It has been found that when quality programmes are invoked, reliability is usually reported to improve dramatically, often by at least a factor of three, and sometimes by a factor of ten (DTI, 1990). This could be important because safety, as measured by accident rates, also seems to be coupled to the reliability of the production process (Mill, 1992). Clearly if some of the potential for such social engineering can be harnessed for application to PSM improvement this will make a most valuable contribution to the judgements that might be made.

STABILITY OF HUMAN PERFORMANCE

The studies reported by Williams and Willey (1985) on experienced versus inexperienced workers show that a predictable reduction in nominal error probability occurs that is associated with accumulated experience. This improvement is fairly consistent across all tasks (electrical installation, panel wiring, and welding) and amounts to about a three-fold reduction in the likelihood of error for every ten-fold increase in accumulated experience, and has been corroborated by the studies of Spiker et al. (1985). Gitomer's (1988) work on trouble-shooting has also shown not only how much absolute probabilities of failure may be reduced, but it highlights the fact that relative performance variability remains roughly constant, regardless of absolute probability of failure.

From the work of Brown et al. (1991) and some recent research results (Taylor-Adams and Williams, 1992) it is apparent that it may now be possible to predict the amount of human variability associated with most nominal assessments, even if this prediction is somewhat coarse. Some unpublished research in this area suggests that the extent of individual human performance variability is not markedly different to group variability. Within limits, it is also possible to be confident about the extent of such estimates in relation to the ranges of task failure encountered. For example, on the basis of data processed in the studies mentioned above, if a human probability of failure of 4×10^{-2} were encountered, the 5th. to 95th. centile variability that would be predicted could be expressed via an error factor of about three, as 2×10^{-2} to 1×10^{-1}. Any failure to appreciate the relative stability of human performance in relation to process safety assessments may reside more in the analysts' ability to model the influences on human performance, therefore, and their impact on system reliability, than the quantitative control that can be exerted by training, selection or narrowed skill development.

Indeed, if one intends to control/minimise overall performance variability, one should be attacking sources of the task performance variability as well as the absolute probability of task failure. The available evidence suggests, however, that task failure variability is not amenable to major reduction via overall improvement in task performance, but rather by reason of the nature of the task itself, and that this is much more susceptible to variability the more skill-based it becomes. Thus, there seem to be two ways of accommodating human performance variability. One is to attempt to narrow the range for a given task, but this is resource-intensive, and not very productive. The other way is to attempt to change the nature of the task so that the nominal probability of failure is reduced. This has a greater impact on overall system performance, and is why efforts are applied to developing training programs, decision aids, and human engineering improvement. A paradox, however, associated with this approach, is that the more skill-based a task becomes (with a lower concomitant probability of failure), the greater its general variability tends to become. The overall effect of such initiatives is not to reduce variability in practice, but to reduce the overall nominal probability of human failure (which usually is the ultimate objective, of course).

IMPACT OF HUMAN PERFORMANCE ON PROCESS SAFETY

At the Salem nuclear power plant in New Jersey in 1983, circuit breakers designed to initiate automatic shutdown of the reactor failed twice in 3 days. Atkinson et al. (1988), commenting on this event, concluded that the equipment performed between six and 1000 times less reliably than would have been assessed using normal assessment procedures. Such failures are usually traceable to human involvement and have been observed in other process industries. It will be seen that, despite current analytical approaches to process safety assessment therefore, human performance can sometimes exert a strong and immediate influence over the levels of process safety that are achieved in practice.

Process safety assessments also show that assumptions about human performance can have profound effects on the way in which risks would be managed. The work of Samanta et al. (1981), Wong et al. (1990) and Samanta et al. (1990), for example, shows that human performance, and, in particular, human variability, can change process safety assessments in powerful ways. Slight changes in nominal assessments can result in massive differences in the computed risk values that an overall assessment might otherwise produce. Samanta et al.'s theoretical studies show that if estimates of human error are underestimated, the calculated effects on hazardous releases will be very great, whereas if the estimates are over-estimated the effects will be comparatively small.

Recently, it has been shown in a study by Brookhaven National Laboratory (1989) that under certain conditions, there need only be an underestimation of the contribution of human error by as little as an overall factor of four, for this to affect top event frequency computations for the worse by more than a factor of ten. Clearly, as has been illustrated by Brown et al. (1990), such differences are non-trivial, and could result in different risk management decisions being taken simply as a consequence of the computational process.

Individual characteristics, leader behavior and injury occurrence will also affect risk management decisions, and these have been studied by Butler and Jones (1978). As might be expected, individual variability with respect to accident occurrence can be partly explained by differences between individuals, but it is noticeable that these differences are not great in a statistical sense. Studies by Rasmussen (1980), Samanta et al. (1981), Ghertman and Griffon-Fouco (1985), INPO (1985), and Bellamy et al. (1989) suggest that it may now be possible not only to identify the causes of human failure, both individual and system-related, but via safety management measures, find ways to reduce their overall likelihood.

Clearly if some techniques can be devised to assess the overall impact of human beings on overall system safety, it should be possible to target resources at the management failures and reduce the likelihood of major accidents. There are many reports regarding the significance of management to the control of industrial hazards, and it is now generally agreed that the evidence for regarding management variability as a fundamental issue in PSM is substantial (Whalley and Lihou, 1988; Brearley, 1990; Bellamy and Geyer, 1991; Reason, 1991 and Jenkins et al. 1991).

IDENTIFICATION AND ASSESSMENT OF INTERACTIONS AND POTENTIALLY ERROR-PRODUCING CONDITIONS

The Advisory Committee on the Safety of Nuclear Installations' Study Group on Human Factors second report (Health and Safety Commission, 1991) confirms the importance of making a systematic attempt to shed light on the human factor in any potential risk or accident situation. Fortunately there is a well-developed literature associated with the identification of human error potential, e.g. Kirwan (1992a). A great many techniques for human error identification are available such as Barrier Analysis, Human HAZOP, Task Analysis and Sneak Analysis. Although some of the techniques are less developed than others, they can all be used, with varying degrees of success (Whalley and Kirwan, 1989), to identify human actions and the error potential associated such actions in the context of process safety assessment (Kirwan, 1992b).

Techniques such as CADA (Critical Action and Decision Approach), (Gall, 1988) and IMAS (Influence Modelling and Assessment System) (Embrey, 1986) show some promise with respect to the identification of cognitive errors. Some generic issues associated with error producing conditions have been identified and assessed for their overall impact by Williams (1988). Hudson (1992), in a more specific sense, has described an approach to identifying and assessing interactions and potentially error-producing conditions in the chemical process industries, and Hunns (1991) has described on a case-by-case basis the factors that have been seen to affect human reliability in particular accidents.

HUMAN FACTORS REVIEW TECHNIQUES AND SAFETY MANAGEMENT ASSESSMENT

A wide range of review and assessment techniques are available to determine the impact of human performance factors on PSM. Currently these are split into two separate approaches. The human factors methods are derived from the human factors experimental literature, and have found ready application within military, aviation, and more recently, nuclear systems assessment. Such techniques and technology are well-represented with respect to PSM by *A Manager's Guide to Reducing Human Error* (Lorenzo, 1990) and the *Guide to Reducing Human Error in Process Operation* (HFRG, 1991). It is interesting to note that this approach is now moving toward a safety management perspective, as characterised by documents such as the *Management at Risk* guide (Jenkins et al., 1991) and *Engineers and Risk Issues - Code of Professional Practice* (Engineering Council, 1992).

By way of contrast, the safety management literature which is also well-established, is now moving toward a more comprehensive portrayal of human factors technology. Such efforts are characterised by documents such as *Techniques of Safety Management* (Petersen, 1990), *An Engineer's View of Human Error* (Kletz, 1990) and *Human Error Reduction Strategies* (Society of International Gas tanker and Gas Terminal Operators, 1990).

Almost all of these initiatives can be seen to be converging on a shared technical understanding of the impact of human factors on PSM, and it is to be hoped that this trend will continue for some time yet.

In the area of safety management assessment, there is now a growing body of literature (much of it commercially sensitive) on the subject and the methods that are available. It would not be possible to do justice to all these techniques within the confines of this paper, but a few generalisations can be offered. Many of the techniques draw upon the concepts propounded by Bird and Germain (1985), namely that evidence of accidents and near misses can be used proactively to engineer similar failure potential out of safety management systems, providing that the process is seen as an overall management function, rather than an idiosyncratic response to isolated events. In order to achieve this management growth, the user of the safety management system is usually invited to evaluate his safety strategies as a component part of the loss prevention effectiveness that can be achieved. The evidence for the claims made by safety management system initiatives can be seen in the work of Gaunt (1989), Sehdev et al. (1990), Guastello (1991) and Wilpert (1991).

GUIDELINES, TECHNIQUES AND CRITERIA FOR IDENTIFYING HUMAN ERROR POTENTIAL

Human factors technology deals with the actions of people in their working environment, and therefore for analytical completeness, not only must we consider the potential impact of human failure, but we must also acknowledge the individual differences of each person performing a particular task. This will provide general insight into the likelihood of system success, and it will also furnish specific insights into the vulnerability of the assessment to human variability, and the sorts of factor that can contribute to this vulnerability. Such an approach is analogous to that adopted by risk analysts when dealing with the ranges used to characterise the probability of failure of components or sub-systems. The approach can be particularly important when human actions are known to dominate or contribute strongly to effective system performance, or when slight changes in levels of human performance can alter the resulting system reliability in a major sense.

The collected wisdom of many practitioners applied to the identification and prevention of human error is now available from a number of sources such as Seminara et al. (1980), Kinkade and Anderson (1984), Health and Safety Executive (HSE, 1987) and the HFRG (1991). In addition, the two recent papers by Kirwan (1992a and 1992b), already mentioned, consider in some detail the appropriateness of a range of specific human error identification techniques, and the ways in which they may be used to identify the potential for human error. Although the methods exhibit some variability and do not always produce equally detailed insights, as a whole when applied as guidelines, they do nevertheless comprise a substantial body of knowhow for application to this problem area.

MODIFICATION OF GENERIC FAILURE FREQUENCIES

It is clear from the work of researchers such as Paté-Cornell and Bea (1992) that management errors and system reliability are inextricably linked. For some time now PSM researchers have been trying to identify and develop ways of modifying generic failure frequencies on the basis that some plant managements are noticeably "bad" and some are noticeably "good". The argument continues that if a management is verifiably "good", then it should not be penalised with respect to a noticeably "bad" management, when generic plant risk calculations are performed. Likewise, if management performance could be said to be "better than average" then perhaps, it has been argued by some, it might be reasonable for a PSM assessor to utilise equipment failure rates that are somewhat more optimistic than generic values.

Several systems have been developed to assess the quality of safety management or to quantify the human role in risk assessment studies. For example, the International Safety Rating System, ISRS (Bird and Germain, 1985) has been developed as an audit technique to provide an assessment of the quality of safety management, and the Instantaneous Fractional Annual Loss, IFAL, technique has been developed to indicate where there may be potential areas of loss that could be attributable to safety management effectiveness (Whitehouse, 1987).

As mentioned, a number of attempts are being made at present to link assumptions regarding PSM and QRA. These have evolved from the work of Powell and Canter (1985) and Bellamy et al. (1989), for example, on the quantified impact of human failure on industrial accident frequency. Further indications of the general soundness of such ideas in relation to attitude and measured safety performance are given in Canter and Donald (1990) and Ward (1992), and descriptions of various approaches to incorporating safety management influences within QRA are given in Pitblado et al. (1990), Smith (1992), Mallett (1992) and Hurst et al. (1990).

Pitblado et al's approach is to modify generic frequencies in relation to broad indications of safety management effectiveness, whereas Smith's approach is to use the variability inherent in Lost Time Injury data, combined with the respective contributions of the International Safety Rating System's elements' judged to be made to overall top event management and organisational control. Mallett's proposed methodology is to derive a Facility or Unit Risk Factor which could be used to adjust risk by means of judgement based on previous hazardous materials incidents, code violations and active participation in community emergency preparedness and joint emergency response exercises. Hurst et al.'s approach is centered on pipe and vessel failure data, taking account of identified management influences in order to match predicted failure and recovery modes to those observed, and thereby weight each part of the safety management assessment in relation to its specific contribution to component failure. In their various ways each method seeks to relate the generalised belief that a battery of safety management indicators might be used to "measure" an organisation's ability to control its overall risk, and perhaps thereby quantify, and if appropriate, adjust, the influence of its PSM on assessed risk.

Although such notions are attractive at face-value, they are beginning to receive some criticism for a number of reasons. There have been criticisms of the factors used to indicate safety management success, and the weightings which may or may not be applied to such factors. This paper cannot fully represent the various factional views, but simply highlights the debate that is currently underway. On a different, but related subject, Tweeddale (1992) has said that accidents do not result from a poor average condition of hardware, but from particular instances, which usually involve abnormalities. He goes on to say that it is important to undertake risk assessments as if quantitative assessments cover all risks, or that the quantifiable risks are not greatly affected by unquantifiable factors.

Tweeddale says, that "the patently phony features of quantitative risk assessment (such as manipulation of data by unsubstantiated judgment; reliance on the final value of the assessed risk for decision making) will have to be discarded", and he believes that "the path to better management of major industrial hazards is not primarily more and better mathematics, but more widespread and deeper understanding of the hazards and of the means of controlling them, and a stronger commitment to doing so."

Modification of Risk (MOR), it seems, is a subject that will not lie down, and as mentioned, there are various initiatives underway to investigate the feasibility of such an approach to risk assessment. The subject is complex, and affected by a multitude of influencing factors, including, as has been demonstrated by Rushton (1992), the design of protection systems, which, whilst notionally producing the same overall probabilistic effect, are vulnerable to completely different types of management failure. Whether it will prove possible to reduce the influences to simple easily-quantified factors remains unresolved at present.

HUMAN PERFORMANCE IMPACT ON ENGINEERED SYSTEMS

As has been highlighted, there is a growing realisation that there is an inextricable linkage between human performance and engineered system performance. As a minimum such links can be seen to affect accident rates, and the availability of systems both for production purposes, and for safety systems actuation (Guastello, 1991 and Kletz, 1990).

Ellingwood in Blockley (1992) reports that only 10 to 20 percent of structural failures are traceable to stochastic variability in loads or strengths, the remaining 80 to 90 per cent, he says, are due mainly to errors. He goes on to explain that most risk and reliability analyses developed to date have not considered failures due to error, and he notes that human error effects generally cannot be related to the random deviations in the loads and strengths that current risk analyses would take into account. This type of error, Ellingwood explains, may correspond to a different event entirely, one that may change the applicable probabilistic models and the relevant limit state as well, i.e. the design is not as intended, and not covered by the Process Safety Assessment. This is a most important issue, because the phenomenon is well-understood by the public.

HUMAN RELIABILITY IMPACT ON SAFETY CASES

The impact of human reliability on safety cases is now well-established (Health and Safety Commission, 1991), and the means to incorporate such considerations have been considered by a range of authors. Waters (1990) has shown how to match human factors assessment effort against risk management principles, whilst Bridges et al. (1992) have illustrated how to integrate human factors engineering into PSM. The premises for incorporating human factors considerations in PSM have been considered by Bellamy et al. (1986), Kirwan (1988b) and Embrey (1990). Carey and Bennett (1992) and Hashemi (1992) have shown that human reliability assessment should play an intimate role in safety case preparation, and they have given examples of how human factors principles can be applied in an integrated fashion, whilst Williams (1990) and Livingston (1992) have described how the process operates, and is regulated, to achieve the desired results.

Delboy et al. (1991) have shown the sensitivity of process safety to human error, and Cox (1991) has discussed the question of integrating human factors into numerical safety analysis, indicating that an integrated analysis involving human factors considerations represents a tool for creating a very much more balanced and authentic overview of total plant safety. It is clear that organisations are now taking on board the need to give comprehensive consideration to human factors and the ways in which safety assessments and cases may be affected by human behavior, and these principles are outlined in Cox (1992).

IMPACT OF HUMAN BEHAVIOR ON QRA

At present significant difficulties can be experienced by analysts and assessors when interpreting and adapting such limited quantifiable human probability of failure data as are available. As already mentioned, it is now recognised that safety management systems may be expected to have a major influence on failures and the rate at which they occur. Currently, when performing QRAs in order to determine what the major components of risk might be it is normal to examine the principal failure cases for their susceptibility to human failures with respect to the initiation, control, consequence, impact and mitigation of foreseeable accident sequences. This approach has taken on a special urgency in the light of observations by Hurst et al. (1992) suggesting that about 40% of pipework and vessel ignitions are influenced by human and managerial factors which are not modelled explicitly in classical QRA procedures. As a consequence of such considerations, the impact of human factors in relation to QRA is now becoming better established.

CONCAWE (1988) has highlighted the fact that, "QRA should be viewed as a supplement to more established safety techniques rather than a substitute", and it observes that, "difficulties about the adequacy of human factors considerations in QRA have led some to question the utility or even validity of such quantified approaches as a predictive tool to aid risk management decision-making". Williams and Hurst's (1992) work on a comparison of safety managements in relation to measured occupational safety, however, has indicated that not only are human factors and PSM matters important, but that worthwhile judgements can be made with respect to overall occupational safety that match observed data.

It will be clear from the foregoing that not only is there a need to make accurate assessments of human error and their contribution to system unreliability, but there is a growing requirement to characterise the amount of human variability that might be associated with any such assessments in order to predict and control the consequences. The use of human reliability analysis in risk studies in order to determine the impact of human behavior on QRA has been addressed by Whalley (1991).

METHODS FOR MODIFYING SYSTEM FAILURE ESTIMATES

At present, most QRA studies use deterministic failure rates derived from a limited range of data sources, mainly power station and large chemical plant facilities in the Northern Hemisphere. These historical data are normally referred to as generic data. When generic failure rate data are used in QRA studies it implies that the same average value would be appropriate for all sites, regardless of the quality of safety management systems employed. As has been highlighted, however, this assumption is difficult to believe and/or substantiate, and it is clear that QRA studies need to be supported with a reasonable qualitative overview of the effectiveness of PSM, and preferably some quantitative estimate as well.

It was noted in the Bellamy et al. (1989) review of pipe and vessel failure causation that over 80% of such failures might have been preventable or safely recovered. Also, as there appears to be a factor of 1000 variation (Rijnmond Study, 1982), between failure frequencies, it has been suggested by Pitblado et al. (1990) therefore, that a non-linear assessment scheme covering 3 orders of magnitude of management influence might be an appropriate way to provide the necessary means to connect the influence of PSM to modifying system failure rates for use in QRA.

Bellamy et al.'s work on failed recovery or preventive mechanisms associated with pipe and vessel failure suggested that about a third of the failures observed in design might have been prevented via hazard review, and that about a quarter of the failures that are observed during operations might be prevented by means of human factors review. These failures represent assumptions about PSM systems that did not hold true. Methods for ascertaining the extent to which such assumptions might not be borne out in practice hold out the prospect of identifying the amount by which generic failure frequencies perhaps ought to modified. Such an approach also holds out the prospect of considerable risk reduction, if suitable means can be found to identify the overall risk potential represented by a particular management.

HUMAN FACTORS AND PROCESS SAFETY MANAGEMENT IMPROVEMENT

Human performance in industrial tasks is determined by the nature of the tasks personnel are called upon to perform and the circumstances under which such demands are made. It is important, therefore, for engineers and executives to possess a good working knowledge of the factors that are known to influence task failure, how these might be identified, and how their probable effects on the success of the mission might be evaluated.

In order to harness the positive contribution the human being can make to enhance process operation and safety, and diminish the negative contribution he can unwittingly make to otherwise seemingly reliable systems, the process safety manager also needs a working knowledge of man's capabilities and limitations. This, together with an understanding of how operators interact with systems, and an appreciation of the likelihood of error given certain task conditions, and the means to assess the likely contribution of error-producing conditions to prevailing work situations, will enable him to understand the principal components affecting predicted performance.

If an understanding of the key human factors issues throughout a chemical project's lifetime can be given to a wide range of personnel, it is possible that a step-change in PSM effectiveness might be achieved. A workable methodology would be to initiate a human factors sensitisation programme, teach workforces about human performance, its variability, the ways in which human performance can affect process safety and the assumptions that are made in areas that are known to impinge upon the effectiveness of PSM.

The chemical and process industries already apply comprehensive measures to ensure that the levels of safety and availability achieved are of a high order. The means to achieve even higher levels of human performance are now becoming less technically orientated, and much more driven, and limited, by human capabilities. Incident report data, whilst showing gradual signs of improvement, still indicate a large exposure to the impact of human performance, and in order to make sizeable changes in the reported data, it may be appropriate to consider adopting broad measures across industries as well as the more traditional focused approach.

Process Engineering is about emphasising a few key attributes and, by design, ignoring the potential contribution that may be made by others in as scientific a way as possible. However, as Dykes (1976) has observed, "engineering is the art of modelling materials we do not wholly understand into shapes we cannot precisely analyze, so as to withstand forces we cannot properly assess in such a way that the public has no reason to suspect the extent of our ignorance". This activity also takes into account implicit assumptions about how human beings may be expected to interact with engineered systems, and it behoves an engineer to have at least as good a grasp of the users' probable performance, as the hardware's probable behavior under load. Because of the inherent uncertainties in both engineering and human engineering, it should not be too surprising, therefore, that, despite our best endeavors, we may still experience some difficulties with regard to both the engineering of process systems and the assessment of their safety.

It will be apparent that there are a number of ways in which human variability might be controlled and exploited (Rimland and Larson, 1986). Indeed, some appropriate measures are already taken, but because they have become so endemic in our lives, some of us would not recognise them for what they are. The minimum height requirements for entry into the police, the requirement for good colour vision in pilots, the age limits on drivers, judges, and juries, are all indications of a desire to reduce sources of variability that might be prejudicial to appropriate human performance. Such an approach could perhaps be extended to operations personnel to include requirements to demonstrate fault diagnostic skills within prescribed limits, and the more exacting definition of the intellectual capabilities of plant managers, again, within prescribed limits.

DISCUSSION

Whilst it is obvious that human performance is variable, it should no longer be necessary to claim that human performance is so variable that it cannot be predicted in relation to process safety assessment. For process safety assessments, such a position is not tenable, because the in-built analytical uncertainty, when combined with the variability of human performance may be sufficient to render any predictions that may be made, relatively meaningless. Furthermore, it is clear that because reasonably good predictions can be made, these should be stated if possible, and human performance variability control measures applied from the literature as appropriate to the circumstances and consequences of the prediction, in order to support the principal features of the assessment.

Various ways of assessing and accommodating human performance variability have been explored, including the identification and assessment of error-producing conditions, the use of safety management and human factors guidelines and review techniques. These have been considered in relation to the ways in which process safety assessments are made, and it has been shown that there is compelling evidence that human performance influences, in their own right, can have a sizeable impact on QRAs.

The variability of these impacts, however, seems to be embedded more in the diversity of the man-machine system failure modes than in the variability of the personnel involved in design and operation. This, it seems, is the principal reason why there could be a case for modifying nominal QRA failure rates, not because of direct human performance variability, but because the diversity of the failure modes causes the variability of the systems performance to increase to a disproportionate extent.

CONCLUSIONS

This paper has shown that the impact of human performance in relation to process safety assessments can be quite considerable, and that such assessments can be altered in significant ways, even when quite small changes in performance are assumed. It has also shown that, once a nominal assessment has been made, human variability is quite tightly constrained around the nominal value, and is generally little greater than the known variability of components.

Therefore a good way in which to incorporate human variability in process safety assessments will be to utilise the methods described for human error identification, to recognise that the actual task variability, when defined, will be relatively small, and to modify any failure or mitigation frequencies on the basis of the contribution that can genuinely be attributed to human performance influences. Addressing human performance variability issues using these means allows them to be confronted and incorporated in process safety assessments, so that they become an integral part of the safety management process, rather than remaining a somewhat perplexing after-thought.

REFERENCES

Atkinson, A., Davis, M. and Fergusson, M. (1988) *Nuclear Facilities and Emergency Planning in the United Kingdom:A Report to the National Steering Committee of the Nuclear Free Zone Local Authorities*. London: Earth Resources Research Limited, 1988

Bellamy, L.J., Kirwan, B. and Cox, R.A. (1986) Incorporating Human Reliability into Probabilistic Risk Assessment. in *proceedings of the 5th. International Symposium on Loss Prevention and Safety Promotion in the process Industries*, European Federation of Chemical Engineering 337th. event

Bellamy, L.J., Geyer, T.A. and Astley, J.A. (1989) *Evaluation of the Human Contribution to Pipework and In-Line Equipment Failure Frequencies*. Contract Research Report No. 89/15, Health and Safety Executive, Sheffield, ISBN 0717603245, 1989

Bellamy, L.J. and Geyer, T.A.W. (1991), (ed. J.C. Williams), *Organisational, Management and Human Factors in Quantified Risk Assessment, Interim Report 1*, Sheffield, UK: Health and Safety Executive

Bird, F.E. and Germain, G.L. (1985), *Practical Loss Control Leadership*, International Loss Control Institute, Atlanta, Georgia

Blockley, D. (ed.) (1992) *Engineering Safety*. London: McGraw-Hill, p.112

Bodsberg, L. Ingstad, O., Sten, T. (1987) *Alarm and shutdown frequencies in offshore production*, SINTEF Report N-7034, Trondheim, Norway

Brearley, S.A. (1990) *High Level Management and Disaster*, Disaster Prevention Conference, Bradford University, October 1990

Brian, P.L.T. (1988) Managing Safety in the chemical industry. In *Institution of Chemical Engineers Symposium Series No.110, Preventing Major Chemical and related accidents*, Rugby, pp. 615-619

Bridges, W.G., Kirkman, J.Q. and Lorenzo, D.K. (1992) Strategies for Integrating Human Reliability Analysis into Process Hazard Evaluations. *In proceedings of an International Conference on Hazard Identification and Risk Analysis, Human Factors and Human Reliability in Process Safety*, New York: AIChE

Brookhaven National Laboratory (1989) *Risk Sensitivity to Human Error*. NUREG/CR-5319, U.S. Nuclear Regulatory Commission, Washington, D.C., 1989

Brown, W., Crouch, D., Higgins, J., O'Hara, J. and Wong, S. (1990) *The Quantification of Human Variability and its Effect on Nuclear Power Plant Risk*. Department of Nuclear Energy, Brookhaven National Laboratory, Upton, New York, 11973, Technical Report L-1170 (1) 10/90.

Butler, M.C. and Jones, A.P. (1979) Perceived Leader Behavior, Individual Characteristics, and Injury Occurrence in Hazardous Work Environments. *Journal of Applied Psychology*, **64**, 3, 299-304

Canter, D. and Donald, I. (1990) *Accidents by Design: Environmental, Attitudinal and Organisational Aspects of Accidents*, Guildford: University of Surrey

Carey, M.S. and Bennett, G.R. (1992) Addressing human factors throughout the safety lifecycle of an installation. *Offshore Safety:Protection of Life and the Environment*, Marine management (Holdings) 1992., 8-1 to 8-10

CONCAWE (1988) *Quantified Risk Assessment*. CONCAWE, Report no. 88/56, pp. 7-9

Cox, R.A. (1991) Use of Quantitative Risk Assessment in Design of Plant. in Risk and The Energy Industries, *Proceedings of the Watt Committee on Energy Twenty-Seventh Consultative Conference*, 18 September 1990, University of Birmingham, chapter 3, pp. 10-11

Cox, R.F. (1992) The human factor in safety and reliability, *Atom*, May/June 1992, pp. 30-33

Deming, W.E. (1982) *Quality, Productivity and Competitive Position*, Cambridge, Mass: MIT Center for Advanced Engineering Study, 1982

Delboy, W.J., Dubnansky,R.F. and Lapp, S.A. (1991) Sensitivity of Process Risk to Human Error in an Ammonia Plant, *Plant/Operations Progress*, **10**, 4, 207-211

DTI (1990) *The case for quality, and the case for costing quality*. publications 216 and 212, London: Department of Trade and Industry

Doyle, W.H. (1969) Industrial Explosions and Insurance. *Loss Prevention*, 1969, **3**, 11-17

Dykes A.R. (1976) *Address to the British Institute of Structural Engineers*, 1976

Embrey, D.E. (1986) Approaches to aiding and training operators' diagnoses in abnormal situations, *Chemistry and Industry*, **7**, 454-459

Embrey, D.E. (1990) A Framework for the management of human reliability in quality management. In *Proceedings of International Conference of Mechanical Engineers, Quality Management in the Nuclear Industry: The Human Factor*, 1990, pp. 91-98

Engineering Council (1991) *Engineers and Risk Issues - Code of Professional Practice*. London: Engineering Council, September 1992

Gall, W. (1988) Error Analysis. *SRD Human Reliability Course Notes*, UK Atomic Energy Authority, Culcheth, Warrington

Garrison, W.G. (1983) *One Hundred Largest Losses. A Thirty Year Review of Property Damage in the Hydrocarbon-Chemical Industries (6th. Edition)*. Marsh and McLennan Protection Consultants, Illinois, 1983

Gaunt, L.D. (1989) *The Effect of the International Safety Rating System (I.S.R.S.) on Organizational Performance*. Research Report Series No. 89-2, The Center for Risk Management and Insurance, Georgia, 1989

Ghertman, F. and Griffon-Fouco, M. (1985) Investigation of Human Performance Events at French Power Stations. *Conference Record for IEEE Third Conference on Human Factors and Nuclear Safety*, June 23-27, Monterey, California, U.S.A., pp.49-59

Gitomer, D.H. (1988) Individual Differences in Technical Troubleshooting. *Human Performance*, I(2), 111-131

Guastello, S.J. (1991) Some further evaluations of the International Safety Rating System. *Safety Science*, **14**, 253-259

Hashemi, K. (1992) Human Factor Considerations in Safety Techniques and Management. in *Proceedings of a seminar on The Practicalities and Realities of Human Factors in Offshore Safety, 30 Sept-1 October 1992*, Aberdeen, London: Business Seminars International

Hawksley, J.L. (1984) Some social, technical and economical aspects of the risks of large chemical plants. in *Proceedings of the CHEMRAWN III World conference (Chemical Research Applied to World Needs)*, The Hague, 25-29 June 1984, Paper 3.V.2

Health and Safety Executive (1990) *Out of Control*. HSE Guidance Document on inadequacies of Control systems, Bootle: HSE, 7 July, 1990, Draft

Health and Safety Commission (1991) *Second Report: Human Reliability Assessment - a critical overview*. ACSNI Study Group on Human Factors, London: HMSO

Hudson, P.T.W. (1991) *Prevention of Accidents involving Hazardous Substances:The Role of Human Factor in Plant Operations*. Discussion Document, OECD, Paris

HFRG, Human Factors in Reliability Group (1991) *Guide to Reducing Human Error in Process Operations*. London, HMSO, ISBN 0-85356-357-8

Hunns, D. (1991) An Examination of Factors affecting Human Reliability. In *Hazards X: Process Safety in Fine and Speciality Chemical Plants Including Developments in Computer Control Plants*, Symposium Series 115, Rugby: Institution of Chemical Engineers, 347-359

Hurst, N.W., Bellamy, L.J. and Geyer, T.A.W. (1990) Organisational, Management and Human Factors in Quantified Risk Assessment. A Theoretical and Empirical Basis for Modification of Risk Estimates. *Safety and Reliability in the '90s (SARRS '90)*, edited by M.H. Walter and R.F. Cox, London: Elsevier Applied Science

Hurst, N.W., Bellamy, L.J. and Wright, M.S. (1992) Research Models of Safety Management of OnShore Major Hazards and their possible Application to OffShore Safety. in *Proceedings of a Symposium on Major Hazards Onshore and Offshore*, Symposium Series No. 130, Rugby: Institution of Chemical Engineers, pp. 129-148

INPO (1985) *An Analysis of Root Causes in 1983 and 1984 Significant Event Reports*. INPO Report 85-027, Georgia: Institute of Nuclear Power Operations, July 1985

Instone, B.T. (1989) *Losses in the Hydrocarbon Process Industries*. presentation made at the 6th. International Symposium On Prevention and Safety Promotion in the Chemical Industries, Oslo, Norway, 19-22 June 1989

Jenkins, A.M., Brearley, S.A. and Stephens, P. (1991)   *Management at Risk.* SRDA-R4, AEA Technology, Culcheth, Warrington: The SRD Association, 1991

Joschek, H.I. (1981) Risk Assessment in the Chemical Industry. in *Proceedings of the ANS/ENS Topical Meeting on Probabilistic Risk Assessment, Port Chester, NY, September 1981,* La Grange Park, Il1: American Nuclear Society

Joschek, H.I. (1986) Results of a study on causes of accidents. in *proceedings of 5th. International Symposium on Loss Prevention and Safety Promotion in the Process Industries,* Cannes 15-19 September 1986

Juran, J. (1979) *Quality Control Handbook, 3rd edition.* McGraw-Hill, New York, 1979

Kinkade R.G. and Anderson, J. (1984) *Human Factors Guide for Nuclear Power Plant Control Room Development.* EPRI Report NP-3659, Palo Alto, CA: Electric Power Research Institute

Kirwan, B. (1988a)   A Comparative Evaluation of Five Human Reliability Assessment Techniques. in *Proceedings of the Safety and Reliability Society Symposium 1988, Human Factors and Decision Making: their Influence on Safety and Reliability,* edited by B.A. Sayers, London: Elsevier Applied Science, pp. 87-109

Kirwan, B. (1988b) Integrating Human Factors and Reliability into the Plant Design and Assessment Process, in *Contemporary Ergonomics, Proceedings of the Ergonomics Society Conference,* UMIST, 11-15 April, 1988, London: Taylor and Francis

Kirwan, B., Martin, B., Rycraft, H. (1990), Human Error Data Collection and Data Generation, *International Journal of Quality and Reliability Management,* **7**, no. 4, pp. 34-66

Kirwan, B. (1992a) Human error identification in human reliability assessment. Part 1: Overview of approaches. *Applied Ergonomics,* **23**, 5, 299-318

Kirwan, B. (1992b) Human error identification in human reliability assessment. Part 2: Detailed comparison of techniques. *Applied Ergonomics,* **23**, 6, 371-381

Kletz, T.A. (1990) *An Engineer's View of Human Error - 2nd. edition.* Rugby: Institution of Chemical Engineers

Kletz, T.A. (1991) Mistakes Operators Make. *Enhancing Process Operator Skills, Symposium Papers 1991 No.4*, Institution of Chemical Engineers, North Western Branch, Wilmslow, 2 October 1991

Kletz, T.A. (1992) *HAZOP AND HAZAN - Identifying and Assessing Process Industry Hazards, Third Edition.* Rugby: Institution of Chemical Engineers, pp.130-133

Layfield, F. (1987) *Sizewell 'B' public inquiry - Summary of conclusions and recommendations.* London: HMSO

Livingston, A. (1992) Incorporation of Human Factors in Safety Cases. in *Proceedings of a seminar on The Practicalities and Realities of Human Factors in Offshore Safety, 30 Sept-1 October 1992*, Aberdeen, London: Business Seminars International

Lorenzo, D.K. (1990) *A Manager's Guide to Reducing Human Error.* Washington: Chemical Manufacturers Association Inc.

Mallett, R. (1992) Rate your risk management plans. *Hydrocarbon Processing*, August 1992, 111-115

Mill, R.C. (ed) (1992) *Human Factors in Process Operations*. Rugby: Institution of Chemical Engineers. p.43

Paté-Cornell, M.E. and Bea, R.G. (1992) Management Errors and System Reliability: A Probabilistic Approach and Application to Offshore Platforms. *Risk Analysis*, 12, 1, 1-18

Petersen, D. (1990) *Safety Management, 2nd. Edition.* Goshen, NY: Aloray, 1990

Pitblado, R.M., Slater, D.H. and Williams, J.C. (1990) Quantitative Assessment of Process Safety Programs. *Plant/Operations Progress*, 9, 3, 169-175

Powell, J. and Canter, D. (1985) Quantifying the Human Contribution to Losses in the Chemical Industry. *Journal of Environmental Psychology*, 5, 37-53

Rasmussen, J. (1980) What can be learned from Human Error Reports? in *Changes in Working Life*, Duncan, K.D., Gruneberg, M.M., and Wallis, D. (eds), J. Wiley and Sons, Chichester, UK

Reason, J.T. (1991) *On the Nature of Safety and the Art of Managing the Manageable*. Workshop on Control of Safety, Bad Homburg, 2-4 May 1991, Werner Reimer Institute, Bad Homburg

Reber, R.A., Wallin, J.A. and Chhokar, J.S. (1990) Improving Safety Performance with Goal Setting and Feedback. *Human Performance*, 3, 1, 51-61

Rimland, B. and Larson, G.E. (1986) Individual differences: an underdeveloped opportunity for military psychology. *Journal of Applied Social Psychology*, 16, 5656-575

Rijnmond Public Authority (1982) *A Risk Analysis of 6 Potentially Hazardous Industrial Objects in the Rijnmond Area - A Pilot Study*. Netherlands and Boston: D. Reidel, 1982

Rushton, A.G. (1992) Protective Device Faults - Vulnerability to Management Failure. in *Proceedings of a Symposium on Major Hazards Onshore and Offshore*, Rugby: Institution of Chemical Engineers Symposium Series No. 130, pp. 183-194

Samanta, P.K. Swoboda, A.L. and Hall, R.E. (1981) *Sensitivity of Risk Parameters to Human Errors in Reactor Safety for a PWR*. NUREG/CR-1879, U.S. Nuclear Regulatory Commission, Washington, D.C., 1981

Samanta, P., Wong, S., Haber, S., Luckas, W., Higgins, J. and Crouch, D. (1989) *Risk Sensitivity to Human Error*. NUREG/CR-5319, U.S. Nuclear Regulatory Commission, Washington, D.C.

Society of International Gas tanker and Terminal Operators (1990) *Human Error Reduction Strategies*. London:SIGTTO Information Paper No.9

Sehdev, S.K. et al. (1990) *The International Safety Rating System:An Evaluation of Effectiveness*. Industrial Accident Prevention Association, Toronto, Canada

Seminara, J.L., Eckert, S.K., Seidenstein, S., Gonzalez, W.R. and Stempson, R.L. (1980) *Human Factors Methods for Nuclear Control Room Design, Volumes 3 and 4*, EPRI Report NP-1118, Palo Alto, CA: Electric Power Research Institute

Smith, A.J. (1992) *The Development of a Model to Incorporate Management and Organisational Influences in Quantified Risk Assessment*. Sheffield, UK: Health and Safety Executive

Snaith, E.R. (1981) *The correlation between the predicted and observed reliabilities of components, equipment and systems*. Report No. NCSR R18, Culcheth, Warrington, UK: UK Atomic Energy Authority

Spiker, V.A., Harper, W.R., and Hayes, J.F. (1985) The Effect of Job Experience on the Maintenance Proficiency of Army Automotive Mechanics. *Human Factors*, **27** (3), 301-311

Suokas, J. (1989) The identification of human contribution in chemical accidents. *6th. International Symposium on Loss Prevention and Safety Promotion in the Process Industries*, Oslo, 19-22 June 1989

Sweeney, J.C. (1992) Measuring Process Safety Management. *Plant/Operations Progress*, 11, 2, 89-98

Taylor-Adams, S.E. and Williams, J.C. (1992) Characterising, Predicting, and Controlling Human Variability for System Reliability Assessments. *Safety and Reliability '92*, Proceedings of the European Safety and Reliability Conference '92, 10-12 June 1992, London: Elsevier Applied Science, pp. 121-139

Taguchi, G. and Clausing, D. (1990) Robust Quality. *Harvard Business Review*, January-February 1990, 65-75

U.S. Department of Labor (1990) *The Phillips 66 Company Houston Chemical Complex Explosion and Fire, A Report to the President.* Washington, D.C.: Occupational Safety and Health Administration

Ward, R.B. (1989) Towards a Management System to Reduce Risk. *Chemeca 89 Technology for our Third Century*, Broadbeach, Queensland, 23-25 August, 1989

Waters, T.L. (1990) The Development of a Human Factors Strategy for the Design and Assessment of Plant and Equipment. in *Proceedings of the Safety and Reliability Society Symposium 1990*, Altrincham, London: Elsevier Applied Science, pp. 107-113

Whalley, S.P. and Lihou, D. (1988) Management Factors and System Safety. in *Human Factors in Decision Making*, London: Elsevier Applied Science, pp. 172-188

Whalley, S.P. and Kirwan, B. (1989) An evaluation of five human error identification techniques. in *Proceedings of 5th. International Loss Prevention Symposium*, Oslo, June 1989

Whalley, S.P. (1991)   The use of Human Reliability Analysis in Risk Studies. *Proceedings of a Workshop on Risk Analysis in the Process Industries*, 12 -14 March 1991, Chester, London: IBC Technical Services Ltd.

Whitehouse, H. (1987) IFAL - A new risk analysis tool. in *Symposium Series No. 93*, Rugby: Institution of Chemical Engineers, pp. 309-322

Williams, J.C., and Willey, J. (1985)   Quantification of Human Error in Maintenance for Process Plant Probabilistic Risk Assessment. *Proceedings of a Symposium on the Assessment and Control of Major Hazards,*, Rugby, England: Institution of Chemical Engineers, Symposium Series No. 93, pp.353-366.

Williams, J.C. (1988)   A Human Factors Data-Base to Influence Safety and Reliability. in *Human Factors and Decision Making: Their influence on Safety and Reliability*, edited by B.A. Sayers, London: Elsevier Applied Science

Williams, J.C. (1989)   Human Reliability Data - The State of the Art and the Possibilities. in *Proceedings of Reliability '89, Volume 1*, pp.3B/5/1 - 3B/5/16

Williams, J.C. (1990) Quality management of human factors technology applied to Sizewell 'B'. in *Quality Management in the Nuclear Industry - The Human Factor*, London: Institution of Mechanical Engineers, C409/026, pp. 87-90

Williams, J.C. and Hurst, N.W. (1992) A comparative study of the management effectiveness of two technically similar major hazard sites. in *Proceedings of a Symposium on Major Hazards Onshore and Offshore*, Rugby: Institution of Chemical Engineers Symposium Series No. 130, pp. 73-84

Wilpert, B. (1991) *System Safety and Safety Culture*. IAEA, IIASA meeting on "The Influence of Organization and Management on the Safety of NPPs and other industrial systems, Vienna, Austria, 18-20 March 1991

Wilson, K. (1990) The Human Factor in safe plant operation. *Energy World*, February 1990, pp. 8-10

Wong, S., Higgins, J., O'Hara, J., Crouch, D. and Luckas, W. (1990) *Risk Sensitivity to Human Error in the LaSalle PRA*, NUREG/CR-5527, U.S. Nuclear Regulatory Commission, Washington, D.C.

Auditing Performance-Based Process Safety Management Systems

Henry Ozog

Arthur D. Little, Inc., Acorn Park, Cambridge, MA 02140-2390

As companies start implementing their process safety management (PSM) programs in response to regulations, industry guidelines and company standards, they need to first assess the status of their programs. This initial or baseline audit is focussed on identifying gaps in the PSM program elements and associated management systems. After these programs are implemented, maintenance audits will be needed to ensure that programs are functioning effectively. Any PSM compliance audit, whether baseline or maintenance, should assess the program content against both applicable regulations, standards and guidelines, and the characteristics of good management systems to ensure PSM programs are effective and will continue to function effectively.

This paper summarizes the contents of the new CCPS book "Auditing Process Safety Management Systems" published in January 1993.

INTRODUCTION

Companies have long recognized the need for process safety in operations where hazardous materials are handled. However, only recently have guidelines for process safety management (PSM) been recognized as a necessary part of an overall risk management program. In 1989, CCPS published "Guidelines for Technical Management of Chemical Process Safety" (1) which defined twelve elements of a comprehensive process safety management program as shown in Table 1. Since publication of that classic reference document, many other industry groups including the Chemical Manufacturers Association (2), and the American Petroleum Institute (3) have published codes and recommended practices for process safety management. In addition, the US Occupational Safety and Health Administration (4) has published a final rule on process safety management. Each of these documents has recognized the need for verification of PSM programs and includes auditing as one of the elements of a complete PSM program.

TABLE 1
Twelve Elements of Chemical Process Safety Management

- Accountability: Objectives and Goals

- Process Knowledge and Documentation

- Capital Project Review and Design Procedures

- Process Risk Management

- Management of Change

- Process and Equipment Integrity

- Incident Investigation

- Training and Performance

- Human Factors

- Standards, Codes, and Laws

- Audits and Corrective Actions

- Enhancement of Process Safety Knowledge

Source: Reference 1

This paper describes the role of auditing in a PSM program and the approach to auditing against industry guidelines and governmental regulations which for the most part specify performance based PSM requirements.

MANAGEMENT SYSTEMS

In order to effectively audit process safety management systems, the auditor should understand what management systems are and how they are developed and implemented. Process safety management systems are comprehensive sets of policies, procedures and practices designed to ensure that barriers to episodic incidents are in place, in use, and effective (5). Figure 1 shows four functions in the development and implementation of process safety management systems. The process starts with the planning function which defines the goals and objectives. This function is generally performed by senior management of an organization and provides overall direction for the program. The next function is organizing. Here the specific policies and procedures are written which define how the system will work. This function is generally performed by supervisory staff and provides the overall structure to the program. The third function is implementing. Here the management system is put into action and results are generated. This function is generally performed by a wage-roll personnel who have primary responsibility for operating the process. The fourth function is controlling. This function is aimed at measuring performance by verifying that systems are in place and functioning and is generally performed by an independent individual or group to provide an unbiased audit of the overall management system. The results of the audit are then used to revise appropriate components of the other three functions.

Effective development and implementation of sound process safety management systems is the product of strong leadership which provides the necessary commitment and accountability.

An auditor needs to assess how each of these four functions is working and how effective they are. It is typical to find good management systems in any process, otherwise we would be experiencing many serous accidents in our facilities. However, many of these management systems are informal. These systems have been developed and championed by individuals or organizations which have recognized their need. In small organizations, informal management systems can be effective, particularly if they have been in place long enough to

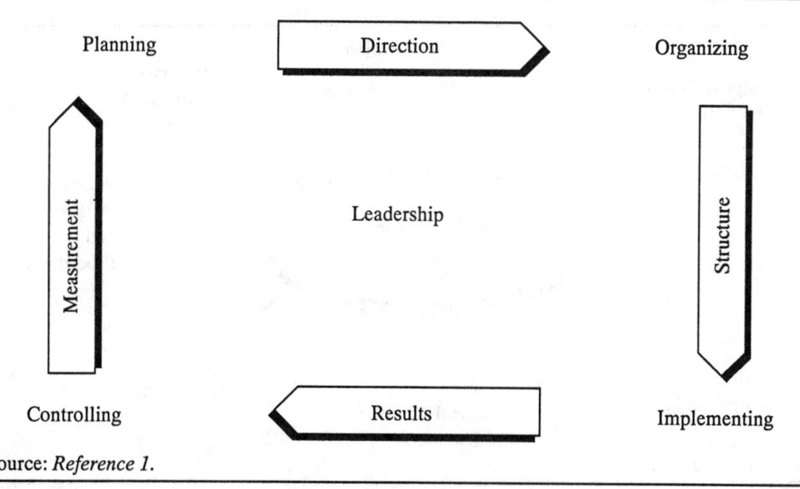

Source: *Reference 1.*

Figure 1. Functions of a process safety management system

establish a "culture" which is ingrained in the individuals who manage and operate the process. Too often, however, informal systems break down when their champions leave the organization, or when personnel are re-organized. The only truly effective way to ensure that management systems are in place and effective, is to make then formal. We have identified seven characteristics of good management systems which are summarized in Figure 2 and Table 2 (5).

AUDITING

An audit consists of three sets of activities:

- planning (pre-visit)

- gathering data and reporting findings (on-site)

- correcting deficiencies (post-visit)

The typical steps within each set of activities is shown in figure 3.

Planning

Planning involves selection of an audit team, development of an audit guide, gathering background information, and scheduling the data gathering process. The audit team should have a balance of expertise which includes familiarity with the process being audited and objectivity. A common approach used to achieve this

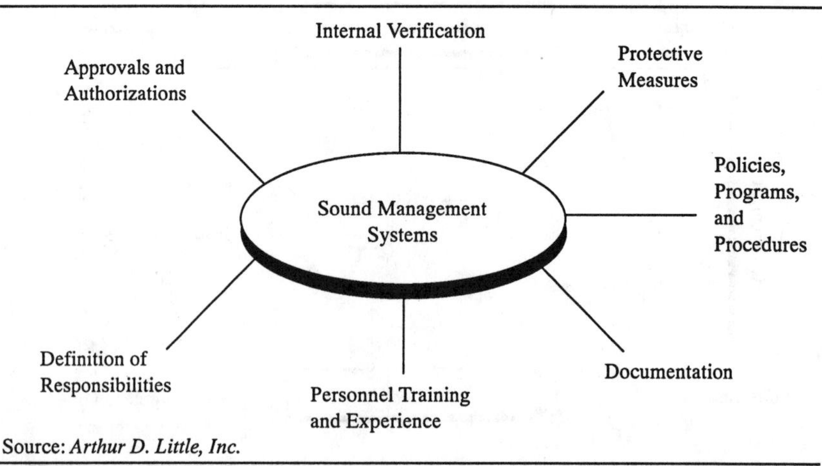

Figure 2. General characteristics of management systems

TABLE 2
General Characteristics of Management Systems

Given the absence of explicit criteria for assessing management systems, the seven characteristics listed below are frequently used to evaluate management systems. These characteristics are not absolute requirements. Facility management systems may vary significantly and still be capable of achieving the desired results.

1. **Policies, Programs, and Procedures:** Goals and objectives have been set. Written corporate and facility plans, policies, guidelines, standards and procedures are available to clearly define PSM program requirements. Management demonstrates commitment and sponsorship for these programs.

2. **Definition of Responsibilities:** Facility personnel understand their roles and responsibilities in achieving the desired level of PSM performance. Appropriate checks and balances have been established to minimize conflicts of interest.

3. **Approvals and Authorizations:** Appropriate delegations have been established, and authorities are clearly established for approval of specific routine operations and non-routine or out-of-specification operations. Approval levels are commensurate with the importance of the task.

4. **Personnel Training and Experience:** Facility personnel have sufficient experience, training, and awareness to accomplish the PSM program or activity. Personnel are familiar with applicable regulatory requirements, and internal standards and guidelines. Employees at all levels in the organization are involved in program development.

5. **Protective Measures:** Safeguards have been established to prevent or control major problems; administrative controls are in place to cross-check completion of critical operations.

6. **Documentation:** Results of PSM activities are documented as are compliance/performance results.

7. **Internal Verification:** Systems or procedures are in place for reviewing performance against standards and milestones and identifying departures from established (external or internal) standards. Exceptions and deficiencies are corrected in a timely manner.

Source: reference 6

---

balance is to select personnel from a similar process, but who don't have supervisory responsibilities for the process being audited. Another important activity is developing an audit guide (see Figure 4). The audit guide is the tool used by the auditor to direct the data gathering process and ensure all requirements are evaluated. Audit guides also help ensure consistency of audits when the personnel

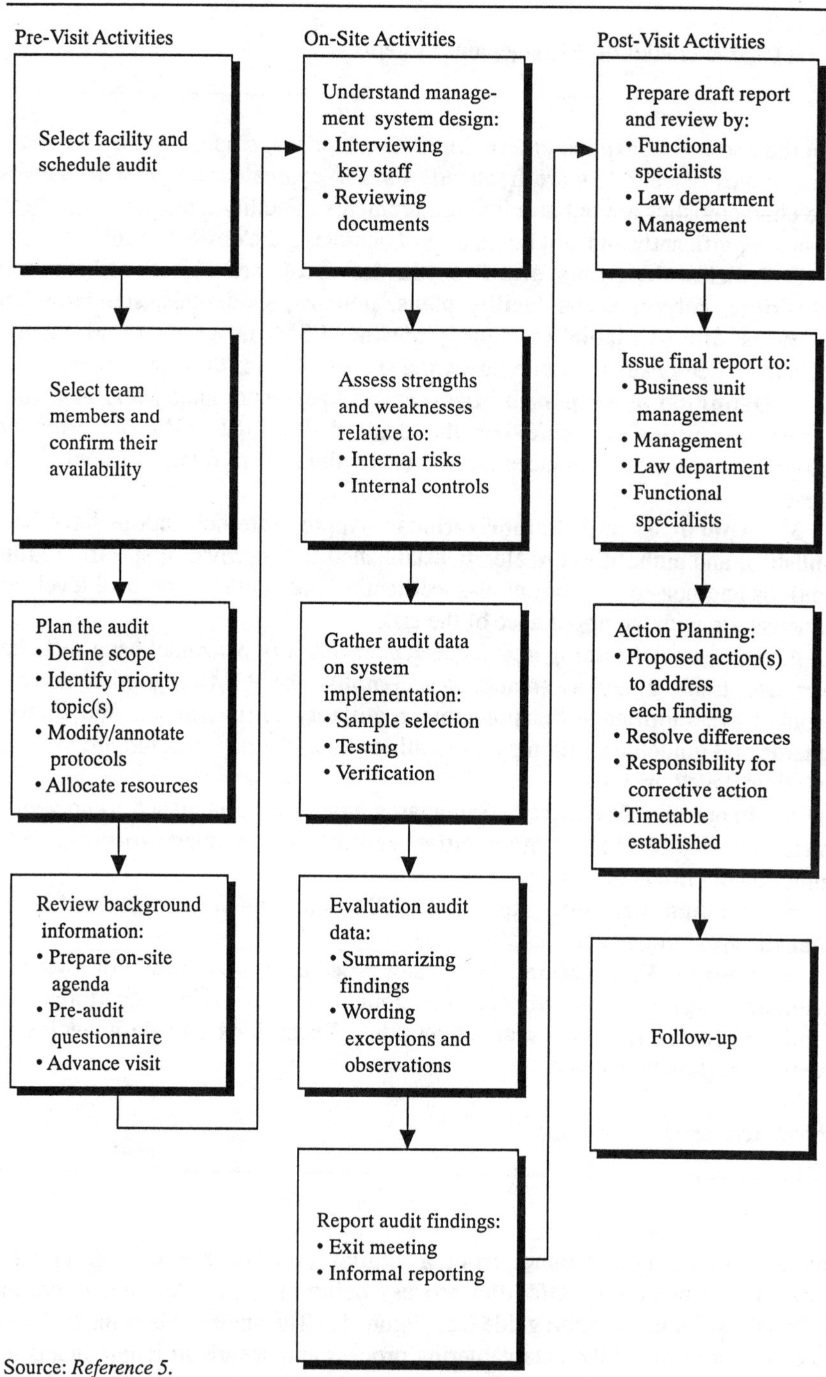

Source: *Reference 5.*

Figure 3. Typical steps in the process safety management audit process

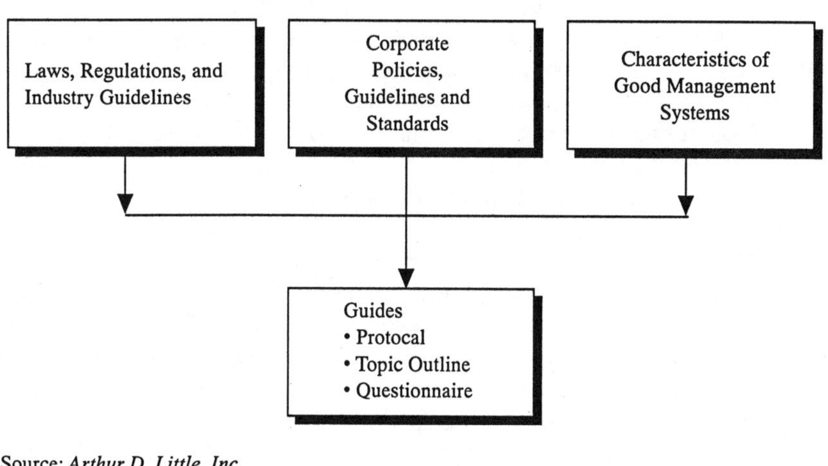

Source: *Arthur D. Little, Inc.*

Figure 4. Development of audit guides

on audit teams change. Common audits guides are protocols, topical outlines and questionnaires (5). We recommend the protocol which provides detailed instructions to the auditor on how to conduct the audit. It is particularly useful for auditors who are inexperienced in conducting audits and/or who are unfamiliar with the guidelines or regulations against which the audit is being conducted. Whichever guide is developed, it is best to organize the guide by PSM element. Background information should be collected to familiarize the audit team with the process, and to plan the time the audit team spends on-site at the process. It is important to identify the key people who need to be interviewed and the key documents which need to be reviewed. Once the key people have been identified, a schedule can be developed which makes best use of the auditor's and facility staff's available time.

Gathering Data and Reporting Findings

Data gathering begins with collection of the background information and continues while the audit team is on-site. In conducting the audit, the first step is to obtain a good understanding of the management system in place for each PSM element (7). This is particularly useful when some or all of the system is informal. Based on this understanding, the auditor should make an assessment of how well the management system satisfies the criteria specified in the appropriate guidelines or regulation as summarized in the audit guide. This initial assessment is aimed at identifying whether a management system is in place and whether it has the basic characteristics of a good management system.

The next step of the audit is to verify that the system is functioning effectively. Here the auditor reviews appropriate procedures and records, inspects facilities, observes operational tasks, and conducts follow-up interviews. Figure 5 shows the distribution of data gathered in a typical organization.

A considerable amount of time can be spent in the verification activity. However, the auditor is not intended to examine every applicable document, interview all personnel involved, or inspect every piece of equipment. The audit is based on reviewing a representative sample of these sources of data in order to reach a conclusion as to whether the system meets the required criteria and contains the characteristics of a good management system. For example, lets look at the requirements for the element compliance audits as specified in OSHA 1910.119 (4). If we review the requirements against the seven characteristics of a good management system (as shown earlier in Table 2), we find that the regulation specifies some of the seven characteristics (see Table 3).

An auditor who focussed strictly on compliance would only be concerned with the items listed above. However, an auditor who was looking for a complete and effective management system would also look for:

- a written audit program

- an audit guide

- an audit team

- definition of responsibilities for scheduling, staffing and conducting audits, as well as responsibilities for developing action plans, follow-up on action items and documentation that deficiencies have been corrected

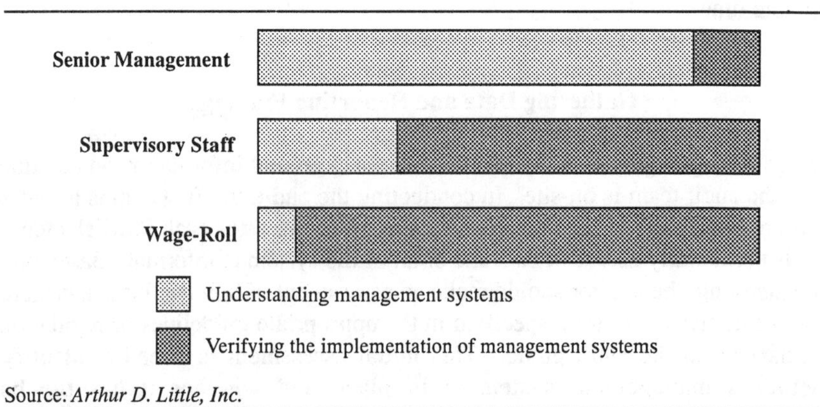

Figure 5. Typical audit information gathering by organizational level

TABLE 3
Evaluation of OSHA Compliance Audits Element vs. the Characteristics of a Good Management System

Requirements	Characteristics*						
	1	2	3	4	5	6	7
Certify evaluation of compliance						X	
Compliance audits conducted at least every three years					X		
Audits conducted by at least one person knowledgeable in the process				X			
Compliance audit report						X	
System for addressing audit findings and documenting that deficiencies have been corrected					X	X	
The two most recent compliance audit reports are maintained						X	

*   1. Policies, Programs, and Procedures
      2. Definition of Responsibilities
      3. Approvals and Authorizations
      4. Personnel Training and Experience
      5. Protective Measures
      6. Documentation
      7. Internal Verification

Source: Arthur D. Little, Inc.

- required approval of action plans

- training in audit skills and techniques

- documentation of current status of management system

- internal audits conducted between compliance audits

It is generally the responsibility of the auditor to determine which characteristics of good management systems are important for a particular element to function effectively in an organization. However, a good management system should contain at least:

- written procedures

- definition of responsibilities

- required approvals/authorization

- training and experience

- documentation

An auditor may gain insight into why a management system is not working by listening to responses from audit interviewees. Table 4 lists some typical audit data which may be an indication of having a deficiency in one of the seven characteristics.

During an audit, data will be obtained through interviews (testimonial data), through review of procedures and records (documentary data), and inspection of facilities and operations (physical data). Whereas each type of data is valuable, an auditor should not rely totally on testimonial data to support a finding. Where possible, testimonial data should be confirmed through either physical or documentary data. However, as discussed earlier, not all sources of data can be or need to be reviewed during an audit. This means that the auditor needs to review a representative sample of the relevant data.

Once the audit data is compiled and evaluated, a list of findings should be developed. Findings should be phrased in the proper context. For example, a finding could be stated as "In a sample of five hot work operations, one did not have an authorized permit," rather than "Permits are not always issued for hot work operations."It is important to present a balanced viewpoint when reporting audit

TABLE 4
Indicators of Missing PSM Characteristics

Characteristics	**Audit Data**
Policies and Procedures	I remember seeing a procedure for that.
Responsibilities	Not by job!
Approvals and Authorizations	Anyone can do it.
Personnel Training and Experience	We learn by doing.
Protective Measures	I don't know how we missed that.
Documentation	Trust me!
Internal Verification	We had a list of action items.

findings. Therefore, a good technique is to include some positive findings. We like to include a section for each element called "Status of Current System" which describes the existing management system. This is particularly important in the initial (or baseline) audit report as it establishes the basis for comparison in future audits. The negative findings (or exceptions) should also be reported. We present our audit findings orally to facility management prior to issuing a draft report.

Correcting Deficiencies

The audit ream's responsibilities are generally completed when the final audit report is issued. At that time many companies destroy copies of draft reports and working papers (i.e., notes taken by the auditors while gathering data). It now becomes the responsibility of the facility to develop an action plan to address the exceptions. As each action item is completed, the action taken should be documented, as well as the rationale for not taking action to address an exception. Sometimes the audit team, or at least the team leader may review the action plan to verify that the planned action fully addresses the audit finding or recommendation. The audit team should verify implementation of action items, during the subsequent audit.

If we compare out audit approach to that proposed by OSHA in their 1910.119 Compliance Directive (8), we notice some striking similarities (see Figure 6). OSHA calls their compliance audits Program - Quality - Verification or PQV Inspections. The Program step is similar to our Understanding step, the Quality step is similar to our Assessing step and both approaches have the third step as Verification.

BASELINE VS. MAINTENANCE COMPLIANCE AUDITS

As companies struggle to implement their PSM programs, conducting an audit should be a top priority. A "baseline" audit, will identify gaps in the PSM program

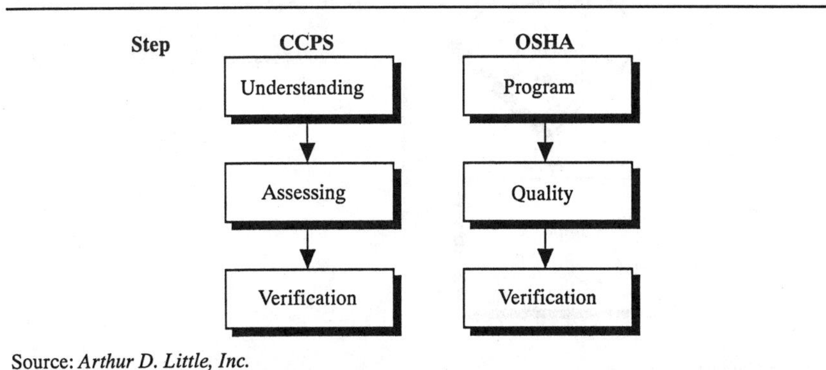

Source: *Arthur D. Little, Inc.*

Figure 6. Comparison of audit approaches: CCPS and OSHA

and aid in planning the implementation process (9). Figure 7 shows where baseline and maintenance audits fit into PSM program development, implementation and improvement. The process for conducting a baseline audit is identical to that of a maintenance audit, but since the expectation is to find gaps in the PSM program, more time is spent in understanding and assessing vs. verification. The results of the baseline audit should be used to develop an Implementation Plan which defines the tasks, resources and timing to achieve compliance. Once a PSM program is fully implemented, a maintenance audit can be conducted to verify that the PSM program is in use and effective.

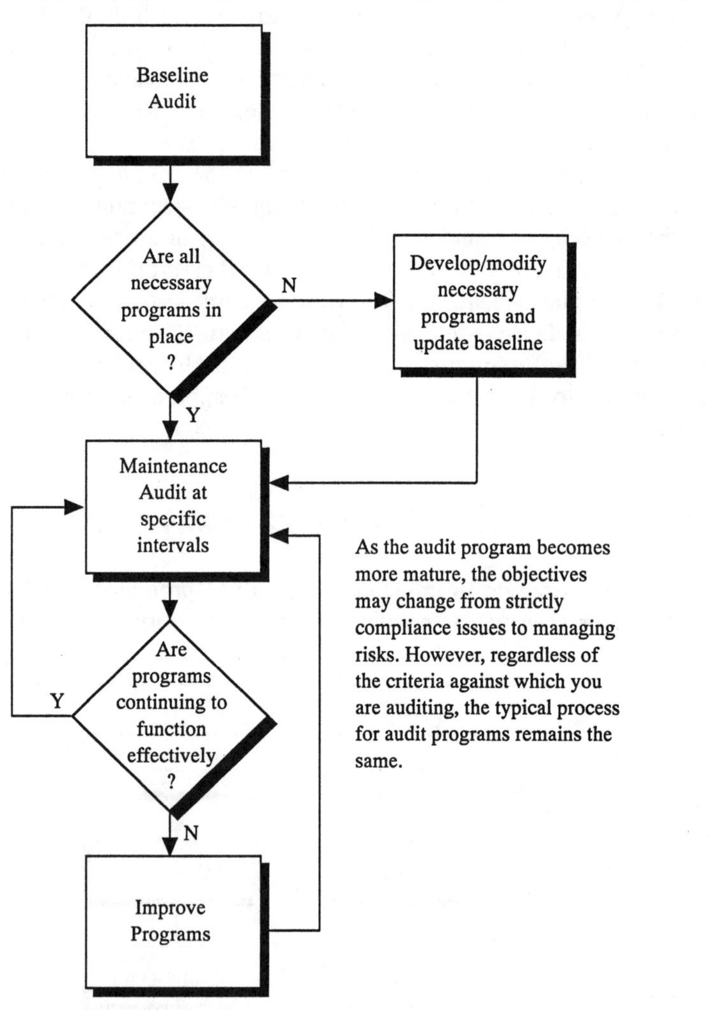

Source: *Arthur D. Little, Inc.*

Figure 7. Typical process for new audit programs

SUMMARY

Auditing is a required element in most industry guidelines and regulations which addresses process safety management. Although each of these documents specifies criteria for individual PSM program elements, the auditor needs to evaluate whether the management system contains the necessary characteristics to be effective. There are a variety of techniques for conducting audits, however, an effective audit program should have clear objectives, a systematic approach, and qualified personnel. The audit approach should include some basic tools, including an audit guide. The audit team should be objective. Finally, to be effective, an audit should be conducted periodically.

REFERENCES

1. Center for Chemical Process Safety (CCPS), *Guidelines for Technical Management of Chemical Process Safety*, American Institute of Chemical Engineers, New York 1989.

2. Chemical Manufacturers Association (CMA), *Responsible Care® A Resource Guide for the Process Safety Code of Management Practices*, Washington, DC, October 1990.

3. American Petroleum Institute (API), *Recommended Practice 750 - Guidelines for Management of Process Hazards*, Washington, DC, 1988.

4. Occupational Safety and Health Administration (OSHA) 29CFR 1910.119, *Process Safety Management of Highly Hazardous Chemicals, Explosives and Blasting Agents; Final Rule*, February 24, 1992.

5. Center for Chemical Process Safety (CCPS), *Guidelines for Auditing Process Safety Management Systems*, American Institute of Chemical Engineers, New York, (1993).

6. Ozog, H. and Stickles, R.P., *What to Do About Process Safety Audits*, Chemical Engineering, September 1992.

7. Greeno, J.L., et al., *Environmental Auditing - Fundamentals and Techniques, Section Edition*, Arthur D. Little, Inc., Cambridge, MA 1987.

8. Occupational Safety and Health Administration (OSHA), Instruction CPL 2-2.45A, 29 CFR 1910.119, *Process Safety Management of Highly Hazardous Chemicals - Compliance Guidelines and Enforcement Procedures*, September 28, 1992.

A Multilevel Approach to Monitoring the Implementation of Safety, Health, Environmental (SHE) Standards in the ICI Group

J. L. Hawksley
ICI Group SHE Department, U.K.

ABSTRACT

In the system for the management of SHE in the ICI Group the basic requirements of mandatory Group SHE Policy and Standards (issued by the Board of Directors) are implemented via local procedures in each operating unit. The principles to be followed in these local procedures are defined in Group Guidelines. The efficacy of the system is monitored by auditing at several levels for each activity within the operating unit (eg production site, distribution operation, laboratory, etc). Firstly, regular "Operational" audits check that what is done in practice conforms to the requirements of the local procedures. Secondly, periodic "Specialist" audits check that each procedure meets the specific requirements of Group Guidelines. The third level, typically on a one to three year frequency, is a "Management" audit to give an overall assessment of the implementation of the Group Standards. The levels of audit interact as a series of overlapping feedback loops to the various levels of management of the operating unit as part of the continuous improvement process. Over and above these three levels of auditing is a requirement for the CEO of each International Business within the Group to report formally to the Board via an annual "letter of assurance" on the extent of implementation of Group SHE requirements within the unit and the actions being taken to correct significant deficiencies.

INTRODUCTION

It has long been recognised that a systematic approach to the management of safety, health and environment protection is necessary in the chemicals and process industries, and others. Most companies have in place more or less formalised systems. In many countries there are now specific regulatory requirements for management systems. For example, the OSHA Process Safety Management Standard in the USA and regulations to implement the

"Seveso" Directive in the member states of the European Economic Community. A fundamental review of the Seveso Directive, currently underway, is including significantly greater emphasis on management systems.

Key elements of a management system are:

- A policy statement of intent, desired performance, etc, which has to come from the most senior manager responsible.

- A definition of the specific requirements necessary to put the policy into practice.

  It can be convenient to define these requirements at two levels. Firstly the key basic requirements - variously called "Minimum Requirements", "SHE Standards", "Guiding Principles", "Elements", etc. Secondly the detailed principles that have to be followed to meet each basic requirement fully.

- Local procedures and instructions defining clearly who does what so that work activities can be carried out in accordance with all the relevant principles.

- An "in-house" auditing process to check that work activities do in fact comply with local procedures and instructions, to verify that local procedures and instructions are in fact in accordance with all the relevant principles and to identify opportunities for improvement.

- A management review process, embracing all levels, to consider performance achieved and to action improvements to both the system and its implementation.

Recently, ICI has brought together the accumulated experience and good practice from its constituent businesses into a unified system for the management of SHE. This includes a common framework for auditing and monitoring the efficacy of the system and the performance achieved. The purpose of this paper is to describe the auditing process, but it is first necessary to outline the overall system.

THE ICI GROUP SHE MANAGEMENT SYSTEM

ICI is a multi-national business-led chemical company comprising five* International Businesses and several Regional Companies covering a broad range of products from paints through various industrial chemicals and materials to explosives. The aim was to establish a unified system in which certain requirements set centrally can be implemented by the individual businesses in a way appropriate to their operations. The CEO's of the constituent businesses are accountable to the Corporate Board of Directors for the management of SHE in their businesses. The system evolved is outlined in Figure 1.

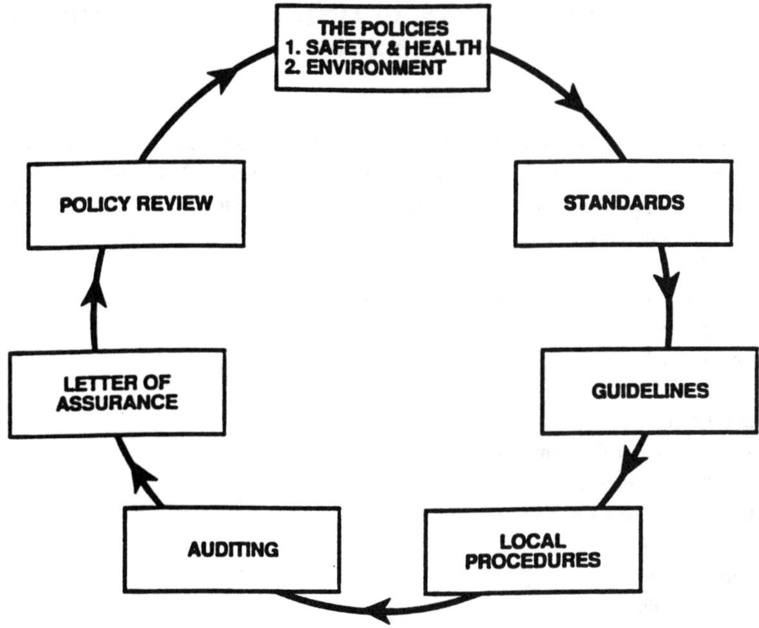

Figure 1 - The overall ICI SHE Management System

* Three other International Businesses covering pharmaceuticals, agrochemicals and specialty chemicals that were previously part of the ICI Group now operate as a separate company; ZENECA.

The key elements of the system are:

a)   Policies

The Board of ICI has set down its Safety, Health and Environment Policies outlining its approach to safety, health and environment improvement as well as its arrangements for ensuring that these are carried through the Company.

b)   Standards

Over recent years the key principles for Safety, Health and Environmental Protection improvement have been "distilled" into a series of management principles which set out succinctly the requirements necessary to ensure effective implementation of the Policies. These are referred to as "Standards". Currently nineteen such Standards cover these principles. They can be regarded as "minimum requirements" and are mandatory throughout ICI. Table 1 lists the Standards.

TABLE 1 - ICI GROUP SAFETY, HEALTH AND ENVIRONMENTAL STANDARDS

The following Standards set out the essential requirements to secure implementation of the ICI Group Safety and Health, and Group Environmental Policies. They apply to all aspects of the ICI Group's activities and their implementation is mandatory. The Standards are set out under the following headings:

1   Safety, Health and Environmental (SHE) Commitment
2   Management and Resources
3   Communication and Consultation
4   Training
5   Material Hazards
6   Acquisitions and Divestments
7   New Plant, Equipment and Process Design
8   Modifications and Changes
9   SHE Assurance
10 Systems of Work
11 Emergency Plans
12 Contractors
13 Environmental Impact Assessment
14 Resource Conservation
15 Waste Management
16 Soil and Groundwater Protection
17 Product Stewardship
18 SHE Performance and Reporting
19 Auditing

Example - Group SHE Standard No 8 : Modifications and Changes

"There shall be arrangements to ensure that no modifications compromise SHE performance. Proposals shall be registered and assessed, and modifications shall be authorised. Necessary hazard studies shall be carried out, appropriate design considerations made and all changes properly engineered and recorded."

---

c)   Guidelines

It will be recognised from the Standards that there needs to be more "detail" if they are to be converted into workable procedures to be used at all locations. This is provided by a series of Guidelines. Since any local procedure must ensure compliance with local legislation as well as take account of national culture, these Guidelines for the preparation of local procedures are of an advisory nature. Individual locations are allowed the opportunity to draw from the best information available to them. The Guidelines can be thought of as setting down good management practice.

The nineteen SHE Standards embrace over one hundred distinct elements and a Guideline has been prepared for each element. Guidelines relevant to each Standard are identified by a "map", part of which is shown in Figure 2. The Guidelines follow a consistent format which covers:

- purpose and scope
- key definitions
- relevant Group documentation
- specific legal requirements
- principles to be followed
- advice on how to apply the principles
- personal responsibilities
- local documentation requirements

The principles to be followed are the "heart" of the Guideline. Table 2 gives an example of these from a Guideline relevant to the Standard given in Table 1.

ICI Group SHE Policies, Standards and Guidelines
Part 2 : Section 0
June 1992

Group SHE Standard (GS)		Group SHE Guideline (GG)		
GS No.	Subject	GG No.	Subject	Issued
		.5.8	Biological hazards	June 92
6	Acquisitions & Divestments	6.1	Procedures for divestment	June 92
		6.2	Procedures for acquisition	June 92
7	New Plant, Equipment & Process Design	7.1	Capital projects	June 92
		7.2	Site development plans	June 92
		7.3	Hazard Studies	June 92
		7.4	Natural disasters	June 92
8	Modifications & Changes	8.1	Control of modifications	June 92
		8.2	Temporary repairs	June 92
		8.3	Modifications to programmable electronic systems	June 92
9	SHE Assurance	9.1	SHE assurance programmes	June 92
		9.2	Establishment of a health assessment programme	June 92
		9.3	Fire management	June 92

Figure 2 - Extract from the Guidelines "Map"

TABLE 2 - SUMMARY OF "PRINCIPLES TO BE FOLLOWED" FROM A TYPICAL GUIDELINE (GG 8.1 CONTROL OF MODIFICATIONS)

Principles to be followed are:

a)   Changes which constitute a Modification should be clearly defined.

b)   There should be a procedure for the control of all Modifications requiring documentation to ensure a systematic approach to the consideration and installation of the Modification.

c)   A statement of the purpose and benefits of a proposed Modification should be approved by a nominated person.

d)   A technical/SHE assessment of the Modification should be carried out in a systematic manner and approved. The design, installation, training and commissioning should be subject to review by the manager responsible for the operation of the equipment or process.

e)   Technical/SHE assessments should be approved by a nominated manager with the appropriate knowledge and experience.

f)   Line diagrams should be updated and operating and maintenance instructions should be written or amended. People likely to be affected by the Modification should receive the appropriate training and validation.

g)   There may be certain exceptional circumstances where work may be undertaken without following the prescribed Modification procedure, but these should be specified and may include:

   1 Emergency work to protect people, plant or the environment where rapid steps may have to be taken to minimise or bring under control the effects of an event. A full assessment should follow as soon as possible by the appointed persons if the plant is to continue operation with the emergency repair in situ.

   2 Certain temporary repairs, which may be an acceptable temporary standard in certain, clearly specified duties. (There is a separate Guideline for procedures to be adopted for temporary repairs to plant (see GG 8.2)).

h)   Each plant/site should keep a register and file of all Modifications in the plant dossier.

d)　Local Procedures

All businesses, sites, locations and works are required to have in place their own SHE local procedures which incorporate the good practice as provided in the Guidelines. These procedures set out the local arrangements and requirements by which work is to be carried out in a safe, healthy and environmentally sound way and are locally mandatory.

e)　Auditing

The process of auditing SHE activities is a vital step in safeguarding people, property and the environment and in ensuring that continuous improvement takes place. It involves systematic checking and in-depth examination of activities and systems at several levels to assess the effectiveness of implementation of ICI's SHE standards and to identify areas of improvement.

One of the Guidelines covers the auditing requirements and is described later.

f)　Letter of Assurance

The auditing process feeds into an annual "letter of assurance" in which the CEO of each operating unit within the Group reports formally to the ICI Executive Director with responsibility for SHE on the extent of implementation of Group SHE requirements within the unit and the actions being taken to correct significant deficiencies. The process is described later.

g)　Policy Review

The ICI Board is provided with an annual summary of how all the businesses are performing. This review shows the Board how the policies are being fulfilled, and helps provide them with the basis for further decisions with regard to the overall management of SHE.

THE AUDIT PROCESS

Principles to be followed

The principles to be followed are set out in a Guideline which defines the requirements of the audit process to be implemented within each unit from which a letter of assurance is required with regard to the extent and adequacy of conformance with the Group SHE Standards.

The first principle is that there should be three basic types of audit. These are defined as follows:

a) Operational

Systematic checks to establish whether all activities are being carried out in accordance with local procedures and instructions.

b) Specialist

Periodic in-depth examinations of the adequacy of particular aspects of local procedures, instructions and arrangements against SHE Standards and recognised good practice.

c) Management

An overall assessment of the effectiveness of management implementation of SHE Standards.

The place of these three types of audit in the SHE management system is illustrated in the diagram in Figure 3 and each is described more fully later.

It is necessary to distinguish between the objectives of Operational, Specialist and Management audits in order that effort is not duplicated. Each is intended to make a distinctive and additive contribution, although there is some overlap. Specialist auditors use the findings from Operational auditing, and Management auditors use the findings from Operational and Specialist auditing and they also do some selective auditing in the other audit categories. In some situations, for example where there are few operations to be audited, it may be appropriate for more than one type of audit to be carried out as a single exercise.

Other important principles are summarised as follows:

- Each unit prepares and implements audit programmes for those activities within its direct control.

- Auditors record and report their conclusions and any recommendations for improvement. Where appropriate, they assist managers in developing action plans.

- Management formulates a written action plan for effecting any improvements, including responsibilities and completion dates. The reasons why any recommendations are not to be implemented should be recorded.

- Auditors should have appropriate experience and be trained in the techniques of auditing.

- Individuals should not audit their own personal activities.

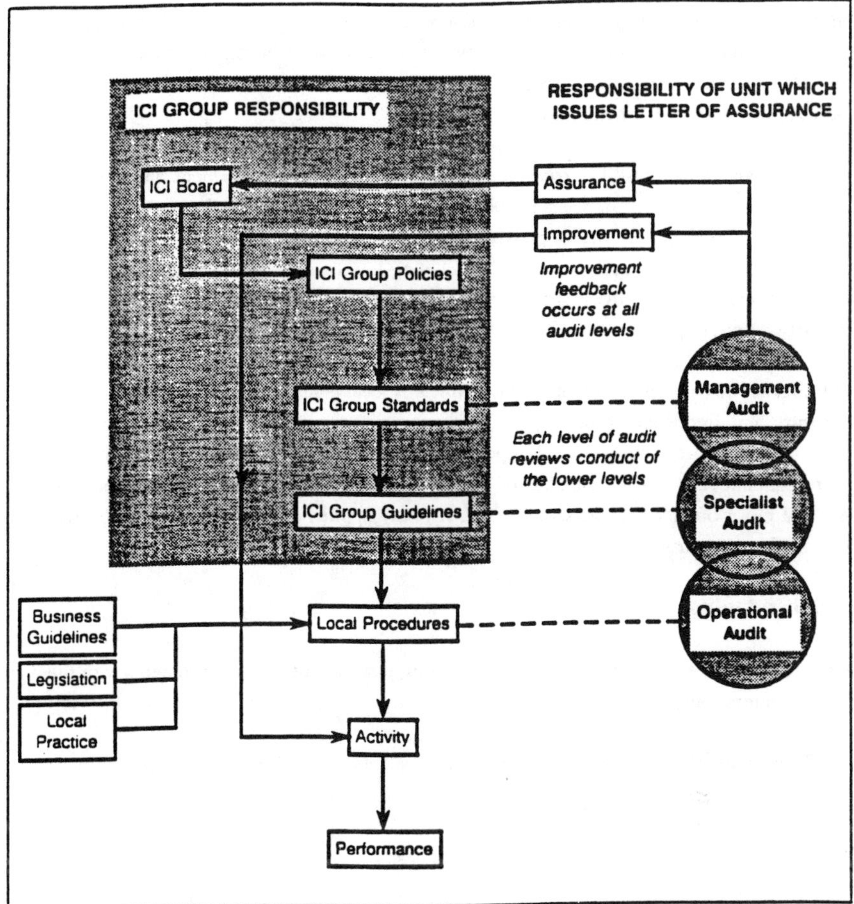

Figure 3 - The ICI SHE Audit Model

Operational Audits

These are regular systematic checks to an annual plan with the object of determining:

- the understanding of those involved of the requirements of all those procedures which are relevant to SHE performance;

- the degree of conformance with those requirements;

- the steps necessary to achieve further improvement.

In quality assurance terminology, these audits correspond to "Systems Compliance Audits".

Conformance has to be judged not only by examination of critical records and collection of other verbal and written evidence, but also by visual examination of physical conditions and methods of work. Typically checklists of, for example, the key steps of a particular procedure are prepared for the audit.

Operational audits are mostly carried out by members of local management and staff and need to be accepted as a normal operational work feature. The audits of a particular work area should be conducted primarily by personnel not directly associated with that work area and also by others more directly involved, including line management and operating staff to ensure ownership, objectivity and sharing of experience. All management levels in the organisation should play a part.

The frequency of Operational audits is determined by local management, taking into account the nature and frequency of the activity, the consequences of non-conformance and the track record of conformance. Mostly operational audits should be carried out frequently. It is recommended that each element of the system should be checked at least every two years and more often where necessary for frequently applied systems or those associated with significant hazards.

Specialist Audits

These are in-depth examinations of local procedures, instructions and arrangements with the object of determining the extent to which they:

- fulfil consistently the intent of the Standard, related Guidelines and recognised good practice;

- adequately take into account local circumstances;

- achieve further improvement.

The audits range from checks that a specific local procedure (eg for plant modifications) properly incorporates all the principles set down in the relevant Guideline (eg Table 2) to wide-ranging examinations of broader aspects, eg fire safety, engineering standards, healthcare arrangements, environmental protection.

The auditors need to have specialist knowledge and experience of the aspects to be covered by the audit. They should be independent of the activity being audited, but the audit should be conducted in conjunction with local personnel. The audit may involve a team of auditors.

As with Operational Audits the frequency of Specialist Audits is determined by local management taking similar considerations into account, but in addition, noting any appropriate specialist advice. It is recommended that the procedures and arrangements associated with each Standard are subject to Specialist Audit at least every five years or more frequently when judged to be necessary.

Management Audits

The objective of these audits is to provide an independent overall assessment of the arrangements for the management of SHE on an activity by examining:

■ the manner and extent of implementation of the relevant SHE Standards;

■ the appropriateness and adequacy of the arrangements for implementing the policy of continuous improvement of SHE performance, taking into account the track record of performance, the nature of the hazards managed and local circumstances.

The audits are carried out by experienced managers independent of the activity being audited (but not necessarily independent of the business). They need to have familiarity with auditing techniques either from experience or training. Within the audit team there needs to be recognised and relevant experience of the type of activity being examined.

It is recommended that each activity is subject to management audit at least once every three years and a "rolling programme" covering all the activities (ie manufacturing plants, warehousing, distribution operation, laboratories, etc) within a business should be drawn up.

Management Audit is conducted primarily by inquiry, with presentations on special topics requested as and when necessary. The audit normally will include tours of the activity to obtain an overview of conditions and the general approach to SHE matters across the activity.

Typically, the audit will use a series of structured questions in order to:

■ determine whether appropriate procedures are in place to address each of the applicable Group SHE Standards, local regulations and business standards;

■ determine whether Specialised and Operational Audits have been conducted;

■ review progress against Improvement Plans.

A model set of Management Audit generic questions is given in Table 3. More specific questions may be developed to apply these to each Group SHE Standard, based on the key principles from the relevant Group Guideline(s).

For each of the SHE Standards considered during a Management Audit, a rating of conformance is generated, based on the auditor's evaluation of conformance with the principles set out in the relevant Guidelines. The rating is based on a descriptive range of categories or a numerical scoring system related directly to the requirements.

The Management Audit is seen as a cornerstone of improvement and is the audit element referred to in Figure 1. It feeds directly into the "letter of assurance". Currently a "cross-business" element is being introduced into the Management Audit whereby a proportion of

Table 3 - Generic Question Set for Reviewing Implementation of Group SHE Standards

Apply to each Standard, using selected key elements from Guidelines, to verify specific aspects:

1 Has all legislation, Group Guidelines, etc, relevant to this Standard been identified and understood?

eg Group Guidelines
Business Guidelines
Regional Guidelines
Legislation and regulations
Other, eg industry codes of practices, incident history

(Check by noting availability of relevant information, awareness of and familiarity with it).

2 Are management systems in place to ensure continuing conformance with the requirements of this Standard in all areas of activity?

2.1 Are there written procedures? *(confirm by inspection)*

2.2 Is there a system for review and update of procedures? *(when was last review? who did it)*

2.3 Are responsibilities for managing and carrying out procedures assigned? *(check nominated persons are "in-date")*

2.4 Have training needs in relation to this Standard and associated Guidelines been identified and satisfied? *(check by inspection of sample of training programmes and records)*

3 Are systems in place to verify conformance with this Standard?

3.1 What Operational Audits are carried out and by whom? *(confirm by inspection of audit plan and records)*

3.2 What Specialist Audits are carried out and by whom? *(confirm by inspection of audit plan and records)*

3.3 What is management's own assessment of the degree of conformance with the Standard and conformity with key principles of Guidelines? *(how assessed? see examples of results)*

3.4 Are local Improvement Plans in place; how are they progressed; what is senior management involvement? *(confirm by inspection of documentation)*

audits involve auditors from another business or a corporate function. This is to avoid any potential disparity of auditing standards between businesses and to provide added objectivity to support the internal arrangements within a business for compiling its letter of assurance.

LETTER OF ASSURANCE

The letter of assurance requirement was introduced about three years ago, initially for safety, but now, with the move to integrated SHE Standards, covering safety, health and environmental protection. Increasingly it is seen as powerful element of the improvement process, because it makes the CEO of a business well aware of its current SHE performance and how any deficiencies are being addressed. CEO's have to have processes in place by which the results of all management audits within their business are collated. This mostly takes the form of a "cascade", with senior managers of activities within a business providing the CEO with a regional or local "letter of assurance" which in turn derive from similar letters from managers responding to them. The process is thus a total management programme of improvement starting now at the lowest level and working through to the Corporate Board of Directors. Not only does it develop a view of the status of SHE management systems, **based on the results of audits supplemented by "self-assessment" by managers, but it also enables a total annual business improvement plan to be prepared.**

The results of these letters of assurance are summarised in an annual report to the Board of Directors giving a view of performance and overall improvement plans.

CONCLUDING COMMENT

The concept of the multi-level approach to auditing that has been described was established in principle some years ago, but only partially implemented in some parts of the Company. Now it is specified as a requirement of the unified Management System which is now being progressively implemented throughout the Company to improve SHE performance. The implementation process involves staff from every level of the organisation and so meets one of the key criteria of any continuous improvement process, ie the involvement of everybody. Importantly, also, full implementation of the system will ensure total conformance with all Management Codes of Practice, Guidelines, etc, of "Responsible Care" programmes to which ICI has pledged commitment.

ACKNOWLEDGEMENT

Developing the management system and audit process described in this paper involved contributions from many SHE and line managers within ICI.

Managing Safety Through Controlling the Causes of Human Error

John Wreathall
James Reason
James Wreathall & Company, Inc., 4157 MacDuff Way, Dublin, OH 43017

It is increasingly recognized that major accidents, whether in the process, petrochemical, transportation, or nuclear industries, are very much the consequence of human errors. The very different accidents at Flixborough (UK), Zeebrugge (Belgium), Chernobyl (Ukraine), North Sea (Piper Alpha), and Channel View (USA) reveal that, while facilities have their own technological defenses, these defenses can be overcome by human errors in common ways. The purpose of this paper is to present ways to achieve the active management of safety through controlling the causes of human errors that underlie the occurrence of most major industrial accidents.

Most reviews of accidents focus on the errors of individuals involved at ''the sharp end'' of the accident-what the maintenance technician or operator did wrong on the day. Numbers such as ''60% of all equipment failures are the result of human error'' are often quoted. These data, however, are misleading in their message. For example, people are a great influence on equipment failures simply because they interface with equipment so often-they design it, install it, test it, repair it, and modify it! It would be astonishing if human error were not a major contributor to equipment failures.

What is important is not the individual percentages of equipment failures, or the percentages of plant shutdowns (or lost production or quality). From a safety perspective, what is important is the way that the facility is operated that make its defenses less vulnerable to multiple failures due to human errors. For example, any process plant has multiple defenses against major accidents. The use of relief valves, vent/flare stacks, design code vessels, fire protection systems, and the like are hardware defenses. Standard operating procedures, training of process workers, permits-to-work systems, and so on are administrative defenses. A catastrophic accident occurs when multiple barriers fail. Reason has observed that most accidents result from latent failures that lie dormant until some triggering event causes these failures to be revealed by the safety challenge. Latent failures could comprise failed standby safety equipment, a relief valve set to the wrong pressure, or an error in an emergency procedure, for example. The extent of latent failures through a process plant define the safety ''health'' of that plant. However, the latent failures are often the final consequences of weaknesses in plant programs such as procedures management, maintenance training, or other programs including those implicit in the OSHA Process Hazard Management regulations.

Such programmatic weaknesses can be discovered easily following a major accident. Increasingly these are the focus of regulatory or other post-mortem investigations. A greater challenge is to have ways of uncovering weaknesses before an accident occurs. Tools and procedures are being developed to provide proactive means for uncovering these deficiencies in everyday work. This paper describes the some of these tools and procedures presently being developed and implemented by the authors.

INTRODUCTION

The history of accidents and their analysis is also the history of human contribution to accidents. Increasingly through the industrial revolutions in the USA and Europe, "man-made" accidents started to become more widespread and with greater consequences. Rolt, for example, has provided accounts of British rail disasters (many of which were human error-related) dating from the beginning of the industry in 1830 [1]. However, more formal consideration of the human contribution to accidents was reflected in the growth of the study of ergonomics in aviation during the Second World War. As technology has become more complex, often in parallel with a greater potential for harm to the public, consideration of the potential problems in human performance has increased. One of the first major process plant accidents that identified human errors as a principal cause was the Flixborough explosion in England on June 1, 1974 [2], where a design modification was inadequately developed because of a rush to continue production. Similar situations have shown themselves since in the petrochemical and process industries [3].

The accident at Three Mile Island, Unit 2, on March 28, 1979, was perhaps the first in the nuclear power industry that led to a critical review of all immediate aspects of the human factors environment: procedures, training, and the human-machine interface [4]. (A review of the human-machine interface issues in nuclear power control rooms had, however, been published shortly before the TMI-2 accident [5].) These three human-factors aspects mirrored the influences (the "performance shaping factors") analyzed in the human reliability technique used almost exclusively at that time: the Technique for Human Error Rate Prediction (THERP) [6].

A series of accidents took place in the 1980's that pushed the frontiers of human performance issues back from the immediate human causes and influences to include more of the organizational framework. These included the release of methyl isocyanate from Bhopal (1984), the destruction of the Space Shuttle Challenger during launch

(1986), the accident at Chernobyl (1986), the capsizing of the passenger ferry Herald of Free Enterprise off Zeebrugge (1987), the fire in the London Underground station at Kings Cross (1987), the explosion on the North Sea oil platform Piper Alpha in 1988, and the British Rail accident at Clapham Junction, London, also in 1988. In all of these cases, much of the discussion of human failures has been directed at organizational issues. In fact, for the most recent cases, the official investigations have dwelt on the failures of the organization rather than of the individuals involved. For example, in the case of Zeebrugge, the formal report states '' ... a full investigation of the circumstances of the disaster leads inexorably to the conclusion that the underlying or cardinal faults lay higher up the company. The Board of Directors did not appreciate their responsibility for the safe management of their ships.'' [7]. Most recently, in its report on the explosion and subsequent fires at the Phillips 66 Company facility in Houston in 1989, OSHA states simply ''The primary causes of the accident were failures of the management of safety systems at the Houston Chemical Complex.'' [8]. Probabilistic risk analysis has, to date, not actually addressed these organization factors, though work is proceeding in this area [9].

However, it is becoming increasingly recognized that the human contribution to disasters is only partly specified by the human factors and organizational issues. One part missing is the context in which the actions take place. This is the area in which work is now starting.

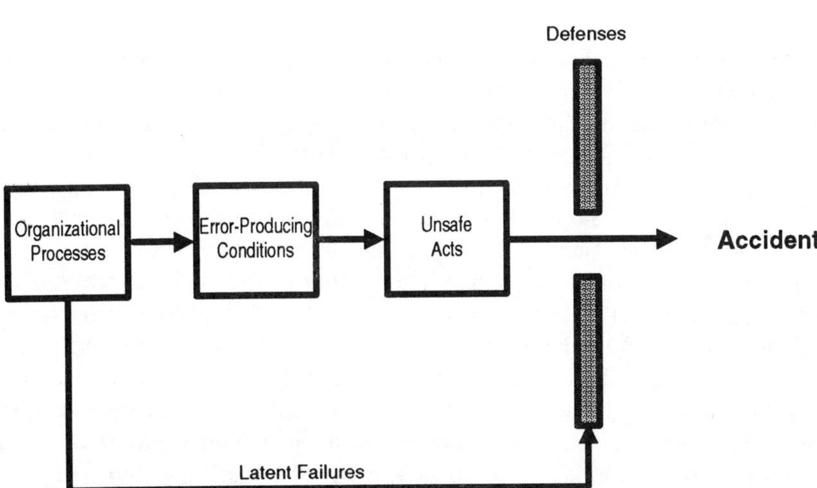

Figure 1. Framework of Accident Causation.

FRAMEWORK OF ACCIDENT CAUSATION

Figure 1 presents the framework of accident causation that forms the core of this discussion; it is heavily influenced by that given by Reason in [10]. The essence of this framework is that when humans perform unsafe acts that breach defenses (or occur in the absence of defenses), accidents can occur. These unsafe acts are those actions (or even lack of actions) by people that are the operators and maintainers of the equipment, that cause a potentially dangerous condition. In well-protected systems, defenses exist to prevent these unsafe acts from becoming what Perrow called incidents and accidents [2]. These defenses can include hardware (safety systems, interlocks) and administrative controls (procedures, rules of conduct).

The unsafe acts do not take place in a vacuum. Rather the characteristics of the ergonomic environment play an important role in shaping the forms and probabilities of the unsafe acts. These factors can be job- and task-specific. However general principals for the effects of these factors have been developed and presented in various sources (discussed below).

At the upstream end, playing an important role in providing both the ergonomic environment and the effectiveness of installed defenses, lie the organizational processes, such as allocation of resources, and the setting of work policies and practices. Here are the wellsprings of good or bad safety potential.

Unsafe Acts

As far as the plant hardware is concerned, unsafe acts can be failures to perform actions to maintain the defenses-errors of omission-such as failing to start emergency equipment, or actions that cause or exacerbate the abnormal event-errors of commission-such as initiating a sequence of events. These unsafe acts can be active (that is, their consequences are immediately revealed as with an initiating event) or latent (that is, their consequences lay dormant in the system until triggered by some event as with deficiencies in maintenance). From the human point of view, there are a variety of unsafe acts: slips/lapses, mistakes, and circumventions. Slips and lapses are unsafe acts where what was performed was not what was intended, as with mis-selecting a control or skipping a step in the procedure. Mistakes are failures where the intentions are erroneous, but are purposefully executed. For example, a misdiagnosed failure in a component will result in a repair that is irrelevant to the failure mode; that is a mistake. The third category, circumventions, are deliberate but non-malicious of safety rules. These are often done for ''good'' reasons like performing a task quickly or to overcome some organizational barrier. These categories are summarized in Figure 2, which provides examples of different kinds of each; they are discussed further in [10].

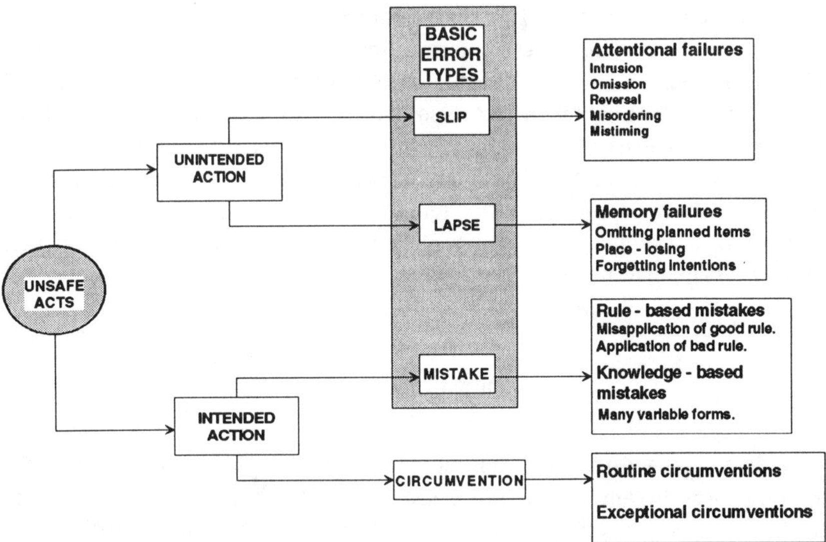

Figure 2. Categories of Unsafe Acts

In terms of human reliability analyses performed in U.S. industrial settings, the emphasis has been on slips and lapses resulting in errors of omission; this is the principal emphasis of the THERP technology and its derivatives. The reason for this focus may be that it is in these failure modes that humans can be modeled as analogs of machines, with simple failure modes. These failure modes can be incorporated simply and directly into engineering frameworks like fault trees. Other human error forms (like errors of commission) have no mechanical counterpart; pumps cannot spuriously perform unplanned and novel actions. New ways of incorporating the more human-like error modes are evolving, particularly those using simulations of human errors.

Error-Producing Conditions

Referring to Figure 1, the improvement in understanding in the middle part of the 1980's was in the area of the error-producing conditions-the so-called performance shaping factors of THERP. In that method, the principal factors include format and content of procedures, adherence to administrative rules, layout and style of displays, item labeling and tagging, stress, and the effectiveness of checking. Alternatively Williams developed an extensive database for quantifying the effects of various error-producing conditions based on a review of published data [11].

An alternative structure of the error-producing conditions has been presented by Wreathall, et al [12]. This structure, termed the onion model, has been developed to portray a more diverse set of influence on job performers that includes traditional ergonomic issues such as human-factors engineering, training, and environment. It also includes psychological and sociological factors, such as team structure, professionalism, policy consistency, and the rewards/punishment structure of the work unit. No formal attempt has been made to quantify the strengths of influences, since it is recognized that the factors are not well understood and their influences are contingent on many factors both internal and external to the work unit.

Organizational Processes

It is recognized that only a limited amount of work has been performed on evaluating how organizational processes influence safety. Noteworthy as early contributors are the work by Westrum on identifying the characteristics of safe organizations [13], and that by the University of California, Berkeley, group on high reliability organizations [14]. Westrum describes the characteristics of organization behavior as generative, calculative, or pathogenic. Generative organizations are those that accomplish high levels of success by apparently unconventional mechanisms and expectations of exceptionally high standards. Hazards are identified and removed by "lower level" personnel empowered to seek out and eliminate problems quickly. "Large lessons" are learned from "small events". Calculative organizations perform functions by the book in conventional ways, often meeting regulatory requirements but rarely exceeding them. Cost-benefit is typically a byword for decision-making. The final group, pathogenic organizations, typically consider safety regulations as a barrier to production. Messengers bearing safety warnings are typically shunted to one side, ignored, or fired.

The work on high reliability organizations describes examples of how organizations fitting Westrum's generative category practically accomplish their performance. These include case studies from aircraft carrier flight-deck operations, air traffic controllers, and electrical power dispatchers. Issues identified in this work include: empowerment of junior staff to stop production, with a reward-based support system for those who do; a transformation process of responsibilities such that the most experienced personnel take control when things go wrong; and that the safety "boundary" is being monitored continuously by several people ready to declare an incident.

A view of organizational activities have been presented by Reason [15] in terms of organizational "core processes"; that is, those activities that must be accomplished within the organization to manage safety. Reason has described the four basic processes of a technical system as comprising Design-Build-Operate-Maintain (D-B-O-M), with management and communication acting between them. (While these apply at the working level, there are two additional processes that contain the technical processes. These are Goal Statement and Organization. Beyond the organization is an additional process: Regulation. Using this concept, Reason then develops a set of eleven Organizational Failure Types (OFTs) that would comprise evidence of inadequacies in these processes. These include: incompatible goals, inappropriate organizational structure, inadequate communications, poor planning and scheduling, poor procedures, poor training, and so on. Once identified and interpreted in an operational context, these indicators could be used to monitor the safety ''health'' of the organization processes.

DEFENSIVE MEASURES AND EVALUATIVE TOOLS

Based on this framework of accident causation, the authors are developing a series of tools to assist in managing safety at several levels, including the error-producing conditions and the organizational performance. These tools are developed on the principle that the development of proactive defenses are more effective than the development of reactive defenses. That is, it is better the prevent the next accident than to prevent the last one.

Defenses Against Error Producing Conditions

Sufficient knowledge of the types of errors induced by deficiencies in such aids as procedures and information displays has led to the development of what may be called a ''Human Factors HAZOP'' of these aids. This approach is much like a traditional HAZOP procedure in that key guidewords are applied to the aids and unacceptable hazards resulting from the process deviations are identified. For example, in the case of procedures, the most typical errors associated with performing the steps (providing the technical contents are appropriate--not always a valid assumption!) are:

> - a step is omitted (*error of omission*);
> - two steps are reversed (*error of reversal*);
> - one inappropriate step is substituted for an appropriate
>   one (*error of substitution*); and
> - one inappropriate step is inserted between two
>   appropriate steps (*error of insertion*).

Errors of omission typically represent the largest fraction of these classes.

Using this knowledge, the procedures are reviewed with the HAZOP team of process engineers, operators, and so on, identifying where one of these errors may lead to an unacceptable outcome. Errors of omission and reversal can be reviewed for each procedure directly. However, the errors of substitution and insertion require additional consideration of other activities at the facility. Like most human errors, there is usually a logical basis for errors of insertion and substitution. These usually represent <u>capture errors</u>, where some similarity in the task being performed with another task induces the task performer to switch to a step in that other task. Therefore it is important to know what other tasks could lead to such capture errors occurring with the procedure under review. Once such potential opportunities have been identified, the HAZOP review team can identify the potential for unacceptable outcomes.

Once such opportunities have been identified, modifications to the design and format of the procedure can be made if necessary to eliminate or reduce the opportunity for these critical errors. Human-factors guides commonly present guidance on ways to reduce these kinds of errors.

Organizational Processes

As mentioned above, one of the characteristics of a high-reliability (and highly safe) organization is its ability to learn large lessons from small events, and monitor continuously the organization's state of safety. This does not mean data that are gathered reactively, as with numbers of lost-time accidents or OSHA citations. Such data come too late for preventing accidents in the first place. Rather, data on the causes of errors are needed if the organization is to implement a comprehensive error management program. The authors are actively developing the means for such programs, to provide early and extensive feedback to the organizational management about the current status of the "organizational failure types" described above. These include deficiencies in procedures, deficiencies in human-machine interface design, deficiencies in training, deficiencies in communications, deficiencies in planning, and so on. Information on the rates of these failure types can be obtained simply by asking the people directly affected on a daily basis "at the sharp end". For example, in one installation, the computerized weekly performance log for the operations supervisors or foremen asks how many times work that week was affected by each of the failure types. That information is summarized and aggregated by the MIS system, with data being sent to the management groups with responsibility for the various functions and the senior corporate management. Data will be trended to indicate adverse changes before any change in outcomes (accident rates, lost production, lost quality, and so on) are seen. Data can be weighted by an assessment of the importance of the various jobs to safety (or productivity or quality), and the various failure types to each job; this can make the results more sensitive.

The U.S. OSHA Process Safety Management requirements [16] are aimed at the existence of management programs rather than on standards or technical requirements.

Such an approach will require interpretation of what is an effective program, and will (ultimately) require demonstration of measures of effectiveness. For example, do the programs, once installed, lead to reductions in near misses or "small" accidents? Is there an effective means of providing feedback from experience at plants or within the corporation? Such measures will differentiate an effective management program from a purely nominal one. The techniques described above are developed from a sound basis of understanding of human errors. It is largely through the management of the causes of human error that safety will ultimately be assured.

REFERENCES

1. Rolt, L.T.C., Red for Danger. London: Pan Books, 1978.

2. Perrow, C., Normal Accidents: Living with High-Risk Technologies. New York: Basic Books, 1984.

3. Kletz, T. A., What Went Wrong?: Case Histories of Process Plant Disasters. Houston: Gulf Publishing Co., 1985.

4. U.S. Nuclear Regulatory Commission, NRC Action Plan Developed as a Result of the TMI-2 Accident (NUREG-0660). Washington, D.C., May 1980.

5. Seminara, J.L., Gonzalez, W., and Parsons, S., Human Factors Review of Nuclear Power Plant Control Room Design (EPRI NP-309). Palo Alto, CA: Electric Power Research Institute, March 1977.

6. Swain, A.D., and Guttmann, H.E., Handbook of Human Reliability Analysis with Emphasis on Nuclear Power Plant Applications (NUREG/CR-1278). Albuquerque, NM: Sandia National Laboratories, 1983.

7. Department of Transport (UK), mv HERALD OF FREE ENTERPRISE, Report of Court No. 8074, Formal Investigation. London: HMSO, July 1987, p 14.

8. U.S. Department of Labor, The Phillips 66 Company Houston Chemical Complex Explosion and Fire. Washington, D.C., April 1990.

9. See, for example, Apostolakis, G., On the Inclusion of Organizational Factors into Probabilistic Safety Assessments of Nuclear Power Plants, Fifth IEEE Conference on Human Factors and Power Plants, Monterey, CA, June 1992.

10. Reason, J., Human Error. New York: Cambridge University Press, 1990.

.11. Williams, J.C., "A Data-based Method for Assessing and Reducing Human Error to Improve Operational Performance," in <u>1988 IEEE Fourth Conference on Human Factors and Power Plants</u>, E.W. Hagen (Ed.). New York: Institute of Electrical and Electronics Engineers, 1988.

12. Wreathall, J., Schurman, D.L., and Anderson, N.A., "An Observation on Human Performance and Safety: The Onion Model of Human Performance Influence Factors." in <u>Proceedings of the International Conference on Probabilistic Safety Assessment and Management (PSAM)</u>, Apostolakis, G., (Ed.) New York: Elsevier Science Publishing Co. Inc., 1991.

13. Westrum, R., Organizational and Interorganizational Thought. <u>World Bank Workshop on Safety Control and Risk Management</u>, Washington, D.C., October 1988.

14. See, for example, Roberts, K.H., "Some Characteristics of High Reliability Organizations," <u>Organization Science</u>, vol. 1 no. 2, pp 160-177, 1990. Also Rochlin, G., LaPorte, T.R., and Roberts, K.H., "The Self-designing High-Reliability Organization: Aircraft Carrier Flight Operations at Sea," <u>Naval War College Review</u>, vol. 40 no. 4, pp76-90, 1987.

15. Reason, J., "Disasters and Human Failures," in <u>Psychological Aspects of Disasters</u>, A. Taylor, D. Lane, and H. Muir (Eds.). Leicester (UK): British Psychological Society, 1991.

16. Occupational Health & Safety Administration, Process Safety Management of Highly Hazardous Chemicals, Explosives, and Blasting Agents, 29 CFR Part 1910.109, U.S. Department of Labor, Washington, D.C., February 24, 1992.

The Smart Project: A Framework and Tools for Addressing and Improving Management of Safety

Jacques F. J. Van Steen
Louis J. B. Koehorst

*Department of Industrial Safety, Netherlands Organization for Applied
Scientific Research TNO, P.O. Box 342, 7300 AH Apeldoom, The Netherlands
The Netherlands*

ABSTRACT

Analysis of recent industrial disasters has shown that these were not simply a consequence of technical failure or human error. The underlying causes were deeply rooted in management aspects of the organization, such as company policy, management style, communication or procedures. The SMART Project on Risk Management is concerned with investigating the relationship between such management factors and safety. Two lines of development are distinguished: the SMART framework and SMART tools. Whereas the framework describes the causal relationships between management factors and safety, the tools are of a more instrumental nature. These consist of an assessment guideline and associated instruments which will give confidence in both the completeness and the effectiveness of an organization's management of safety.

A first version of the SMART framework has been developed, and was applied in two industrial case studies on existing problems; developing the tools is currently in progress. The paper summarizes the framework and is mainly concerned with presenting the results of one of the case studies.

INTRODUCTION

Extensive analysis of recent industrial disasters, such as those at Three Mile Island, Bhopal, Chernobyl and Zeebrugge, has shown that these were not simply a consequence of direct technical failure or operator tasks which were carried out incorrectly. The underlying causes were deeply rooted in management aspects of the organization, such as company policy, management style, communication or procedures (see for example [1,2]). The SMART Project on Risk

Management is concerned with investigating the relationship between such management factors and safety; it is carried out jointly by TNO in the Netherlands and SRD in England, and partly sponsored by the Commission of the European Communities.

Two lines of development are distinguished: the SMART framework and SMART tools. The *framework* describes the causal relationships between management factors and safety. It is intended to improve awareness at all levels of company management with respect to the impact of decisions on safety. The *tools* are of a more instrumental nature, consisting of an assessment guideline and associated instruments which will give confidence in the completeness and effectiveness of an organization's management of safety. Thus, these are intended to improve the quality of safety management systems. Both the SMART framework and SMART tools are to be developed in close co-operation with industry.

A first version of the SMART framework has been developed. It was applied on existing problems at Dutch establishments of Dow Benelux and GE Plastics Europe. The SMART tools line of development is currently in progress and is focused on the development of an assessment guideline for safety management systems. As part of this guideline, instruments will be developed in order to assess specific aspects of safety management, such as commitment and communication.

This paper continues the discussion in earlier contributions [3,4], and is mainly focused on the case study at GE Plastics. Prior to discussing this case study, the SMART framework is summarized. The paper concludes with sketching a perspective.

THE SMART FRAMEWORK

As noted above, the underlying causes of industrial accidents and disasters may lie deeply rooted in management aspects of the organization, such as company policy, management style, communication or procedures. This observation motivates the necessity to investigate and identify the relationships between such management factors and safety. The purpose of the SMART framework is to characterize and visualize these relationships, in order to create the opportunity to recognize and eliminate management factors, thus improving safety and contributing to accident and disaster prevention.

This section summarizes the SMART framework (a more detailed discussion is presented in [4]). It starts with describing its structure and some of its fundamentals, followed by a general introduction to applying the framework in practical situations.

Structure of the SMART framework

The SMART framework combines existing insights from various disciplines, such as organizational theory and accident analysis. It consists of the following major building blocks:

1. Fundamental organizational requirements. There is a limited number of fundamental organizational requirements with respect to safety, which should be taken into account in order to achieve a high level of safety. These requirements apply to every organization. Examples of such requirements are commitment of management to safety, allocation of tasks and responsibilities, communication, and organizational learning.

Fundamental organizational requirements which have not been taken care of in a sufficient way are assumed to be strongly related to type failures, a concept which originates from the literature on accident analysis. There, a clear distinction is made between so-called token and type failures [5]. Token failures are related to symptoms, of which there are an infinite number. On the other hand, there exist a limited number of type failures, and these are related to more fundamental mechanisms (sometimes called 'root causes'). Type failures usually exist for a long time in an organization before they are noticed. Token failures are the symptoms of the underlying type failures, mostly directly visible and recognizable.

2. Organizational characteristics. Each organization is unique, and differs from other organizations. Three major characteristics are distinguished which determine such differences: organizational structure, organizational culture, and the history of the organization. Organizational structure is to a large degree a matter of design. The culture of an organization, associated with 'the way things are done here', is less visible than its structure. The history of the organization is associated with the way in which it developed into its current condition. In order to achieve (and maintain) a high level of safety, the way of implementing the fundamental organizational requirements and the way of implementing changes must match the characteristics of the organization.

3. The management circle. Since safety is considered to be an integral part of all business activities, it should be managed in the same way as all other areas of concern. Consequently, the management circle:

... ► Policy ► Decision ► Actions ► Control ► Policy ► ...

also applies to the field of safety. It refers to the basic process which each manager has to deal with daily. The management circle appears in the centre of the SMART framework.

4. External pressures. There are a number of external pressures which can influence the managerial decision making process. These may affect decisions of management with respect to resources, design, expectations, standards, and priorities. Consciously or unconsciously, the decision making process may be influenced in such a way that the eventual decisions cause the introduction of additional risks. Examples of external pressures are commercial and financial constraints, and legal and political constraints.

The relationships between the above-named building blocks are visualized in the SMART framework (see Figure 1; a more generic version is given in [3]). Its key characteristic is that it shows the connection between managerial decision making and safety. The SMART framework distinguishes two kinds of failures in managerial decision making, which are associated with the distinction between token and type failures: (1) decisions are focused on resolving token failures, or characterized by an inadequate balance between resolving type failures and addressing other considerations or external pressures ('what' is decided, is wrong), and (2) the way of implementing decisions is characterized by an insufficient balance between the organizational requirements for safety and the characteristics of the organization, either when managers are not aware of the necessity of an appropriate balance between these two aspects or when they are not able to find the right balance despite being aware of its necessity ('how' the decision is implemented, is wrong).

Application of the SMART framework in practice

Summarizing, the SMART framework groups management factors into two kinds of failures which are associated with managerial decision making: (1) 'what' is decided, is wrong, and (2) 'how' the decision is implemented, is wrong. This being the case, one is in the position to recognize and eliminate management factors, thereby improving safety and contributing to accident and disaster prevention.

Applying the framework entails that for an undesirable or unsatisfactory situation the various building blocks of the framework are assessed. The first result of this process will be that managers are provided with recommendations to improve the particular situation itself. Use of the framework guarantees that the proposed recommendations are attuned to the characteristics of the organization and therefore effective, which is a rather concrete and quickly visible result. This result corresponds with resolving token failures in an adequate way.

A more abstract but also more important result is that applying the framework helps in identifying the underlying management factors. Thus, management can be advised on how to initiate a process to eliminate these factors. Such a process takes time, and concrete results will not be directly visible to everybody. However, addressing underlying causal mechanisms is essential in order to gain structural improvement.

Thus, it appears that the focus of applying the SMART framework is on *improvement*. More specifically, as a consequence of the distinction between 'what' and 'how', the focus is on improving the quality of management decision

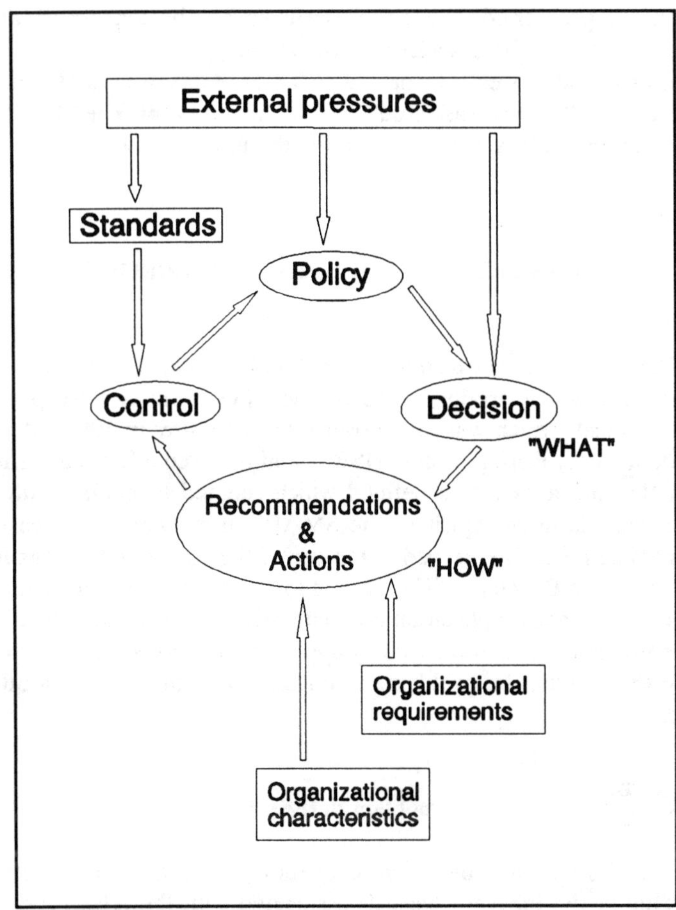

Figure 1. Structure of SMART-1 framework

making ('what'-perspective) as well as the quality of implementation ('how'-perspective).

In order to validate the SMART framework and to test its applicability and usefulness in practical situations, a number of companies in the chemical process industry were asked to participate in the project by allowing on-site investigation of a case. Three companies responded positively to this request: Dow Benelux, GE Plastics Europe, and Hydro Agri Sluiskil. A number of criteria were defined for practical situations to be appropriate within the scope of the SMART Project. These criteria are as follows. First of all, it must concern an existing problem. Furthermore, this problem must be persistent: investigation has led to recommendations which were implemented, but the result is unsatisfactory ('it does not work'). The problem does not necessarily have to be concerned with an accident, although it must be associated with safety.

Two case studies were performed so far, at plants of Dow Benelux and GE Plastics Europe. The Dow case study is summarized in [4]. The GE Plastics case study is documented in a report [6], and is discussed below.

CASE STUDY AT GE PLASTICS EUROPE

The GE Plastics site in Amsterdam, Netherlands, is one of three sites within GE Plastics Europe which produce ABS powder. (These sites were part of Borg-Warner Chemicals which was taken over by GE Plastics in 1989.) It applies the International Safety Rating System (ISRS) and received a five star rating several times. At this site a case was defined which seemed to contain sufficient elements for analysis in the spirit of the SMART framework. This case involved problems which recently emerged in the use of the correct filter bags in the dust collectors in the ABS plants. The case study was concerned with the question how a particular undesirable situation could arise and whether it would be possible to identify management factors by applying the SMART framework. Before discussing the results, the problems in question and the approach adopted are described.

Problem description

The GE Plastics site in Amsterdam consists of two plants manufacturing ABS powder. One of the inherent hazards associated with the ABS production process is the occurrence of dust explosions as a result of static discharge. This risk is particularly known to arise in the dust collectors. Each plant has a drier to which a large dust collector is connected. In these dust collectors, termed d.c. A and d.c. B, filter bags are used to separate the ABS powder from the air discharged by the drier.

Prevention and elimination of static charges is explicitly considered in the design of dust collectors and the selection of filter bag types. D.c. A and d.c. B are equipped with different filter bag suspension systems, and originally filter bags of different materials were used. Since its start-up in 1984, bags of antistatic or conductive material have been used in d.c. B; these are called epitropic. Nonconductive bags with a ground wire were used in d.c. A until recently. A study in 1990 led to the recommendation to use only filter bags which are conductive but at least antistatic. In 1991 it was decided to modify the suspension system of d.c. A in such a way that epitropic bags could also be used there. Moreover, an investigation covering the filter bags used in d.c. B was begun during that period. This revealed a number of uncertainties. First of all, the specifications to be satisfied by the filter bags were unknown. Furthermore, it was not clear whether the filter bags were made conductive/antistatic by adding stainless steel fibres or carbon fibres, and it also was not clear what was meant by 'epitropic'. In the light of these uncertainties it is possible that the wrong filter bags have been used in d.c. B which may have increased the likelihood of a dust explosion.

Meanwhile, GE Plastics drew up the correct specifications and also found a supplier who can provide filter bags to satisfy the prepared specifications. That was not the focus of the TNO case study (although the above was not yet completed at the time of the study); rather, it was intended to provide insight into the circumstances which resulted in the above uncertain situation and to answer the question of how to prevent the occurrence of comparable problems in future.

Approach

In order to obtain the necessary information, extensive interviews were held with representatives of production, maintenance, safety, engineering and purchasing at various levels in the organization. Moreover, a few people from GE Plastics Europe (headquarters in Bergen op Zoom, Netherlands) were also interviewed who have a 'historical' and/or topical involvement with the problems in question. Altogether, 18 people were interviewed.

The interviews were based on a list with points to examine; a matrix with tasks/responsibilities relating to filter bags was also used. The interviews not only focused on the problems around the filter bags; several other, more general issues, such as company history, communication, and the use of the ISRS, were also addressed during most of the interviews. Each interviewee was given two questionnaires at the end of the interview and asked to fill them in and return them. These questionnaires aimed to give insight into the organizational culture of the company and the roles played by the interviewee within a team.

The interview round was followed by a verbal presentation of initial findings to those most directly involved with the problems of the filter bags: five persons, four of them from the management team. This presentation was the first opportunity to talk to more than one person at the same time, and formed part of the study.

Results (1): filter bags

This section is focused on the results concerning the actual problems around the filter bags. First, the background and history of these problems and the present state of affairs are discussed. Next, the analysis of tasks and responsibilities relating to filter bags is summarized. Finally, a description is given of how the interviewees themselves assessed the severity of the problems.

History and state of affairs

To gain some insight into the problems with the filter bags it is necessary to examine its development. This development started during the design stages of the B drier and its related dust collector B. At the time, selection of the filter bags was examined comprehensively. Conductive filter cloth was selected. The conductive nature of the cloth can be realized in different manners: by stainless steel addition, carbon addition or chemical treatment. Partly based on discussions with Bayer ('never use filter bags with stainless steel addition'), carbon addition was selected. Moreover, the choice was presented to and approved by management; this approval was required because the choice was made for a different type of filter cloth than the type used till then.

So the type of filter cloth to be selected was entirely clear during the design phase: it was the intention from the start to have *no stainless steel addition* and to use filter bags with *carbon (C) addition*. The filter cloth was among other things specified as 'epitropic', a protected tradename of ICI who supply the carbon fibre. The filter bags were ordered from the supplier of the dust collector, in accordance with the specifications valid at the time ('epitropic'). A considerable number of interviewees were convinced that filter bags with C fibre addition were in fact fitted during the installation of d.c. B.

The takeover of Borg-Warner Chemicals by GE Plastics in 1989 led to several changes and developments. One of these is worth mentioning in this context: a risk assessment which also considered dust explosions. The risk assessment has answered the question about the safest way to prevent ignition. This was also laid down in GE Plastics Engineering Standards which indicate, among other things, that metal fibre addition is undesirable and that epitropic filter bags (= with C fibre addition) are preferred. As a result of this development, the question has arisen of which filter bags are actually in d.c. B. That question has not proved easy to answer due to the lack of labels or certificates with the filter bags used. Moreover, it appeared that clear specifications were not available and that confusion existed about the meaning of the word 'epitropic'. Then it became evident from discussions with the supplier that the filter bags used were made conductive by stainless steel addition!

So more or less by accident it was discovered that the filter bags used for years did not meet the original intention of not using stainless steel addition, whereas detailed data were not available at GE Plastics and had to be specified by a supplier. Everyone thought everything was properly organized; 'epitropic' filter bags were ordered each time on the assumption that this was an unambiguous term. There was confidence in the supplier, yet it is not clear what this confidence was based on in view of the impossibility of a proper inspection of delivered material. So nobody ever went to the effort of investigating whether the filter bags were in accordance with the order specifications.

The situation at the time of the case study was still characterized by uncertainty. The small group of interviewees who definitely knew which filter bags were in d.c. B said: filter bags with stainless steel addition. Most of the interviewees, and that certainly applied in production, did not know (or were not certain) what was in it. Moreover, the question about the meaning of the word 'epitropic' also produced various answers.

The above discussion shows that knowledge has been lost with time. This seems astonishing in the context of a production process characterized by the risk of dust explosions and considering the fact that such explosions occur regularly. In 1987, for example, GE Plastics (at the time still Borg-Warner Chemicals) experienced various dust explosions at two of its establishments.

Tasks and responsibilities

The responsibility for sub-tasks regarding filter bags was discussed with almost all interviewees. The following sub-tasks were addressed:

1. Preparation of the specifications relating to the type of filter bags.
2. Selection of the bags to be ordered.
3. Ordering the bags.
4. Inspection of the ordered bags after delivery.
5. Installation (and maintenance) of the bags.
6. Safe operation of the dust collectors.
7. Determining the time for replacing the bags.

Since several interviewees preferred to answer in terms of organizational units as opposed to functions, the results of the analysis of tasks and responsibilities are given by stating the responsible unit per sub-task.

The first two sub-tasks, preparation of specifications and selection of the bags to be ordered, presented the greatest spread in answers. For the *preparation of the specifications,* the emphasis of the answers was on engineering, which was also indicated by engineering itself. As regards *selection of the bags to be ordered,* emphasis was on maintenance and this was also indicated by maintenance itself. But in both cases several 'broad' answers were also given on the basis of which the question nonetheless arises as to whether these tasks are in fact arranged clearly enough.

Emphasis among the answers on *ordering* was on purchasing (also indicated by purchasing itself) and on maintenance (also indicated by maintenance itself). Here too there seems to be some degree of uncertainty.

The answers for the other sub-tasks were more equivocal. As regards both *inspection of the ordered bags after delivery* and *installation/maintenance of the bags,* the emphasis among the answers on maintenance was abundantly clear. This was also indicated by maintenance itself. For *safe operation of the dust collectors,* the emphasis was very clearly on production, and this was also indicated by production itself. *Determining the time of replacement* seems to be the responsibility of maintenance (also indicated by maintenance itself) with a clear involvement of production.

Assessment by the interviewees

The interviews revealed a distinct awareness of the hazards of static electricity as the cause of dust explosions, both in ABS production in general and in the dust collectors in particular. But opinions varied on what had to be done about the filter bag problems. Roughly speaking, two opposed groups could be distinguished here. The *first* group, who were mainly in production, did not really see a problem regarding safety. If a problem is not acknowledged, there is obviously no problem. The *second* group were of the opinion that something had to be done immediately/quickly/as soon as possible because stainless steel addition was involved (or: if stainless steel addition was indeed involved). In that case it was not right but wrong (and not in-between). So statements such as 'little decisiveness', 'underestimation' and 'it is treated too casually' emerged from this group. It is remarkable that even this group made statements such as 'it has been going all right for years'.

Thoughts about the origin of the problems and suggestions for future improvement referred to three aspects: knowledge, procedure and responsibility:

• There is insufficient knowledge in the organization: 'We know for sure that we don't know anything for sure', 'There are many people who know little about it and few people who know a lot about it'. This is linked to the fact that external bodies (such as suppliers) are trusted too much and taken for granted.

• There are gaps in the procedure (e.g. unclear specifications, inadequate inspection of delivered material). Improvement is possible by making the entire procedure traceable: proper specifications, requiring certificates and/or labels with the delivered filter bags, and a sound inspection of delivered material.

• Nobody is the owner of the problem: 'Nobody has ever seen this as *his* problem and hardly anyone has ever bothered about it'.

Results (2): general

This section discusses results of a more general nature, which allows the problems with the filter bags to be placed in a wider context, in order to identify other areas of potential danger. First of all, several aspects relating to the filter bags are discussed which were also brought up for discussion in a general sense, such as the perception of communication and of tasks and responsibilities. Then the results of the questionnaires on organizational culture and team roles are discussed.

If the findings on the filter bags are plotted against the way the interviewees see the company in a more general sense, the impression emerges that

a certain degree of contradiction and ambivalence exists:
- Most interviewees stated that tasks and responsibilities are clearly arranged. The analysis of tasks and responsibilities regarding filter bags nonetheless shows that this clarity should be queried for some tasks (preparation of specifications, for example). One of the interviewees also noted that it is not clear whether responsibilities are clear-cut, in connection with the observation that many matters are dealt with as joint action by maintenance, production and safety.
- The interviews suggested that matters are well arranged via procedures. There was nonetheless confusion about the specifications of the filter bags in d.c. B, and ultimately detailed data had to be provided by a supplier.
- The interviewees generally referred to good and open communication. There was nonetheless a large degree of confusion about the state of affairs regarding the problems with the filter bags.
- The interviews showed that safety has a high priority. However, the filter bag problems were something of an issue which drags on and in which little decisive action was taken (whereas the expression 'time bomb' was sometimes used in connection with the possible presence of stainless steel fibres in the d.c. B filter bags).

Organizational culture

Culture is sometimes defined as the entirety of written and unwritten standards and values, and indicates the way of working in a company ('that happens to be the way we do it here'). The culture of a company is not tangible or clearly demonstrable; however, understanding it is of great importance in, for example, implementation of decisions, in order to increase the chances of success.

At the GE Plastics site in Amsterdam the consequences of the GE Plastics takeover of Borg-Warner are still clearly noticeable. So, for the present situation to be properly judged, the two companies as they operated before the takeover are characterized first.

The previous Borg-Warner comes across as a very well organized company. There were many procedures and there was a well-considered but slow decision-making process going through many channels and involving many departments. The competitiveness of the company and the market on which it operated made it necessary to have a very critical look at the cost-benefit ratio of investments. Decisions took time, and the company was clearly characterized by prudence.

GE Plastics has a completely different background. At the time of the takeover of Borg-Warner, the company had nothing or all but nothing to do with competition and enjoyed generous profit margins. There was a lot of money to be spent and there was no hesitation in doing so. The lines of decision-making were short, so quick action was possible.

The present situation at GE Plastics Amsterdam shows a mixture of both cultures. Substantial investments were made after the takeover to adjust and modernize the company and to make up for back maintenance. At the same time it is a fact that many elements from the previous Borg-Warner culture have remained. The takeover by another company and changes in personnel alone just do not lead to a different company culture. That is partly due to the fact that most of the present managers come from the previous Borg-Warner. After all, the effect of management on the culture of a company is very evident because it enhances the prevailing culture to which it has also given shape. Those who cannot agree with the way things go in the company will easily be inclined to leave. This means that, once developed, a company culture can preserve itself very well.

A picture of the present culture in GE Plastics Amsterdam developed during the interviews. In order to give this picture more shape, each interviewee was asked to complete a questionnaire which is focused on identification of the

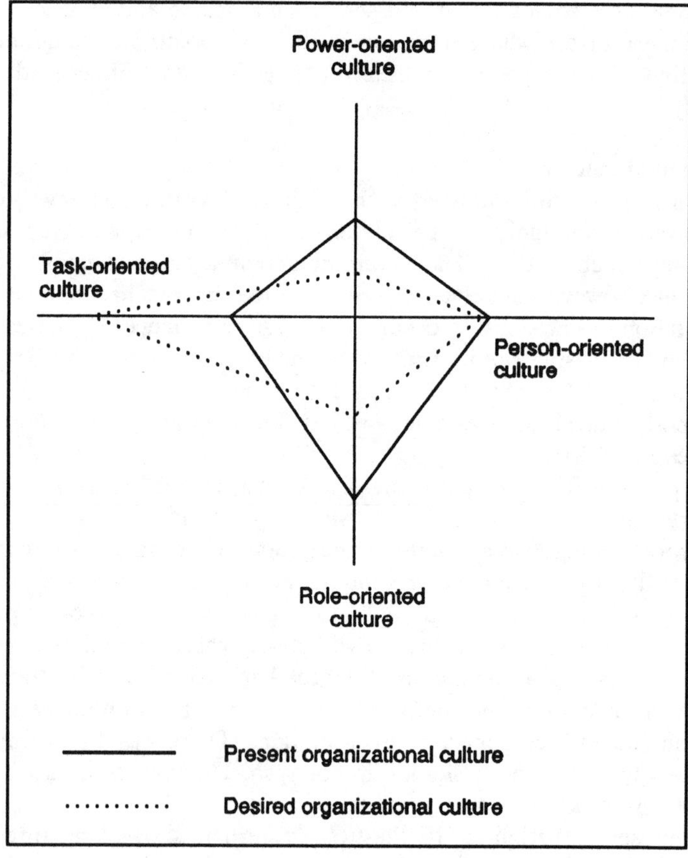

Figure 2. Culture profiles management team (GE Plastics Amsterdam)

present and the desired organizational culture. The questionnaire is based on the culture characterization developed and described by Harrison [7] who distinguishes four cultures:

- power-oriented (personal power is decisive; key persons with delegated power in essential positions; individual performance counts),
- role-oriented (order is paramount; rules and procedures are very important; sound arguments, logic and rationality; correct behavior more important than effective behavior),
- person-oriented (strong emphasis on individual objectives; minimum structure and regulations; expertise and job satisfaction are important), and
- task-oriented (organizational objectives are paramount; the task to be performed is determining; rules are irrelevant; pragmatic way of acting).

Although companies will generally exhibit characteristics of each of the four cultures, one type is dominant in many cases. That culture and its corresponding characteristics are then decisive for the company in question.

The questionnaire produces a score for each of the four possible cultures. These scores can be plotted in a diagram. Figure 2 gives the average scores of the GE Plastics Amsterdam management team, for both the present and the desired situation.

It is evident from Figure 2 that the present culture is characterized by a certain emphasis on role orientation. The desired situation shows a much clearer emphasis on task orientation. This seems a logical tendency in the light of the company's history: the previous Borg-Warner exhibited distinct characteristics of a role-oriented culture, GE Plastics showed distinct characteristics of a task-oriented culture at the time of the Borg-Warner takeover, and the present culture could be characterized as a 'task-oriented role culture', i.e. a role-oriented culture which evolves towards a task-oriented culture and in which the desired task-oriented culture is already recognizably present.

Team roles

Making a company to operate is not a one-man job. Stronger even: the majority of the daily tasks are carried out together with others. People are part of different teams. The management team is special to the extent that it addresses particularly those affairs which could be/are relevant to the entire company. Each person has certain qualities, and a team benefits from as even representation as possible of the various qualities involved with regard to the functioning of a team. In these qualities a distinction can be made between three orientations, where the extremes of each of these orientations are indicated by a specific team role: time orientation (innovator vs preserver), person orientation (entrepreneur vs supporter) and environment orientation (thinker vs practician).

Each interviewee was asked to complete a questionnaire which is focused on determining his pattern of team roles. Figure 3 shows the average scores of

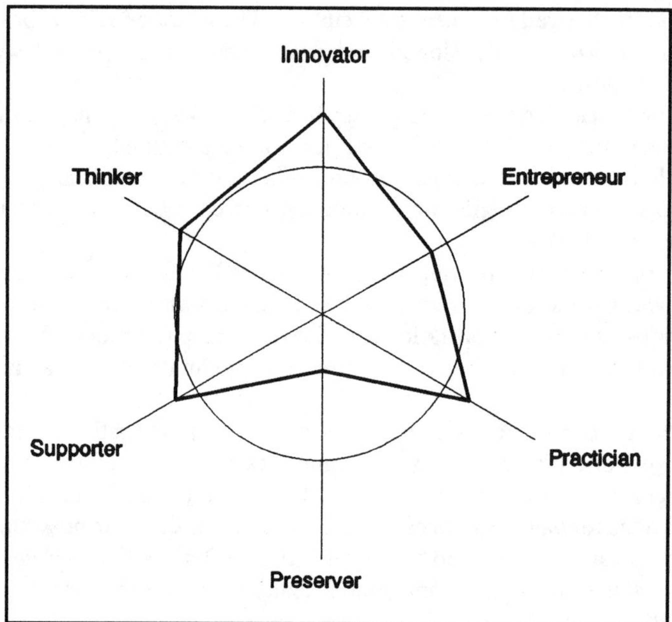

Figure 3. Average team role score management team (GE Plastics Amsterdam)

six persons from the GE Plastics Amsterdam management team. The circle in Figure 3 symbolizes an identical average score in each of the six different team roles and thus corresponds to a balanced team. The main conclusion from this figure is that there is considerable under-representation of the preserving role. This means the team has an inadequate tendency towards 'retrospection' and maintaining results from the past. This finding confirms the observations which were made on this point during the interviews.

Conclusions and recommendations

It was already known at the start of the study that the focus was not on selection of the proper filter bags but on the situation in which such problems could develop and on the factors which contributed to the development of the situation. The most important conclusion in that perspective was that the climate in the company may cause an unjustified feeling of security which is associated with a tendency to think that everything is properly arranged; thus, alertness regarding critical affairs may decrease. This is related to a certain degree of contradiction and ambivalence which was found in the organization and which became evident with regard to tasks and responsibilities, procedures, communication, safety, and decisiveness. This ambivalence is associated with the change in the organizational culture from role orientation to task orientation. The tendency in difficult situations is to return to familiar standards and habits, in this case caution and shared responsibility. In the case of the filter bag problems, this has led to an uncertain situation which was characterized by diverse perceptions, unclear responsibilities, insufficiently decisive action and which dragged on. For example, the involvement of the various persons/departments in the problems considered remained unclear. Nobody seemed to be directly responsible, so progress depended on the degree of involvement experienced individually. It was recommended that efforts be made towards recognizing the unjustified feeling of security, eliminating existing uncertainties by improving communication, and providing a person who is ultimately responsible in situations comparable to the filter bag problems.

The above unjustified feeling of security was concluded to be connected among other things with the use of the ISRS and the level achieved in it (five stars). The impression was that the emphasis has shifted with time from 'safety' to 'system'. Thus, it was recommended that the primary goal of using the ISRS remains: maintaining and improving safety.

Finally, it was concluded that knowledge within the organization has been lost with time. There seemed to be a degree of 'dozing off'. This development relates to under-representation of the preserving role in the management team which ensures that results achieved are maintained for the future. It was recommended that accessibility of the available knowledge be improved and in conjunction with this the preserver role within the management team be enhanced.

CONCLUSION AND PERSPECTIVE

The SMART framework was applied in two case studies so far, at plants of Dow Benelux and GE Plastics Europe. Both companies spend a lot of effort in safety, and also strive for continuous improvement. The Dow case study is summarized in [4], the GE Plastics case study was discussed above. The experiences gained with these case studies clearly show that the SMART framework is useful in practice, and indeed provides the opportunity to recognize management factors. More precisely, applying the framework to a specific problem leads to identifying *generic* underlying causal mechanisms, and thus allows for eventually resolving more than just the problem in question. In most cases, a clearly defined solution to address those underlying mechanisms will not exist. However, applying the SMART framework leads to a possible *route* to a solution which is attuned to the characteristics of the organization. Eventually, it is the responsibility of the company to accept the challenge of initiating a process which will gradually improve the existing situation in a fundamental manner.

The SMART tools line of development is currently in progress, and is concerned with the development of an assessment guideline for safety management systems. This guideline will focus on the question of the quality of such systems. Its purpose is to provide companies and competent authorities with an aid for investigating the quality of a safety management system which is already in place. Within the scope of the guideline, quality is defined in terms of completeness and effectiveness. The guideline will integrate the philosophy of the SMART framework with the lines of thought on which the ISO 9000 series is based, in particular with those of ISO 9004 [8]. Associated with the assessment guideline, instruments will be developed in order to assess specific aspects of safety management, such as commitment and communication. The ideas on which the assessment guideline will be based are introduced in [4].

In addition to the SMART Project, work is in progress on the development of harmonized safety management standards and safety auditing procedures, to be adopted in Member States of the European Community. It is carried out jointly with DuPont de Nemours and Moret Ernst & Young, and is also partly sponsored by the Commission of the European Communities. This research, which is known as the ACRONYM Project, will adopt the SMART philosophy. It will result in proposals for harmonized guidelines on safety management and safety auditing.

ACKNOWLEDGMENTS

The SMART Project is sponsored by the Commission of the European Communities. Furthermore, the authors are grateful to GE Plastics for permission to discuss the case study which was performed at one of its sites, to Willem Renes of Kern Konsult for stimulating discussions and valuable advice on the case studies, and to Marten Brascamp who provided valuable comments on a draft version of this paper. The SMART framework was developed in close co-operation with Sally Brearley of SRD.

REFERENCES

1. J. Reason, Human Error. Cambridge University Press, Cambridge, England, 1990.
2. A.M. Jenkins, S.A. Brearley and P. Stephens, Management at Risk. Report SRDA-R4, SRD, AEA Technology, Culcheth, England, 1991.
3. S.A. Brearley and L.J.B. Koehorst, Hard and soft options to achieve safety: The SMART project. In: K.E. Petersen en B. Rasmussen (eds.), Safety and Reliability '92, Elsevier Applied Science, London, England, 1992, pp. 296-307.

4. M.H. Brascamp, L.J.B. Koehorst and J.F.J. van Steen, Management factors in safety. Presented at the European Safety and Reliability Conference ESREL '93 (Munich, Germany, May 1993).

5. W.A. Wagenaar and J. Reason, Types and tokens in road accident causation. Presented at the CEC Workshop on Errors in Operation of Transport Systems (MRC - Applied Psychology Unit, Cambridge, England, May 1989).

6. J.F.J. van Steen and L.J.B. Koehorst, SMART - Management at Risk: Case study at GE Plastics (Amsterdam). Report 93-073, TNO, Institute of Environmental and Energy Technology, Apeldoorn, Netherlands, 1993.

7. R. Harrison, Understanding your organization's character. Harvard Business Review, May-June 1972, 119-128.

8. ISO 9004: Quality management and quality system elements - Guidelines. International Organization for Standardization, 1987.

Common Documentation Systems for ISO 9000 and Process Safety Management

Alfred W. Bickum
Arnold E. Barnette
The Goodyear Tire & Rubber Company, Akron, OH

ABSTRACT

Successful implementation of ISO 9000[1] and Process Safety Management (PSM) is best achieved by using a systems approach. Certification to ISO 9000 is becoming a market place necessity and often a requirement for doing business. Since many of the ISO elements overlap with those of PSM, there are significant advantages to defining a documentation framework applicable to both. A standard four level framework definition is described which provides a traceable link from policy through records. Details of the framework are described, the overlap issues of ISO 9000 and PSM are presented with references to how the framework serves both, and a mainframe computer based software (PREFERENCE™)[2] which enables on line instantaneous viewing and updating of Policies, System Standards, and System Procedures is presented.

BACKGROUND

Global competition necessitates the adoption of an internationally recognized quality management system. ISO 9000[1,3] certification has been adopted by the European Community as the standard for quality systems, and companies throughout the world are increasing their commitment to implementation. Quality management systems have been around for decades and are the foundation for modern manufacturing. A standard was developed by Government and sophisticated OE manufacturers and audits were conducted by Government, OE or internal quality inspectors to insure conformance for systems and products.

By contrast, for many chemical manufacturers process safety management has only recently come to the forefront as a result of catastrophes like the Bhopal incident in 1984. Comprehensive management systems like those detailed in the CCPS Guidelines books have enhanced the state-of-the-art in Process Safety

™ is a trademark of Goal Systems International, Inc.

Management[4]. Added impetus has been placed through the promulgation of OSHA 29 CFR 1910.119 "Process Safety Management of Highly Hazardous Chemicals"[5].

Both quality and process safety are effectively implemented through adoption of management systems as opposed to a "program" approach which typifies many recent outmoded management styles. ISO 9000 is a clearly defined standard which originated through World War II military quality systems adopted by NATO and assimilated into the British Standard Institution specification in 1979. The standard was reissued as the ISO 9000 series in 1987 after significant discussion by experts in the field. The result was a clearly defined standard through which certification of quality management systems became possible.

APPROACH

Examining the elements of ISO 9000 reveals that in order to satisfy the requirements for certification, complete documentation of the existing quality system is required so that all parts of the system are visible and clearly defined. Documenting the quality system is basically writing down "what is being done" and "doing what you say you are going to do". If deficiencies exist in the quality management system they are identified and a system is designed to satisfy the ISO requirements. As each element of the Process Safety Management model is compared to the ISO elements, commonalties between both are identified and document structure is designed that builds on those commonalties. Once Quality and Process Safety Management systems are examined, overlaps are exposed for management systems beyond the quality and process safety arena. As an example, health and environmental management systems require training as does quality and process safety. The system for identifying training needs, development of a training program, delivery of the material, and assessment of subject understanding is identical for all.

ISO AND PSM MODEL COMPARISONS

With ISO 9000 selected as quality model, the first step is to educate associates ISO 9000 requirements. An understanding of the requirements of the standard is necessary to develop an appreciation for the documentation requirements necessary

PSM Elements → ISO Elements ↓	Account-ability	Process knowledge and document-ation	Capital project review	Process risk manage-ment	Manage-ment of change	Process & equipment integrity	Human factors	Training and perform-ance	Incident investiga-tion	Standards, codes, & laws	Audits & corrective actions	Enhance-ment of process safety knowledge
4.1 Management Responsibility	X											
4.2 Quality System												
4.3 Contract Review												
4.4 Document Control		X										
4.5 Purchasing					X							
4.6 Purchaser supplied product												
4.7 Product identification and traceability												
4.8 Process control		X			X							
4.9 Inspection and testing						X						
4.10 Inspection, measuring and test equipment						X						
4.11 Inspection and test status						X						
4.12 Control of non conforming product												
4.13 Corrective action												
4.14 Handling, storage, packaging and delivery												
4.15 Quality records		X										X
4.16 Internal quality audits											X	
4.17 Training								X				
4.18 Statistical techniques		X										

Table 1:   Common ISO 9002 and CCPS Process Safety Management Elements

for certification.   This appreciation for the role of documentation in defining management systems carries over into the designing and implementation of Process Safety Management systems.

The second step is to select a Process Safety Management model. Many models exist and are used by corporations in their pursuit of compliance under OSHA 29 CFR 1910.119, including API-750[6], CCPS, CMA Responsible Care, etc. The CCPS model consisting of 12 elements represents a management model which contain the planning, organizing, implementing, and controlling characteristics of a management system.

Each CCPS PSM element is compared to each ISO 9000 element to determine where overlaps exist with commonalties shown in Table 1. For these common requirements only one system to manage both quality and safety is necessary. Certainly policy can be written so that both quality and safety requirements can be met. Similarly, descriptions of how the management systems are designed and implemented should be common to both. While some of the specific requirements of procedures used in each system may differ, the system itself can often be identical. A good example of this is the training element. A management system that ensures proper training of associates should be the same whether it applies to quality, safety, process safety, health, or environmental training.

The   advantages of defining a single system includes time savings, documentation reduction, and management system design consistency. How can the documentation requirements be reduced and how can the systems be defined singly while ensuring complete subject matter coverage? A structure for design of a documentation and delivery of this design needs to be specified.

PYRAMID MODEL

The pyramid model would enable common management systems to be defined where appropriate and allow for separate management system definition for differing elements. This model is similar in concept to that proposed by Dowell[7] which describes the process by which values translate into actual practice. This model is used to describe the architecture of the document system (Figure 1).

For the documentation model, the four tier framework in Figure 1 places policy at the top of the pyramid as Level I. Policy describes "why" the policy exists and "what" the policy is designed to do. It does not indicate how this is to be done nor does it indicate who is to do it; however, it does include who is responsible for making or revising policy.

The second level (Level II) is used to describe the management system and is referred to as the System Standard. The System Standard contains information on what is required, who has responsibility for the standard, what criteria are used, and an overview on how it is done.

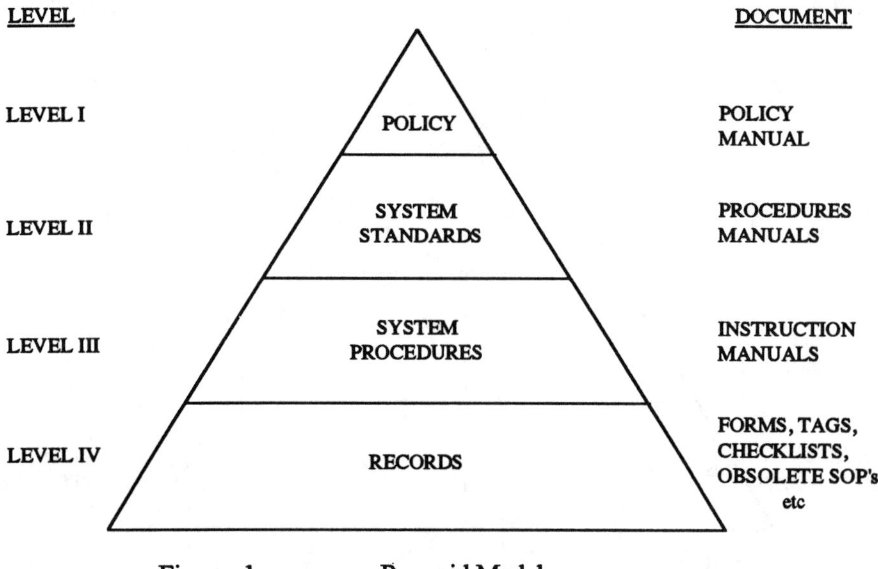

Figure 1:   Pyramid Model

The third level of the model (Level III) is System Procedures, which details work instruction. Included in this level are the operating procedures, training manuals, management of change procedures, etc. Level III may also include references to other systems already in place or other manuals, procedures, or document systems which is described by including reference list or matrix. An example of a matrix is a listing of all records and the corresponding individual or department responsible for their archival. Level III items would also include the forms and checklists which would be used in the normal execution of procedural work.

The final level of the documentation system (Level IV) is records. The information in this level can be used to verify that work is completed in conformance to the procedures described in Level III. Examples of records would be mechanical integrity testing results, incident investigation reports, completed check sheets, old revisions of operating procedures, old versions of P&ID's, etc. A constant flow of records entering Level IV is provided to the individual responsible for filing as prescribed by the Level III record retention procedure.

Each System Procedure, System Standard, or System Policy refers to the preceding or succeeding level. This ties an entire management system together from a documentation standpoint. For example, training policy may indicate that a training system requires the that a needs analysis be conducted, a training manual be prepared, and a trainer to have specific training skills. These requirements are valid regardless of whether Quality or PSM training is conducted. Figure 2 shows how the training policy and the training System Standard would be the same for either Quality or PSM training .

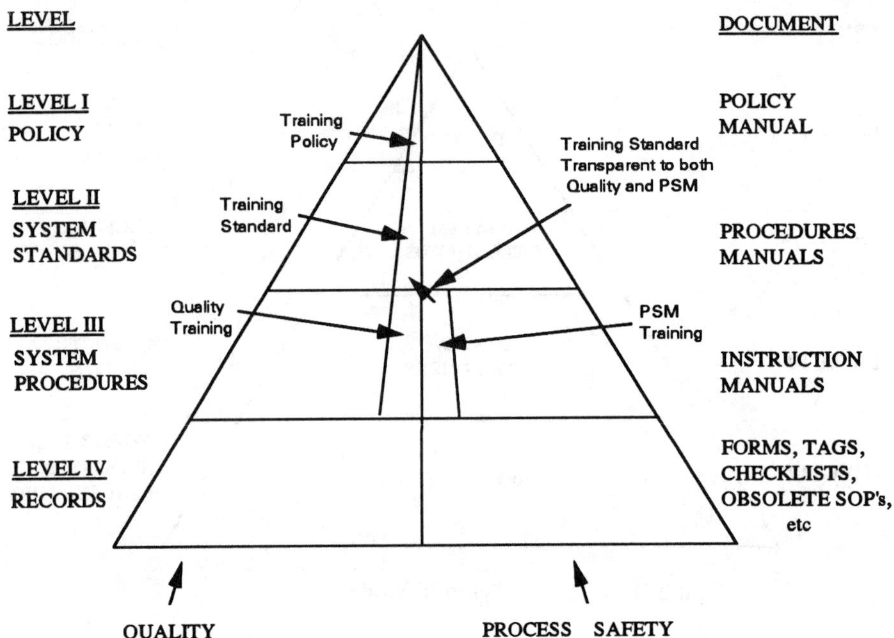

Figure 2:   Example of Common System Standard

For the document system to have traceability a uniform numbering system is required.   A smart numbering scheme effectively differentiates Policy, System Standards, and System Procedures.   The document number shows the quality, health, safety or environmental element, the numerical sequence, revision number, and department, if needed. The format for the numbering scheme is as follows:

TTTLL-EEE-NNN.RR-DDD where,

T = System type, (Q for quality, H for Health, S for Safety which includes
     process safety, or E for Environmental, one or more to be used, i.e.
     HSE)
L = Document level, (SS for System Standard, SP for System Procedure)
E = System Element, (i.e. DOC for document control, II for incident
     investigation, TR for training, AUD for audit, etc.)
N = document number
R = revision number
D = department or area designation (optional)

As an example, HSESS-II-001.00-COR would be the original corporate system standard for all health, safety, and environmental incident investigations.

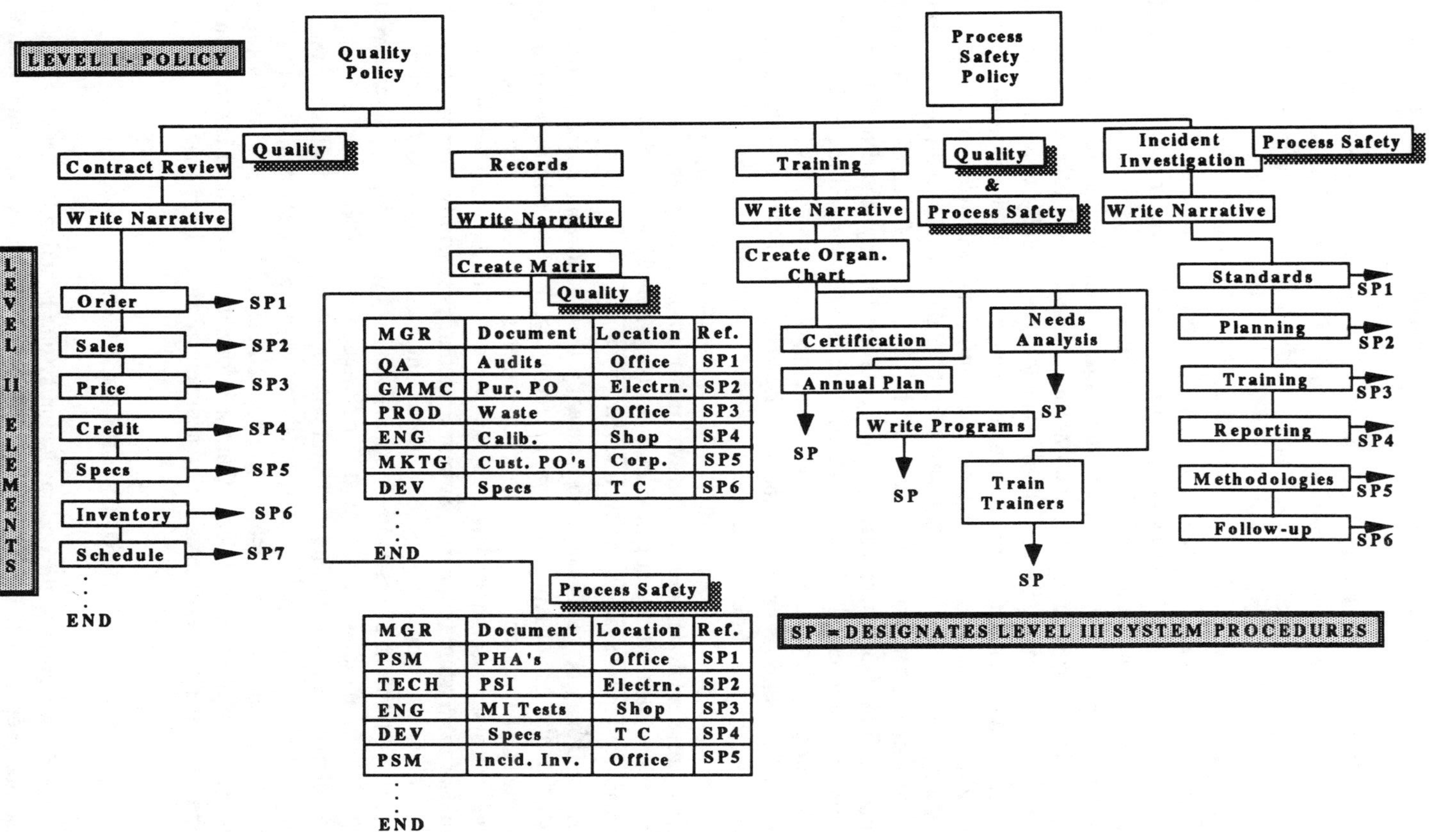

MGR	Document	Location	Ref.
QA	Audits	Office	SP1
GMMC	Pur. PO	Electrn.	SP2
PROD	Waste	Office	SP3
ENG	Calib.	Shop	SP4
MKTG	Cust. PO's	Corp.	SP5
DEV	Specs	T C	SP6

. . . END

MGR	Document	Location	Ref.
PSM	PHA's	Office	SP1
TECH	PSI	Electrn.	SP2
ENG	MI Tests	Shop	SP3
DEV	Specs	T C	SP4
PSM	Incid. Inv.	Office	SP5

. . . END

Figure 3:   Example of Some Common Quality and Process Safety Management Documentation Elements

Similarly, QSP-DOC-001.01-345 would be the first revision of department 345's quality system procedure for document control. The numbering scheme helps identify the function of that document and provides a means of referencing from one document level to another. For ISO 9000 and Process Safety, each system standard is a direct result of the essential elements defined in a policy statement. From the System Standard flows specific procedures required in order to accomplish tasks. Once the tasks are performed, the resultant documentation of those activities generate records which are archived for future use.

It is important that separate documents on Policy, System Standards, and System Procedures be written since the information contained in the policy is less likely to change as frequently and should therefore be carefully written to ensure that whatever management system design is adopted, it will adhere to the policy. As Policy and System Standards are designed and described, wording is selected which defines the requirements that satisfy both quality management and process safety management for those common to both. Examples of this integration are presented in Figure 3 where common elements for Quality and Process Safety Management are presented. Contract review is an ISO element while the definition of a Policy and System Standard for the managing of records covers both. In this case the written policy is used for either Quality or Process Safety and the Level II System Standard has differing matrices for each. For training, the Policy and System Standard would be almost identical. Since any training that is done requires a needs analysis, annual plan, writing of training programs, and trainer elements. The incident investigation element is designated for process safety but would also overlap into safety, health, and environmental investigations.

Level II may be modified more frequently than policy but it should remain somewhat stable. Major reorganization, restructuring, or change in responsibilities would effect the system standards. Assessments on the effectiveness of a management system could lead to modification in its design and would require rewriting of the System Standard. System Procedures constantly change in search of more efficient manufacturing conditions, shipment of materials, testing of equipment, or training of associates. Annual reviews of these procedures reveal discrepancies from practice which must be reflected in an updated procedure.

An important aspect of ISO 9000 and process safety management is document control. The Policies, System Standards, and System Procedures all must be controlled to ensure that the most current, accurate information is available to associates. There must be confidence that the information being used is correct and that when it is changed, associates are immediately aware of the change.

DELIVERY SYSTEMS

Once the structure of the documentation system is defined, a delivery system of these documents needs to be specified as an integral part of document control. The goal of the delivery system is to have ready access, wide distribution, ease of updating, flexibility, compatibility with existing electronic communication systems, and the ability to reference documents common to both Quality and Process Safety Management systems.

With these criteria in mind several systems which are available within Goodyear were investigated and mainframe software called PREFERENCE was selected. The software is designed for computer based training where manuals and help could be easily used by the trainee. The application for this software extends beyond the computer based training realm. Examples of on-line applications of PREFERENCE software include software help, general documentation, manuals, training materials, new product announcements, associates announcements, demonstration materials, and bulletin boards.

Controlling documents to ensure the most up-to-date standards or procedures are available for use by associates was a major concern. The existing main frame document system has a three year life before it is archived. Providing policies, standards, or procedures change during this time period, there is not a problem with updating the document and publishing. However, should one of the documents not be updated, it was deleted after the three year time period.

Following an analysis of advantages and disadvantages of available systems PREFERENCE is chosen software for use in the Chemical Division.

PREFERENCE

In order to describe how the ISO 9000 and process safety management documentation systems are tied together, a brief review of the structure and options that PREFERENCE software offers is appropriate. A single manual in PREFERENCE software is referred to as a volume. Each volume is sub-divided into chapters and each chapter is subsequently divided into topics. A topic can further be divided into sections and each section can contain up to 99 reference items. With this structure (see Figure 4) sufficient flexibility is obtained for the division or classification of most subject matter. In addition, a list of volumes can be presented giving easy access to a variety of subject matter.

Two modes of access exist in PREFERENCE software, one available for the writer of the volume and one for a reader or user of the volume. A single individual is assigned the responsibility of maintaining the integrity of a volume. Revisions, once approved, are made to the volume, and once entered become immediately available to the user. A reader can be granted access to a specific volume or any number of volumes depending upon the readers needs. A volume can also have access to anyone in a Division or the entire company./ This access is determined by policy and controlled by the PREFERENCE software administrator.

This system allows the user to automatically develop a table of contents as the volume is created. Indexing of terms can be done manually or automatically.

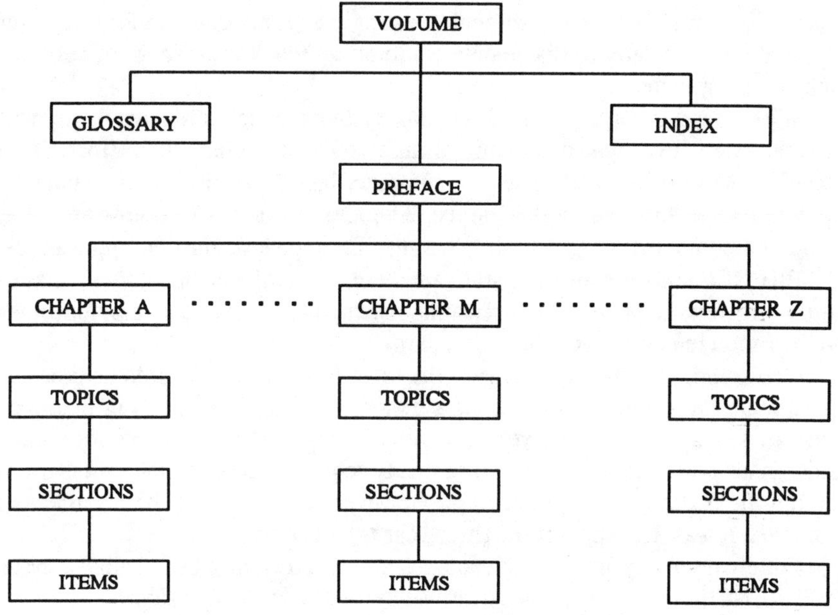

Figure 4:   PREFERENCE Structure

When used with a colored CRT, specific colors can be used to enhance or accentuate specific parts of the item which the writer wants to highlight.

The system is structured to include corporate level policy and system standards. PREFERENCE software has the ability to imbed volumes, topics, or text wherever appropriate into the current manual.   This is an advantage for dissemination of corporate policy. For those policies which are generic for the plants and generic for the ISO or PSM systems, only one policy has to be written and approved. The plant can then incorporate the policy in their manual strictly through an imbedding command.

There is also the option of using a pop-up window to enhance specific parts of the text. This can be done for training or simply to help the reader clarify a particular point in the manual.

One disadvantage of the PREFERENCE software is the inability of the system to provide graphics. The integration of PREFERENCE software with other software is being developed but for now the use of tables, graphs, and drawings is done through separate means.   Another limitation is the inability to perform calculations in the text. Any formulas that are required for operating the process or making decisions on adjustments must be done external to this system. This is not an uncommon disadvantage for a system designed for documentation.

COMMON KNOWLEDGE

The use of the system was designed to provide easy access to a wide variety of information for all associates within the Chemical Division. PREFERENCE software volumes include operating procedures, safety manuals, lab procedures, shipping/receiving procedures, policy, system standards, utility procedures, and reader and writer PREFERENCE software information. Each plant provides a PREFERENCE software volume list that contains all the manuals for the facility and provides a common entry point to the information.

REFERENCES

1. ISO 9000, "Quality management and quality assurance standards - Guidelines for selection and use", International Organization for Standardization, Switzerland, 1987.
2. PREFERENCE, Goal Systems International, Inc., 1990.
3. ISO 9002, "Quality systems - Model for quality assurance in production and installation", International Organization for Standardization, Switzerland, 1987.
4. AIChE-CCPS, Guidelines for Technical Management of Chemical Process Safety, New York: Center for Chemical Process Safety, 1989.
5. OSHA, Subpart H of 29 CFR 1910.119, "Process Safety Management of Highly Hazardous Materials; Explosives and Blasting Agents; Final Rule", February 24, 1992.
6. API-750, "Management of Process Hazards", American Petroleum Institute, January, 1990.
7. Dowell, A. M., III, Getting from Policy to Practices: (or, What is this Standard Really Trying to Do?), 1992 Process Plant Safety Symposium, Houston, TX., February 18-19, 1992.

HAZOP Studies under ISO 9000

Philip Charsley
Director of Technology, DNV Technica, London, England

Bill Brown
CEO, DNV Technica, Inc. Houston, TX

ABSTRACT

HAZOP studies are now probably the preferred method of Process Hazard Identification worldwide. They normally combine the steps of Hazard Identification and Corrective Action.

Senior management and Authorities in many countries, are asking for HAZOP studies, but many have no idea whether the product they receive is up to scratch. They now need an answer to the question *'Has the team done a good enough job?'*

ISO 9000 contains many aspects of value in answering this question. The principles of Responsible Care require us not only to use our best endeavours to identify and cure problems in relation to safety, but also to use the best available technology to assure the quality of those endeavours. However, as with many specialist areas, the application of ISO 9000 to HAZOP is not straightforward. The authors of this paper, both experienced HAZOP leaders, have applied their knowledge of what makes a good study to the formal requirements of the ISO 9000 standard, to ensure that the quality of studies is truly assured.

INTRODUCTION

ISO 9000 is a powerful quality standard, but it is possible to follow it in many areas, including HAZOP Studies, without more than a marginal improvement in the quality of the activity. The basic difficulty that we face in trying to design a quality system to match the HAZOP activity is that the study procedure relies on brainstorming. Even though the guidewords used in HAZOP methodology provide structure to this brainstorming, they do not of themselves ensure that the brainstorming process will produce quality results.

Before DNV Technica could use ISO 9000 to assure the quality of HAZOP Studies, we needed to define a measure of the quality of a study. This in turn required a definition of the objectives of a HAZOP Study. The definition we chose is as follows:

'The objectives of a HAZOP study are:

1   To identify all potential causes of process upset scenarios which could lead to significant safety or operability consequences.
2   To decide whether the current design ensures that the risk from each identified scenario is at a suitably low level.
3   If not, to recommend modifications which will reduce the risk to a suitably low level, or specify a further study to investigate the issue, with the objective of identifying a suitable modification.

Unfortunately, this objective set may be sufficient for a consulting company, only responsible for the conduct of the HAZOP Study itself, but it is incomplete for the client company because it does not put HAZOP in the context of a Safety Management Program. We decided to look beyond the HAZOP itself, to the reasons for carrying out the study, and the objectives for the process unit under study.

This then introduces more objectives, which may be the following:

1   To maximise the value of the facilities to the company by reducing process related risks to tolerable levels and to improve operating efficiency.
2   To recommend cost effective risk mitigation measures in line with company targets.
3   To recommend cost effective measures which will improve the profitability of the operation.

These two sets of objectives then allowed us to put forward quality criteria for a specific HAZOP study, bearing in mind that the client company will require a framework policy to regulate when HAZOP studies will be carried out on what facilities. The HAZOP Study quality criteria we have adopted are:

1   All potential scenarios leading to significant safety or operability issues will be identified.
2   Decisions will be made on the acceptability of the risk level from each of these scenarios.
3   Where the risk is intolerable, recommendations for change will be made which will bring the risk level within tolerable limits.
4   Recommendations will be implemented as rapidly as feasible, when justified against client company criteria.
5   An authorised rejection notice will be produced to explain the background to the rejection of any safety recommendation.

It was then necessary to decide which factors could influence these quality objectives, and how our quality system could ensure their achievement.

In addressing these quality objectives, DNV Technica identified the following factors of importance:

- team composition, and qualifications
- methodology followed (ie brainstorming method)
- guidewords used
- completeness of coverage
- judgement criteria applied
- categorisation of recommendations
- testing of the recommendations against their objectives
- decision making and follow-up of recommendations
- continuous improvement of the HAZOP process

Each of these factors is discussed below, with the solution that we have adopted.

Team Composition and Qualifications

In order to meet our quality objective of identifying all significant safety and operability issues, we need a HAZOP Leader well versed in the HAZOP study technique, and capable of assisting the team to meet this objective. DNV Technica expects this qualification to be gained by a combination of training and experience of studies. It is unlikely that the required skills will be gained without in-depth study of the principles of the technique combined with sitting in on sessions run by several skilled HAZOP Leaders. Our internal qualification scheme includes certification of a new Leader after observation by an existing skilled Leader.

The majority of the team members also need experience in previous studies, but we accept that where this is impractical, training may be given to the group before the study proper starts. This may be done by an appropriately skilled Leader.

The brainstorming technique works best with a minimum core team of four participating members (excluding Leader and Recorder), with a broad range and depth of experience relevant to the study object. It is important that the qualifications of these members include the disciplines necessary to understand the potential hazards of the process under study, and the industry standard methods (including Codes) in use to control these hazards. At least one team member needs a good background knowledge of relevant past accidents on similar facilities, and of the lessons learned from these accidents.

The quality of the team is best judged by the HAZOP Leader, and he must be in a position to require a change of personnel if he is not satisfied. The strongest tell-tale sign of an unbalanced or underqualified team is the proportion of issues raised which require information unknown to the team to draw conclusions either about seriousness, or about suitable recommendations.

It is clear from the above that the procedure for a HAZOP Study must include allowance for the HAZOP Leader to arrange for changes of team members if he is not satisfied with their capabilities. Even if the choice of the team has been made without his input, he should retain this right. He should also be able to postpone a study if the team is not effectively identifying the serious issues. This requires a Leader with enough experience in the industry to recognise some deficiencies himself. These issues cover part of the responsibility of the Leader for the quality assurance of the study under ISO 9000. Naturally exercise of these powers could seriously strain a client/consultant relationship, but the future will show us whether insistence will cause us serious problems.

It goes without saying that a balanced team will be out of balance if a member is absent. Hence our HAZOP procedure states that if a core team member is absent the study will stop until his return or replacement by a suitable substitute. Again the future will show us how much of a problem this gives.

HAZOP Methodology and Guidewords

As explained earlier, the HAZOP method relies on brainstorming. Brainstorming as a technique is very powerful, but it is difficult to ensure that it is sufficiently rigorous to meet our objective of not missing any significant safety or operability issues. One requirement to meet this need has already been discussed, that is the size and capability of the team. The other requirement is a suitable set of 'prompts'.

HAZOP methodology relies on a set of seven guidewords, which should be applied to all the important parameters of the process under study, to identify deviations from design intent which could lead to the issues we seek. The problem of completeness lies in the identification of the 'parameters of importance'. The method adopted by many proponents of the HAZOP technique, including DNV Technica, is first to ask the team to identify this set of 'parameters of importance', then to apply to them the HAZOP guideword list to generate a much longer checklist of guidewords to be used as our 'prompts'.

In practice, for most oil or chemical industry process systems, a single, generic set of some 20-30 guidewords is used, as for example the list of Table 1. The Leader should always check before uising a generic guideword list that the team believes it to be sufficient. When non-standard equipment is involved, or batch operations, or a high level of operator involvement in the process, a special exercise will be required to generate an appropriate guideword set. The use of a guideword set to identify scenarios is illustrated by the flowchart of Figure 1.

The completeness and appropriateness of the guideword set is a key to the completeness of scenario identification, and the choice of guidewords is a mandatory step in our HAZOP procedure.

The leader will divide the equipment under study into 'study sections', and will ensure that these sections are not too large or complex, so that the team does not become confused when applying the guidewords, leading to missed scenarios.

In order to meet our requirement to identify all causes of problems, all guidewords will be applied to each 'study section', and their use will be recorded to aid later quality control. During this process it is vital that the action of all safety systems is ignored by the team, ie that the identification process *assumes the failure of safety systems*. The reliability and effectiveness of these systems is considered when judging the seriousness and likelihood of each scenario (see the section on Judgement Criteria below).

TABLE 1 Generic HAZOP Guidewords for Continuous Chemical Processes

NO FLOW	MORE TEMPERATURE	SAMPLING
REVERSE FLOW	LESS TEMPERATURE	CORROSION/EROSION
MORE FLOW	MORE VISCOSITY	SERVICE FAILURE
LESS FLOW	LESS VISCOSITY	ABNORMAL OPERATION
MORE LEVEL	COMPOSITION CHANGE	MAINTENANCE
LESS LEVEL	CONTAMINATION	IGNITION
MORE PRESSURE	RELIEF	SPARE EQUIPMENT
LESS PRESSURE	INSTRUMENTATION	SAFETY

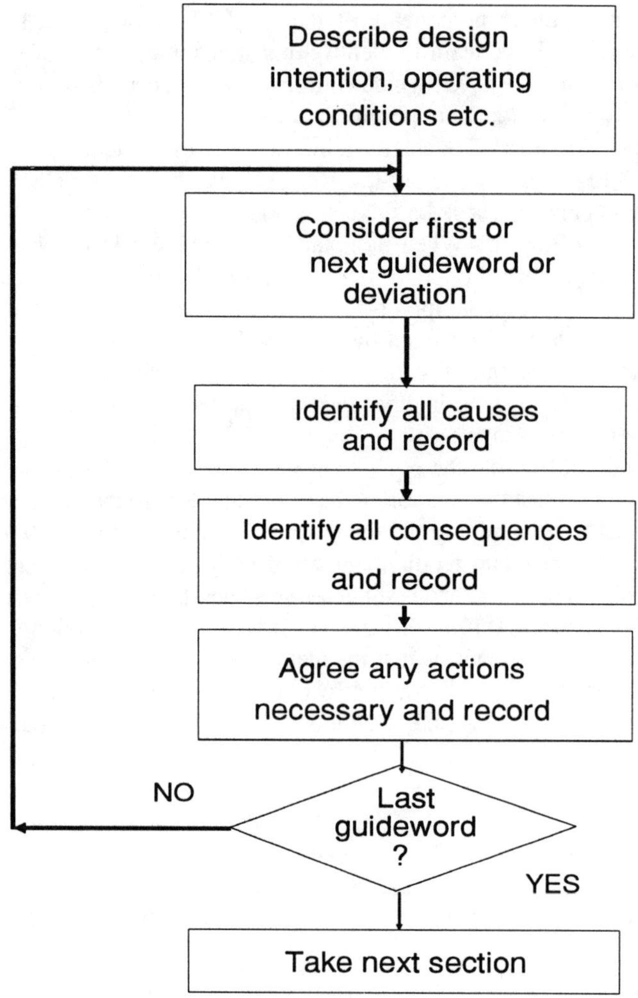

Figure 1   Scenario Identification in HAZOP

Completeness of Coverage

We may follow all the procedural recommendations given above, and still not identify all issues within the scope of the study. This can occur if all the Process and Instrumentation Diagrams (P&IDs) in the scope have not been studied, or if not every section of the P&IDs are subjected to study. It is also possible if the P&IDs are out of date, and omit recent changes. Our procedure therefore calls for a statement of scope listing all P&IDs to be studied, and a requirement that only if the team specifically agrees should particular P&IDs not be studied, and that this will require a suitable explanation.

It is normally the Leader's responsibility to ensure that all equipment on each P&ID has been studied, and if this is done by marking each study section on a master file drawing, the completeness of cover is documented, and can be checked later by a QA reviewer.

The Leader in the study of an existing plant who is informed that a P&ID is not up to date (by a team member) must put that drawing aside until it can be accurately marked up or updated. He should not accept that a team member mark up the drawing in the study: physical checking is essential. This is a mandatory part of our procedure.

For batch systems, it is also necessary to ensure that all sections are studied for each process step. This can normally only be checked against the log sheets.

Judgement Criteria

Before the HAZOP team makes a recommendation for change, it should decide whether the current situation is acceptable. Not to do so can mean a very productive team measured by the number of recommendations produced, but we do not consider this to be a true measure of quality! The process of making the decision is often a difficult one, as there are rarely suitable criteria against which to make a judgement.

For safety issues the question 'is this consequence too serious to tolerate?' can normally generate a simple answer, but the following question 'is the likelihood too great to tolerate?' is sometimes very difficult to answer (see Flowchart of Figure 2).

This is the area where the assistance of a company safety specialist may be required to help the team to determine suitable criteria. A Risk Index approach can sometimes be used to advantage, but is normally too time consuming to be used more than occasionally.

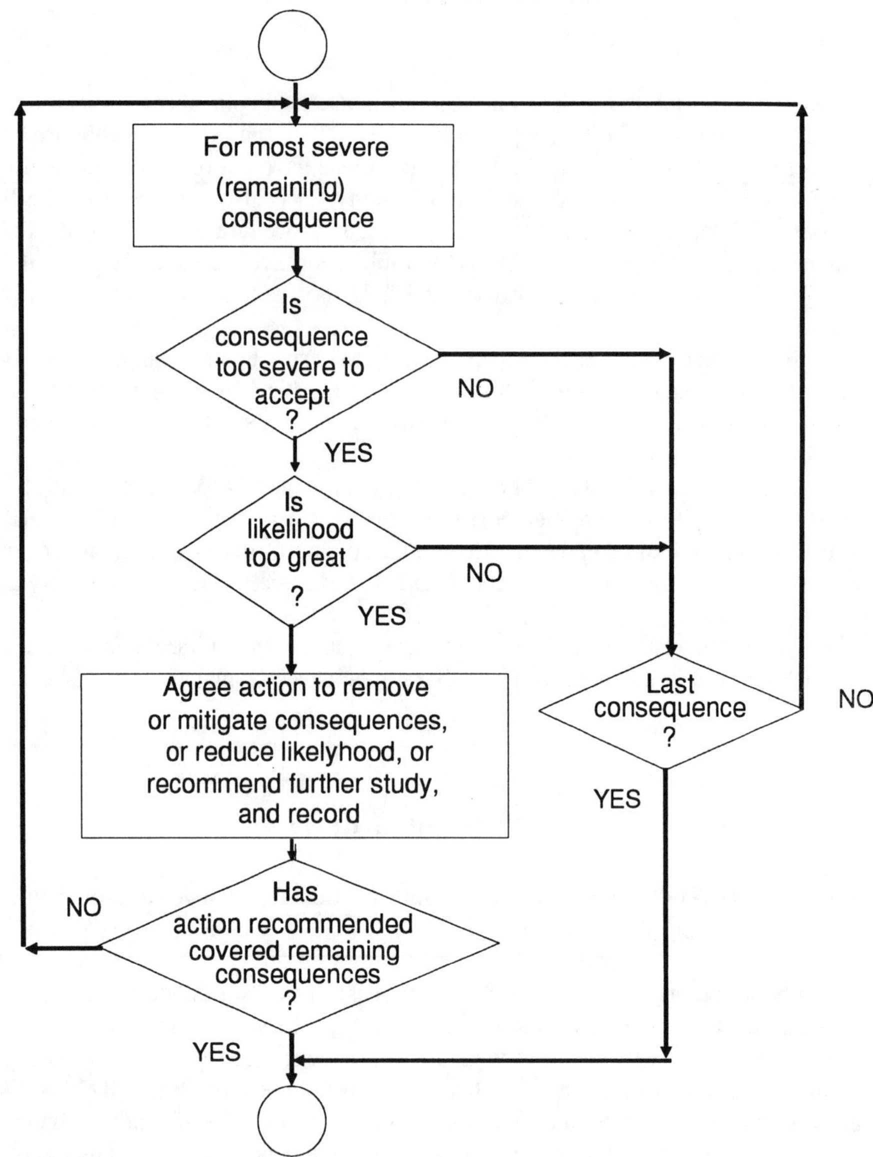

Figure 2   Judgement of Scenario Acceptability

For operability issues the criteria are much simpler, being based on cost/benefit principles. Again judgement is required, but often as long as the benefit looks worthwhile the team will be justified in recording the recommendation for management consideration.

As far as procedures are concerned, we have chosen only to include that the question of acceptability is addressed, and that a judgement is made by the team.

Categorisation of Recommendations

Our objectives made it clear that we expected the decision making on recommendations to be influenced by whether the consequence identified was a safety issue or not. In many situations identifying environmental consequences can also be of importance in decision making. Our procedure therefore has to ensure that the correct categorisation is applied to each recommendation, and passed on to management. We also require our HAZOP Leader to carry out 'off-line' checks for this while checking the recommendations against their objectives.

Testing of Recommendations Against Objectives

While it should be part of the thought process of all team members, Leader and Recorder, it is quite difficult during a study to remember to check that each recommendation does match with the objective of reducing the risk from the scenario under study to an acceptable level. DNV Technica have therefore incorporated into their HAZOP procedure an 'off-line' check by the Leader at the time when he and the recorder are expanding the record to make the recommendations stand-alone (ie understandable in isolation from the log sheet). This is normally done immediately after each HAZOP session, and allows the Leader to come back at the beginning of the next session to resolve any problems. Again this is exercise of the Leader's QA responsibility under ISO 9000.

Decision Making and Follow-up

This activity is part of the framework within which the HAZOP procedure operates, and therefore the responsibility for it falls upon the client company in a consultant/client relationship. In order to achieve the objectives 4 and 5 of our list, the framework procedure needs to address time guidelines for decision making and further study activities. It will also need to specify the requirements for rejection of safety recommendations, and place responsibility for all these activities on particular job-holders within the company.

Continuous Improvement of the HAZOP Process

From the Leader's standpoint there needs to be continuous improvement in the methodology and the effectiveness of the brainstorming activity. Many strides have been taken over the years to improve these areas. Examples include the use of LCD panels and computerised recording software to allow the team to interface with the record on-line; the use of checklists of possible causes of problems; the use of a qualified leader as Technical Recorder to improve team efficiency and help the leader concentrate on the brainstorming activity; the attention given to human factors issues.

From the company management standpoint there is a continuing need to help HAZOP teams to make better risk and business decisions. Defining and periodically upgrading risk criteria that are well understood by team members is crucial to reaching this goal. There is also an obligation to have in place continually improving management systems to address the HAZOP recommendations in a timely and responsible manner. In order to make significant improvement to the overall process, it is often necessary to provide additional expertise to the teams either from other areas of the company, or from outside.

Conclusion

DNV Technica has set up a procedure for HAZOP Studies designed to ensure their quality under ISO 9000. However, to meet all the desired quality objectives, a client company will also require to have in place a Process Safety Management framework procedure into which this procedure fits. The major components of our HAZOP Procedure are as follows:

- A HAZOP leader will be qualified to take the role by in-depth study of HAZOP principles, experience in studies lead by others, and assessment by an experienced leader.
- The majority of team members on a study team will already have experience of previous studies, or will have had training in HAZOP techniques.
- The core team will have a minimum of four participating members (excluding Leader and Recorder), with a broad range and depth of experience relevant to the study object. At least one will have a good background knowledge of relevant past accidents, and lessons learned.
- The HAZOP leader must be in a position to request a change in team personnel, or to halt the study, if he considers this necessary to ensure quality.
- If a member of the core team is absent, the study will stop until his return or replacement by a suitable substitute.
- The team must select a suitable set (or sets) of guidewords to use in the study, chosen to ensure complete coverage of the parameters of importance.
- The leader will divide the equipment under study into 'study sections' which are not too large or complex, to avoid the team being confused when applying the guidewords and hence missing scenarios.
- All guidewords will be applied to each 'study section', and their use will be recorded. When identifying problem scenarios safety systems will be assumed to have failed.

- A statement of study scope is required, listing all P&IDs or other drawings to be studied. Every drawing on the list will be formally studied unless the team agree that it is not necessary, and record their reasoning.
- The HAZOP Leader will ensure that all equipment on each P&ID has been studied, and for batch systems that all equipment has been studied for every step.
- If the HAZOP Leader is informed that a drawing is not up to date, he will put it aside until it has been updated and formally checked.
- Before the team makes a recommendation for change or further study, it will agree that the current situation is unacceptable. A company specialist may be required to advise on suitable criteria.
- The team will decide, and the record will state, whether a recommendation relates to a safety issue, an environmental issue or an operability issue. The HAZOP Leader will carry out a check of this categorisation after the session.
- The HAZOP leader is responsible for ensuring that each recommendation matches the intention to reduce the risk of the scenario under study.
- The Leader and his organisation will strive continuously to identify and implement means to improve the HAZOP process.

The client company's Process Safety Management framework will need to address:

- The responsibility for requesting HAZOP studies, and the situations in which they must be carried out.
- The responsibility and procedure for follow-up of HAZOP recommendations, including time available for decision making, and conditions under which rejection is acceptable.
- The provision to HAZOP team members of continually improving criteria for the assessment of the acceptability of risk, and of business related criteria.

Method for Building Performance Measures for Process Safety Management

Edward M. Connelly
Paul Haas
Kent Myers
Concord Associates Inc., 1625 Autumnwood Dr., Reston, VA 22094

MEASUREMENT OBJECTIVES

Increased competition and reduced budgets have forced corporate and plant management to seek performance assessment tools which can: increase efficiency, increase effectiveness, help define priorities, and be cost-effectively applied. In this environment which stresses efficiency and cost effectiveness, the process safety manager is at the same time tasked to maintain, and in some areas, improve the level of plant process safety. To accomplish that job, process safety managers need performance measures for process safety management systems that will:

1. Permit evaluation of the performance of each element of the process safety management (PSM) system and the overall PSM system,

2. Support defense to higher management of the needed process safety management system resources,

3. Facilitate the communication of the process safety priorities to plant personnel,

4. Permit establishment of process safety criteria identifying superior, satisfactory, and unsatisfactory process safety management system performance (and the translation of that criteria to the task performance necessary to support that system performance.), and

5. Be a practical, easy-to-use tool that can be adjusted to the particular needs of each plant.

BUILDING PERFORMANCE MEASURES IS DECEPTIVELY DIFFICULT

While the need for performance measures and their usefulness is clear, most attempts at measure building result in unsatisfactory measures. This is because the measure development processes typically used do not allow specification of criteria for an acceptable measure, and consequently, the method justification is an appeal to the plausibility of the procedures used, instead of the quality of the performance measures produced. Without criteria for an acceptable measure, there is little guidance as to how to build a performance measure. Without criteria and guidance, the measure development process can drift from one formulation to another without direction and without reaching completion.

In this paper we describe the method used to build performance measures for process safety management (PSM) systems. We believe the measures developed will achieve the goals cited above. The method first establishes the basis of the performance assessment by identifying a test for accepting the measure. Based on that test, the method builds criteria identifying the performance discrimination requirements for the measure, provides feedback to the measure specifier to help insure the specification quality, and finally produces a measure that satisfies the performance discrimination specifications. This method has been used successfully to build many different types of performance measures in a number of technologies. It was used in the application reported here to provide a continuous assessment of a process safety management system (PSM) performance for two of the twelve CCPS PSM elements.

The measure building system presented here relies on subjectivity and is quite different from other techniques for capturing subjective judgement or preference. Some of these differences are cited here while others will be apparent as the method details are described. We do not attempt to force a consensus among experts and we do not average over experts' inputs. Instead, a measure is built for each expert. The differences are identified and used to discover unstated assumptions or different documentation interpretations. Experts are not forced to use a pre-defined measure mathematical structure, because we don't know beforehand what structure is needed to discriminate performance. And we don't pre-specify a psychological rating scale, because, again, the rating scale needed to discriminate performance is not known beforehand.

In the discussion that follows below, the measure basis is described first, including a brief discussion of potential traps for the unwary when building a measure of performance (MOP). Then, and finally, the measure building procedures are described. A companion paper describes the actual measures produced.

MEASURE BUILDING FUNDAMENTALS

MOPs Exist and are Presently Operational

Consider the individuals presently making or influencing process safety decisions; these individuals include plant operators, supervisors, trainers, plant and corporate management and staff, regulatory and other public officials, and public interest groups. These individuals, and perhaps others, taken as a whole, govern the performance of process safety management systems. In doing so they exercise their own concept of what constitutes process safety. These tacit measures of process safety management are already operational; they determine the present level of plant process safety, as well as the cost of that level of safety. In this very real sense, process safety MOPs already exist.

The problem, of course, is that these MOPs are not in an explicit form. We can't document these important MOPs. We don't know all the variables they incorporate or how those variables are combined to provide an overall assessment of process safety management. Further, we don't know if these various experts' MOPs are the same or different, and if different, how they differ. And we don't know if the experts are working toward the same precise goals - even if they think they are.

A Test of Performance Measure Acceptability

A MOP defines how performance is to be assessed. The MOP that mimics an individual's performance rating is said to be acceptable to that individual. A <u>test</u> of a MOP, therefore, is that it rates performances the same way as the individual does, as suggested in Figure 1. What the individual thinks is good or poor performance must be consistently rated by that individual's MOP. Thus, the MOP can be thought of as capturing an operational version of the individual's definition of what constitutes performance measurement. By the term "operational definition" is meant that the definition (i.e., the MOP) uses measurable variables called "indicators" that other individuals can use successfully to collect the same data. Once the MOP is converted into mathematical form, its indicators become explicit specifying the data to collect and how to combine the data into the overall performance rating. *A MOP, then, is a mathematical formula relating indicators to performance.*

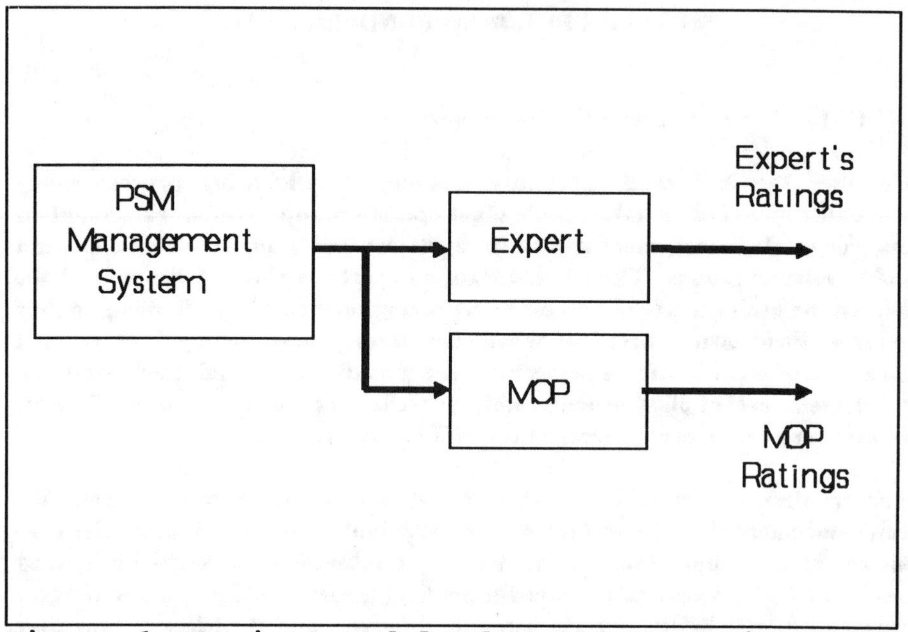

Figure 1 MOP is a Model of Expert's Rating

Experts Have Difficulty Communicating How They Rate Performance

A major difficulty that must be addressed when building measures is that even highly experienced individuals cannot be relied on to accurately and completely explain how they assess performance. These highly experienced individuals will often say "I know it when I see it" or "the plant just talks to me" when discussing their area of expertise. When asked how they make that performance assessment, they will offer some factors and indicators they believe they consider when assessing performance, but often we have found that variables offered as important initially are later found to be not as important (sometimes insignificant) as compared to other variables.

Why is it that highly experienced individuals who are relied on to design, operate, maintain, and manage potentially hazardous plants have difficulty in verbalizing how they assess process safety performance? Apparently assessing process safety and describing how that assessment is accomplished are two different types of thinking. The former is the application of an assessment method to a specific system, while the latter is a description of the method itself. The former focuses on one or more particular systems, while the latter is broader, providing rules for applying the method to all systems. Highly experienced individuals perform the former frequently as part of their jobs, while they may have never considered the general method application. An experienced individual can be an internationally recognized expert in the former, but have little expertise in the latter. Thus, we concentrate our investigation with the expert on the area in which he/she truly is an expert, namely, on assessing the performance of specific systems (either real or hypothetical).

Psychologists Suggest Why Experts Can't Always Tell How They Assess Performance

In a classic study, Bruner (1956) describes how individuals can develop and use a concept (such as performance assessment) before being able to understand and describe that concept. This helps explain the difficulty in extracting information from experts as to how they make performance assessments. Psychologists tell us that as an individual develops experience, that experience is stored in the brain as "schemas". Schemas apparently trigger the sequences of thoughts and actions we use to do work, including assessing performance. Reason (1990) describes schemas: "Although their processing lies beyond the direct reach of awareness, their products- words, images, feelings and actions - are available to consciences". Whatever the underlying explanation may be, there is abundant evidence that it is difficult for experienced individuals to verbalize their performance assessment strategies.

This is important to understand because our method for building measures is designed to overcome an individual's limited awareness of how they assess performance. Our approach to solving this apparent dilemma lets the individual work in a familiar environment — assessing performance of PSM management systems in the context of a familiar plant environment.

Starting with a description of a process safety management system in a familiar plant environment, the management system characteristics are changed one at a time, producing a somewhat different hypothetical system. In this way, a set of example management systems is constructed for the individual to evaluate. The performance of each example system is rated, using whatever rating scale the individual wants to use. Concentrating on an example management system allows raters to work in familiar territory and, most important, to do the type of performance assessment thinking in which they are truly expert. Since each system example is described by its indicators and also provided a performance rating, the inputs and associated outputs (the performance ratings) required to build the mathematical function for the performance measure are available.

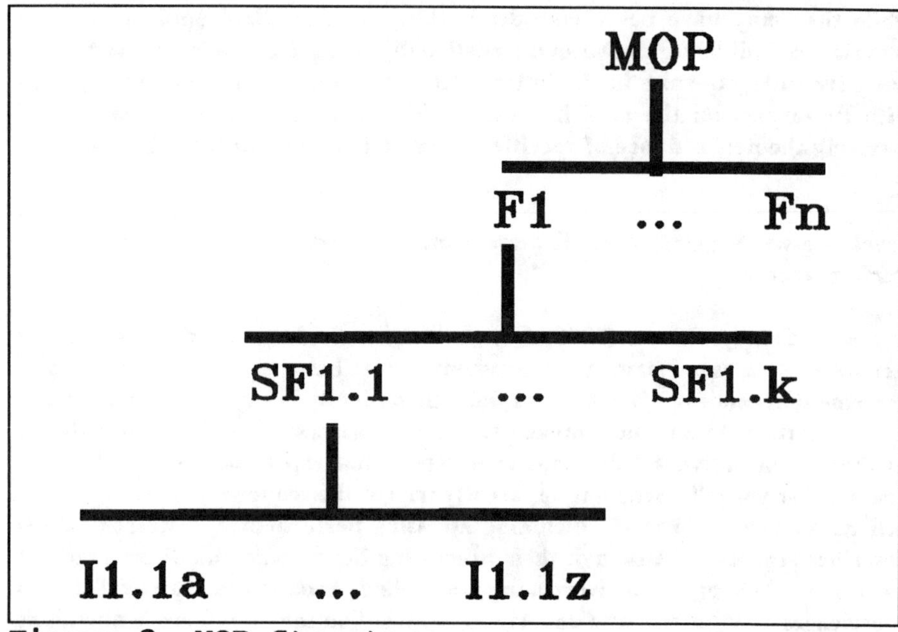

Figure 2   MOP Structure

MEASURE DEVELOPMENT METHOD

The measure building method is described in this section. Rationale and theory of building measures is found in Connelly (1987). A performance measure is a mathematical function of factors, subfactors, and indicators, arranged as shown in Figure 2. Performance is a function of factors (F1,..., Fn). Each factor is a function of subfactors (subfactors for F1 are SF1.1, ..., SF1.k). Subfactors are thought of as measurement objectives of their respective factors. The measurement objectives are implemented by a set of indicators. In the figure indicators for subfactor 1.1 are shown as I1.1a, ..., I1.1z.

Elements of the Measure

Obtaining the terms of the measure and their definitions results from interaction between a subject matter expert (SME) experienced in the relevant PSM system and a facilitator. The terms consist of a set of factors, and indicators associated with those factors. Subfactors may intervene. The factors are not measured directly, but rather are summary dimensions. The indicators are stated as operational definitions, for which data could be collected from instruments, observations, or data that is derived from these sources. There are no firm rules concerning the number of factors and indicators that are used. Several techniques are brought to bear in this phase. Brainstorming identifies many dimensions, from which selection and editing may begin.

A technique for reducing errors of omission is to organize the items (factors, subfactors and indicators) into categories and then to consider whether there are additional unstated items that belong in the categories. Next, a new set of categories is selected and the existing items placed in the new set. Voids in the new set suggest additional items need to be added. This process of selecting a new set of categories is repeated several times to reduce errors of item omission.

Once the initial measure structure is identified, the relationships among the measure elements must be determined. For instance, if the measure structure consists of five factors each with three indicators:

Measure of Performance

Factor	Indicators
1	1a, 1b, 1c
2	2a, 2b, 2c
3	3a, 3b, 3c
4	4a, 4b, 4c
5	5a, 5b, 5c

then six mathematical functions must be determined. One mathematical function relates the Factor 1 to its indicators 1a, 1b, and 1c. Similarly, Factors 2 through 5 are related to their respective indicators, requiring four additional mathematical functions. Finally, the MOP must be related to the five factors, requiring the sixth mathematical function. In mathematical notation this produces six equations:

$$\text{Factor } 1 = F_1(1a,1b,1c)$$

$$\text{Factor } 2 = F_2(2a,2b,2c)$$

$$\text{Factor } 3 = F_3(3a,3b,3c)$$

$$\text{Factor } 4 = F_4(4a,4b,4c)$$

$$\text{Factor } 5 = F_5(5a,5b,5c)$$

$$\text{MOP} = F_{mop}(F1,F2,F3,F4,F5)$$

The subsequent sections provide a description of the method and tools used to build these mathematical functions. Specifically, building the function for Factor

1 is described. The same process is used repeatedly to build the remaining
mathematical functions.

Tools & Procedures for Building the Measure Mathematical
Functions

Baseline Case. The SME is asked to consider a particular implementation of the
PSM system he or she is familiar with or can conceive of and to identify the
numerical data values of the indicators which describe that system.   The
objective is to consider a system implementation that is not the best or worst
possible, but is a baseline system to serve as a reference for constructing other
somewhat different system examples.

System Implementation Variations. The SME is asked to identify indicator data
values for each indicator which would describe a significantly improved system
implementation, near the limit of what would be observed in the field. Also, the
unfavorable indicator data values describing a system implementation with
substantially reduced performance are identified. This provides three values for
each indicator, describing three system implementations resulting in three levels
of performance: baseline, favorable, and unfavorable. Although three values for
each indicator is used here, a greater number of values can be used where
complexity demands greater detail. Three are illustrated here because it is the
simplest to use and explain.

If an indicator has only two values (yes or no, true or false) then obviously three
different values cannot be offered. In these cases, either the favorable or
unfavorable data value will be equal to the baseline data.

After the indicator data are collected, a software tool called MAP (<u>M</u>easurement
and <u>A</u>nalysis of <u>P</u>erformance) is used to generate hypothetical variations of the
baseline system performance. The software tool forms variations of hypothetical
implementations according to the pattern shown in Table 1. In that Table, each
row contains the three data values for an indicator arranged so that:

> Column 1 contains the set of all favorable indicator values,
> Column 2 contains the set of all unfavorable indicator values,
> Column 3 contains the set of all baseline indicator values.

In the remaining columns, 4 through 9, one indicator value is varied from the
baseline value. In the columns numbered 4-6 each indicator in turn is changed
from baseline value to the favorable indicator value. Columns 7-9 show the
unfavorable implementations. The total number of sets (columns) will always

Table 1 Pattern for Generating A Ratings Worksheet

Ind	Fav	Unf	Base	System Variations					
	1	2	3	4	5	6	7	8	9
1	F	U	B	_F_	B	B	_U_	B	B
2	F	U	B	B	_F_	B	B	_U_	B
3	F	U	B	B	B	_F_	B	B	_U_
R	10	0	5	6	7	6	2	1	2

equal to twice the number of indicators plus three. Each column of indicator data, except the third column, describes a hypothetical PSM system implementation which is a variation from the baseline system. System implementations described by columns 4 and above differ from the baseline by the value of a single indicator.

Performance ratings. A printed copy of Table 1 is called the "ratings worksheet" which is given to the SMEs so they can enter their ratings. For the system example illustrated above requiring rating of 5 factors plus the overall performance, six worksheets would be provided to rate each factor and the overall system performance. Although the same work sheet is available on the terminal screen, most SMEs prefer to use the worksheet. They use it to enter ratings but also to make notes, calculations, and list assumptions. When completed the ratings are as listed in the last row of Table 1.

Process of Developing Factor and PSM System Performance Ratings. For each worksheet, the SME are asked to provide a factor performance rating. The first three performances (favorable, unfavorable, and baseline) are rated first, providing an overall scale. SMEs are told to use any rating scale they need to define the necessary factor performance discriminations. Then the SMEs are asked to rate the factor for each of the other data sets. In Table 1, a 10 means "extremely good" and 0 means "extremely bad", but there are no a priori restrictions placed on the scale chosen. And, SMEs often modify their own scales as the exercise proceeds.

Since each of the system implementations, described by the indicator data in columns 4 and greater, differ from the baseline by the value of a single indicator, the SME must consider what the indicator data value means regarding factor performance. For that implementation to occur, the SME might judge that additional unstated variables may have changed. Thus, the SME draws from his or her knowledge of the baseline system modification as specified by the indicator change. This is difficult work requiring detailed knowledge of the system functioning and the ability to conceive of the impact of the indicator value changes.

As the factor ratings are being developed, "war stories" are told along with justifications of the ratings. Inadequacies of the set of indicators are readily offered — sometimes resulting in the need to revise the factors and their indicators. These often provide insights into the underlying basis for the judgments that would not be accessible using highly structured formats. This stage of the process seems to force a critical assessment of the indicator adequacy, since indicators are added or refined by the SME during this activity.

The ratings task also provides an opportunity for the facilitator to assess the SME's depth of knowledge and experience. An SME who knows the functioning of the baseline system that results in various levels of performance described in the worksheet will typically analyze a system implementation (identified by the change in indicator value) by describing the those modified PSM system functions. Then, the factor performance rating is given and its rationale provided.

In contrast, an "SME" without the requisite PSM system performance knowledge and experience will seek a mechanical way to the ratings. For instance, that SME will ask " How do I do this? Do I weight each indicator and multiply the indicator values by the weighting and then sum the products to obtain the rating?" That SME cannot be used to build the measure.

Once the SME completes the worksheet, the ratings are entered into the computer program, which supports three main types of feedback showing the implications of the ratings. In each review, the ratings, i.e. the "preference specifications" are tested by transforming the indicator data into a different form for reconsideration by the SME. The transformed data forces the individual to reconsider the system performance from a different viewpoint, reducing the original biases, mind sets, generalizations, errors, etc. This technique has been shown to reduce errors-of-commission in specification building (Connelly 1984).

Reviewing the Ordered List. Performances are ordered according to the rating assigned, as shown in Table 2. This feedback lets the SME consider the performances ordered by ratings and in context of all the indicator data values.

Table 2 Implementations Ordered by Rating

Ind	Fav	Unf	Base	System Variations					
	1	2	3	5	4	6	7	9	8
1	F	U	B	B	F	B	U	B	B
2	F	U	B	F	B	B	B	B	U
3	F	U	B	B	B	F	B	U	B
R	10	0	5	7	6	6	2	2	1

With it the SME can ask the question, "Is this the ordering of the PSM system performance that I prefer?" In addition, the SME can now consider identifying a criterion level of acceptable performance. A criterion level corresponds to a vertical line drawn on the ordered worksheet indicating that "all system performances to the left are acceptable and those to the right are not acceptable." Other performance categories can be similarly identified, such as performance rating levels: superior, good, fair, and poor.

A final use of the ordered list is the comparison of the ratings of multiple SMEs. Each SME's ratings are characterized by their ordering of the system performances. Comparison of these orderings permits a direct comparison of the SMEs' rating strategies. Wide discrepancies are of particular interest, such as when performances are rated favorably by one SME and are rated unfavorably by another SME. Often these differences are resolved by asking the SMEs to explain their ratings.   In the responses, previously unstated assumptions are frequently offered as the basis of the rating. Another reason is that the SMEs have simply interpreted the indicators differently. When these assumptions are agreed upon or are incorporated explicitly as a new or revised indicators, or when a common indicator interpretation has been established, the rating differences frequently disappear.

Reviewing Graphs. Indicator data describing system implementations provide three variations in which an indicator data value is varied and all other indicator values held constant. For instance, in Table 1, columns 3, 5 and 7 use three different values of indicator number 1 while the values of all other indicators are held at their baseline values. The corresponding three ratings for those performances define a curvilinear relationship that the SME is apparently quite sure of. A very common relationship is that of "diminishing returns,", such as that shown in Figure 3, where favorable indicator increments are less valued after a certain level has been achieved. Or, increasing increments of an

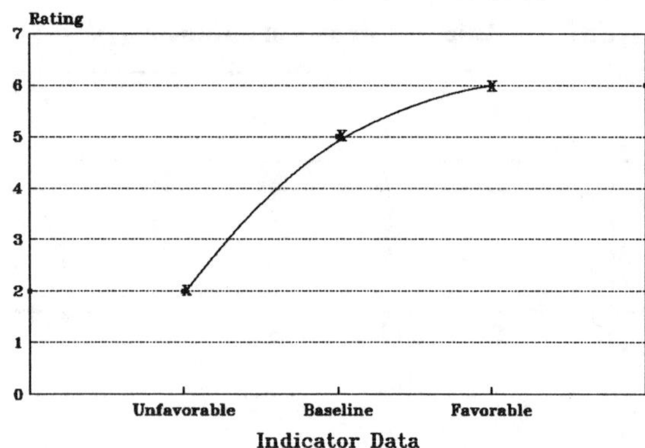

Figure 3 Possible Relationship Between Indicator and Subfactor

indicator may have little importance, until a threshold is reached and contribution to overall performance can begin to be counted. The software graphic tool shows the relationships between these variables, and the SME can confirm that the relationship derived from the rating inputs is as intended.

By allowing for non-linear curves, and not the forced use of only linear functions, the SME is not forced to think in terms of "weights" which in many cases are not meaningful. Figure 3 shows a possible factor rating plotted against the three values of the one indicator that is varied among the performances. The SME can view the graph for each indicator and determine if that relationship is the one intended. If not, the rating and/or indicator data can be modified.

Combining Multiple Indicator Changes

In addition to the transformation of each indicator, described above, the transformed indicators must be combined into an overall factor rating (and subsequently factor ratings combined into the overall PSM performance ratings). A family of possible strategies for combining transformed indicators has been developed and are applied based on the individual needs of the measure. They include methods of accounting for: common variance among indicators, alternative ways of achieving high performance, finite state markov relationships among indicators, logical relationships among indicators, and common indicator weighting.

Review using MOP Calculator. The software calculator allows the whole MOP to be exercised using data that differs from the original data used to generate the MOP. The calculator indicator data are initially set to baseline data values, and the ratings (which are now calculated through the MOP rather than copied from the SMEs inputs) are calculated. When the user changes any of the indicator data values, the MOP can be recalculated, and the effects on a factor score and overall MOP score shown.

MOP Refinement. A facilitator will generally work with one SME initially, to work out an initial "strawman" MOP structure. The resulting MOP reflects the assessment concepts of that individual SME using a single specific baseline system and variations of that system.

Measure refinement process consists of multiple (three is usually the best number) SMEs working with the facilitator to review the "strawman" measure elements: factors, subfactors, indicators, baseline indicator data values, and ratings. The purpose of this refinement is to add, modify, or adjust the measure elements to have a broad application. The SMEs consider a common set of indicators and indicator data values and then individually rate each performance. Next the ratings are compared and where differences are found the reasons for the differences are determined. Often, when asked, the SME will say " I assumed that ..." - and an unstated assumption is revealed. Another reason for differences is the different interpretation of the terms used in the indicators. Once a common definition is developed, the rating differences typically disappear. In some cases SMEs will differ on ratings but agree on outcomes. This is a disagreement on what measurement to use - some like to emphasize process while others outcomes. If the processes lead to those outcomes, then the two SMEs concur as to the result but differ on what the measurements should be. A measure that uses both process and outcome measurements is usually an acceptable accommodation and provides a more robust measure.

SUMMARY

The measure building methodology described above takes advantage of an expert's ability to evaluate and compare performances while accounting for their limited ability to completely and accurately explain *how* they assess performance. Methodology to build performance measures:

> 1. Identifies the basis for performance assessment as the performance preference of the experienced individuals presently working to insure a high level of plant process safety.

2. **Recognizes that experts will have difficulty in accurately and completely describing how they assess performance.**

3. **Lets the experts work in the environment in which they are familiar and perform tasks for which they are accepted as being expert to specify the performance discrimination requirements of the measure.**

4. **Uses examples of PSM implementations which the experts can rate.**

5. **Uses feedback to help reduce errors of omission and commission in producing the measure specification.**

6. **Permits non-linear relationships as needed.**

7. **Builds the mathematical functions for the performance measures.**

8. **Provides a calculator for computing performance ratings from plant data.**

This approach produced measures of performance that permit evaluation of CCPS PSM elements and can be used to defend needed resources to higher management, effectively communicate process safety priorities to plant personnel, and establish performance categories, such as satisfactory and unsatisfactory. Finally, the measure is a practical, easy-to-use tool adaptable to the particular needs of each plant.

REFERENCES

Bruner, J.S., Goodnow, J.J., & Austin, G.A. <u>A Study of Thinking</u>, New York: Science Editions, 1956.

Connelly, E.M. <u>Transformations of Software Design and Code May Lead to Reduced Errors.</u> Paper presented at INTERACT '84 International Federation for Information Processing, London, England, September 1984.

Connelly, E. M. <u>A Theory of Human Performance Assessment.</u> Paper presented at the Human Factors Society 31st meeting New York, October 1987.

Reason, R., <u>Human Error,</u> Cambridge University Press, New York, NY, 1990.

Performance Measures Developed for CCPS's Elements of Process Safety Management

Edward M. Connelly
Paul Haas
Kent Myers
Concord Associates Inc., 1625 Autumnwood Dr., Reston, VA 22094

Performance measures have been developed for two of the CCPS Process Safety Management (PSM) system elements, namely, Management of Change (MOC) and Training & Performance (T&P). The work to build the measures was a corporative effort of CCPS staff, CCPS Technical Management Subcommittee (TMS) members, Concord Associates, Inc., and JBF Associates, Inc.

This paper is a status report on the present state of this new development. MOC and T&P performance measures have been built by a development team and reviewed independently by TMS Committee Teams, the full Committee, and consultants. Overview presentations have been given to the CCPS Technical Steering Committee. Plant evaluations are planned to assess the acceptability and utility of the measures in plant environments, to gather data on work load required to install, operate, and refine the measures, and to identify possible correlations among measure elements. The later could lead to the simplification of in-plant data collection necessary to use the measures.

The paper describes: the activities required to build the measures, the factors and subfactors for the MOC measure to illustrate the performance measure concept, the factors for the T&P measure, ongoing measure work load analysis, plans for in-plant testing of the measures, and implications for building measures for the remaining management system elements. A companion paper in this document, titled "Method for Building Performance Measures for Process Safety Management", describes the methodology used to produce the measures and documents the mathematical structures employed.

To introduce the terminology used here, a performance measure is a mathematical function relating indicators to the performance rating. Indicators are measurable variables, such as the "percent of documents processed correctly" or "are the procedures written". Typically the performance measure, as shown for the management of change system in Figure 1, is organized into factors which are organized into subfactors, and finally indicators. Factors do not always have subfactors but sometimes use indicators directly.

A Short Project History, Its Participants, and Concept Development

The project "officially" started August 15, 1991, but prior to that time, several presentations describing the methodology had been given to the Technical Management Subcommittee. Initial working meetings were held, August 17, 18, 19, at Concord's Office in Knoxville to attempt the first cut at overall measure organization and identification of critical elements for all the CCPS PSM elements. A development team consisting of Dr. Paul Haas, Steve Arendt, and Sandy Schreiber served as subject matter experts during these early meetings. Ed Connelly served as facilitator at those meetings and throughout the project, later assisted by Dr. Kent Myers. Steve Arendt served as continuing subject matter expert throughout the project.

A major product of the early meetings was the evaluation of several alternative measure organization schemes starting with the one found in Appendix A of the <u>Guidelines for Technical Management of Chemical Process Safety</u> (i.e., Organizing, Implementing and Controlling) and evolving to the workable "life cycle" organization (i.e., Design, Development, Installation, Operation, and Test and Evaluation). In addition, the critical features of each of the 12 CCPS Process Safety Management Systems were identified as an aid for selecting the initial two management systems for measure development.

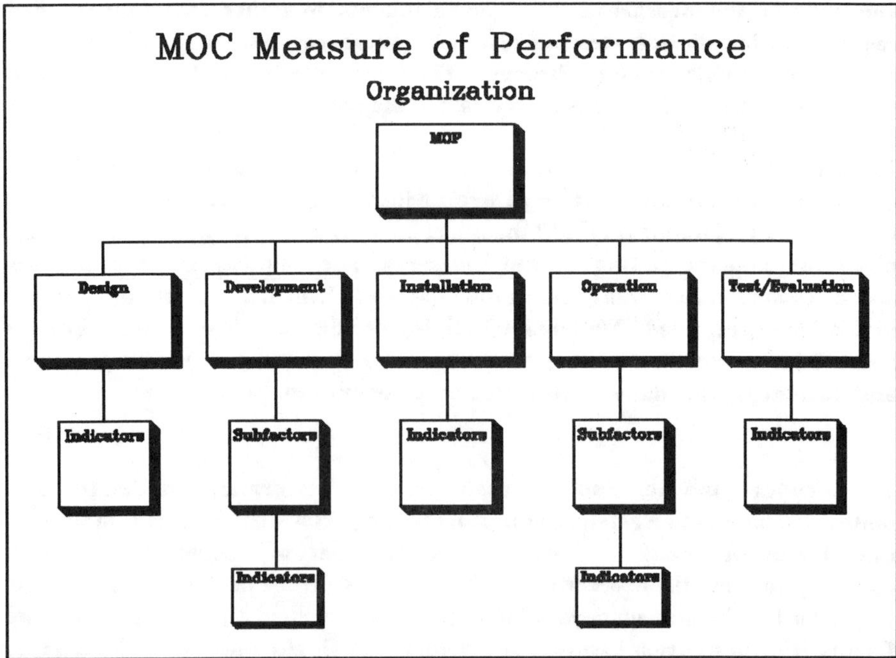

Figure 1   MOC Performance Measure Organization

Preliminary results were presented to the Technical Management Subcommittee in Atlanta on October 15, 16, 17, 1991. Based on the discussions at the Atlanta and subsequent TMS meetings, a common understanding of the measurement needs and acceptable types of indicators began to evolve. Later, the development team held a subsequent interactive meeting with Joe Sweeney in Washington DC in which the concept that a "continuously useful" measure was crystallized.

A presentation and demonstration of the methodology for building measures was given to the Technical Management Subcommittee (TMS) in Houston, Texas on February 19, 1992. A part of the presentation focused on the need to test initial measure constructs for errors-of-omission. This resulted in the suggestion to use a methodology familiar to the committee members, namely the HAZOP process to examine an existing, in-use, exemplar MOC system to identify areas critical to its high level of performance, and thus, to gain additional insight for selecting measure indicators.

These meetings were thought of as preliminary meetings allowing a two-way communication where general concepts of both how to build measures and the needed measure characteristics were exchanged. In addition to those meetings specifically mentioned above, multiple meetings were held by the development team to form what is known as the strawman measure (the initial measure construct). The strawman measure is a measure developed to reflect the management system performance at a specific plant implementation. These meetings resulted in specification of indicators designed to implement the continuously useful measure concept, including: historical data records, analysis of trend deviations from norms to act as warning flags, the use of human observers judgment to obtain rapid feedback of current MOC functions, the detection of important plant changes previously considered constant or not important, and the general work safety culture at the plant. It was recognized that a continuously useful measure employs a spectrum of indicators, some requiring frequent update and others, whose data values don't change often, requiring less frequent updates.

The Technical Steering Committee participants were briefed on the overall project progress and goals at a meeting in Philadelphia, Pennsylvania on April 22, 1992.

The more formal part of the project initiated on March 17, 18, and 19, 1992 at Williamsburg, Virginia. During these meetings the strawman MOC measure was presented to the TMS. Then, TMS members with expertise in MOC systems formed task forces consisting of Bill Helmer, Bob West, Anne McGuiness, Tom Selders, and Jack Dowbekin to provide an indepth review of the MOC Measure Factors, Subfactors, and Indicators. The meeting continued later in Dallas, Texas and Los Angeles, California on June 24 and 25, 1992.

In a parallel effort, another group of CCPS consultants, consisting of Homer Richardson, Lou Perfetti, and Ray Witter formed a task force to review the T&P Measure. A strawman measure for training systems was already available from a previous Concord project. Consequently, this task force reviewed the organization of that measure and specified the desired measurement objectives at the Williamsburg, Virginia meeting. Then, in two later meetings on June 18, and 19, 1992 and July 16 and 17, 1992, at Concord's office in Reston, Virginia, the task force refined the T&P measure indicators, and provided performance ratings and other data required for the measure.

A TMS review meeting was held at the IBM TeamFocus facility in Southbury, Connecticut on July 30,1992. An after action meeting was held the next day to review the results. TeamFocus is the IBM version of a groupware facility consisting of a meeting room containing 15 computer terminals interconnected so that each participant can enter responses simultaneously, each being recorded in a central computer. Using the TeamFocus system the TMS members initially provided ideas on how to assess a performance measure in general and then reviewed the strawman MOC measure, providing an evaluation of its indicators, and performance ratings. An interesting feature of the facility is that meeting results are immediately available. Results were reviewed with the committee the next day. Suggestions and corrections were then incorporated into the measure after the Connecticut meetings.

A further analysis was conducted with the CCPS Steering committee on October 13, 1992 at the IBM TeamFocus in Palisades, New York. A brief presentation was given at the Steering Committee meeting the next day. The Palisades facility provides 25 computer terminals accommodating 50 participants with two individuals per terminal. During the TeamFocus session, the participants input data on how the MOC and T&P Process Safety Management Systems interacted with the other CCPS PSM Systems. These data are needed to help define the structure of the measures for the remaining PSM Systems.

A presentation of the Draft Final Report was given to the TMS in San Diego, California on March 9, and 10, 1993.

Important Features of The CCPS Measures

Continuously Useful Performance Assessment. The performance assessment concept built into the PSM performance measures is analogous to the instrumentation of a chemical process. Measure indicators (which are analogous to chemical plant instrumentation) are scheduled for update according to the dynamics of the variable being measured and when the data are needed. The measure property is noted by describing the measure as a "continuously useful" assessment of PSM system performance.

Multiple Evaluation Points. Several levels of performance ratings are available from the measure. The measure provides an overall PSM element performance rating, ratings for each factor and subfactor, and processed indicator data. Factor ratings for the MOC measure, for example, are design, development, installation, operation, and test & evaluation ratings. Thus, for instance, when the system is being designed or redesigned, the design factor rating is important. At other times other factor ratings will be of primary interest. Subfactor ratings provide performance assessment of a portion of the factor. For instance, Factor 4, of the MOC measure, Operations, has a subfactor assessing the overall operation performance, another assessing the change processing historical performance of a specific type of change, and yet another assessing the historical outcome performance for a specific type of change. With these subfactor assessments, an ongoing change, for instance, can be assessed with historical data as well as current indicator data. If that type of change has a history of difficulty, indicators assessing the on going change can be updated at a more frequent schedule. A deviation from normal indicator values will flag management attention to possible change processing deficiencies. Finally, some indicators will flag significant deviations from normal operations. These flags are meant to trigger management attention without waiting for the normal management review of the measure data.

Adaptable to Each Plant. Some of the measure indicators require building data bases for tracking performance trends. Data bases are plant specific and tuned to the measure to the specific plant needs.

Inputs from Industry Experts. Process safety experts presently working in industry both in consulting and corporate staff contributed to the measure development. The measure's credibility is ultimately attributed to that source of expertise.

Performance/Cost Effectiveness. Performance effectiveness is defined as the change in PSM system performance rating per unit change in that system cost. The method for calculating effectiveness has been developed for both MOC and T&P systems.

Exchange of Information Among Plants. The performance measure indicators provide a "state space" for describing the status of PSM at a plant. As plant personnel improve process safety, the PSM state will change and the performance rating will change from one level to another. Of course there is an investment or cost associated with that change. Of common interest among plant managers will be strategies that result in improvement ratings at least cost. One basis for the exchange of that information is the PSM state as defined by the state variables. Thus plants with dissimilar chemical processes but similar "PSM states" (i.e., similar indicator data values) will be able to use each other's success strategies to predict results in their plants.

Factors, Subfactors, and Indicators for the
Management of Change Performance Measure

As shown in Figure 1, the MOC factors are:

1. DESIGN

2. DEVELOPMENT

3. INSTALLATION

4. OPERATION

5. TESTING & EVALUATION

The Design (Factor 1) assesses both the process of developing a design specification as well as the product of that activity. Factor 1 does not use subfactors; it is assessed directly from indicators. The Development factor assesses the MOC development process for the current MOC written program. To illustrate some indicators the following are indicators for Factor 1. Space limitations and the possible refinement of indicators prevent listing of indicators for all factors.

Factor 1 Indicators are:

1.1a   Does a formal written MOC program design specification exist? (Y/N)

1.1b   How many of the following items are included in the MOC design specification. (no. of items)
- technical basis for the change
- safety/health considerations
- authorization protocol
- emergency change provision
- temporary change provision
- internal auditing provision
- consensus target completion data for MOC review

1.1c   Were the resources used and the process followed in developing the MOC design specifications adequate as judged by a PSM coordinator? (Y/N)

1.1d   Does the design specification require that the MOC program be reviewed and upgraded when the facility changes? (Y/N)

1.1e   Is the rationale and technical basis for the MOC design specification adequately documented and determined to be understood by all concerned? (Y/N)

Development (Factor 2) takes an MOC design specification and turns it into a written program that is ready to be installed. Factor 2 uses subfactors to further decompose the factor into categories for measurement. These subfactors are summarized in the following; however, note that subfactors are a set of measurement <u>objectives</u> for their factor, not the measurements themselves. Indicators which were illustrated above provide the actual measurements.

Subfactor 2.1: Review/test
> (assesses the review and test of the development process)

Subfactor 2.2: Who
> (assesses the quality of the development staff)

Subfactor 2.3: Communications Quality
> (assesses the communications quality of the developed MOC program)

Subfactor 2.4: Temporary Changes
     (assesses treatment of temporary changes in the MOC system)

Subfactor 2.5: Types of Change
     (assesses the comprehensiveness of the various types of changes in the MOC system)

Subfactor 2.6: MOC Coordinator
     (assesses the MOC coordinator qualifications)

Subfactor 2.7: Review Method
     (assesses the quality of change review method incorporated in the MOC system)

Subfactor 2.8: MOC Process Documentation
     (assesses the provisions for documentation in the MOC system)

Subfactor 2.9: MOC Process Personnel
     (assesses the quality of the change review process developed for the MOC system)

Subfactor 2.10: Variances
     (Checks for provision of variances procedures in the MOC system)

Subfactor 2.11: Training
     (assesses the provision for training in the MOC system)

Installation (Factor 3) provides an assessment of the MOC system installation which creates the infrastructure to support and operate the MOC program and involves training of MOC participants and the education of everyone who might originate, review, or approve a change request, or who might implement a change. Factor 3 does not use subfactors; it is assessed directly from indicators.

Operation (Factor 4) assesses the MOC system in its operating environment. It involves the use of the MOC system to process change requests. Two operating modes are considered: active (a change request is being processed) and standby (the system is at rest and awaits a change request). Factor 4 uses subfactors to further decompose the factor into categories for measurement. The subfactors are summarized as follows:

Subfactor 4.1: Overall Operation
(assesses the MOC operation as a whole)

Subfactors 4.2 through 4.8 assess performance of an ongoing or recent MOC action replacement in kind or a change.

Subfactor 4.2: Process: Approach for a Specific Change
(assesses the overall MOC processing performance for this MOC action i.e., replacement in kind or change)

Subfactor 4.3: Process: Participants
(assesses the qualifications of the authorizers for this MOC action)

Subfactor 4.4: Process: Historical Database
(Assesses the organization's historical processing performance for this type of MOC action)

Subfactor 4.5: Process: Human Observer Estimate
(Assess the MOC process for this MOC action from a human observer's viewpoint)

Subfactor 4.6: Product: Completeness
(Assesses the completeness of the documentation package for this MOC action)

Subfactor 4.7: Product: History for Specific Types of Change
(assesses the organization's history on overall performance for this type of MOC action)

Subfactor 4.8: Product: Human Observer Estimates
(assesses the overall MOC performance for this change)

Subfactor 4.9: Detection of a Change in a Previously Constant Plant Factor
(Assesses the appropriateness of the MOC performance measure in a changing plant environment)

Subfactor 4.10: Assessment of the Degree of Compliance of Plant Personnel to Work   **Safety standards**
(Assesses the holistic plant personnel compliance to work safety standards)

Testing & evaluation (Factor 5) assesses the feedback mechanisms for testing and improving the MOC program. Feedback and adjustments can be made during any stage of the life of an MOC program. Factor 5 is evaluated directly

with its indicators.

**Factors, Subfactors, and Indicators for the
Training and Performance Measure**

The performance measure for the T&P PSM management system uses 7 factors
which are as follows:

1.   **JOB/TASK ANALYSIS**
      (Analysis)

2.   **LEARNING OBJECTIVES**
      (Analysis)

3.   **DESIGN**
      (Design)

4.   **DEVELOPMENT**
      (Development)

5.   **INSTALLATION & OPERATION**
      (Delivery)

6.   **EVALUATION OF TRAINING MASTERY**
      (Evaluation)

7.   **EVALUATION OF TRAINING SYSTEM**
      (Evaluation)

Due to space limitations, subfactors for all the factors cannot be listed. Since
T&P assessment interest is typically centered on the last two factors, both
providing trainee assessment, subfactor s for Factors 6 and 7 are given in the
following; however, note that subfactors are a set of measurement <u>objectives</u> for
their factor, not the measurements themselves. Indicators for each subfactor
provide the actual measurements.

Subfactors for Factor 6 (Evaluation of Training Mastery)

6.1   **Exemptions from training are controlled.**

6.2   **Trainee evaluation is appropriate to job performance
       requirements and training objectives.**

6.3   **Performance of trainees is evaluated regularly during the
       training program, and prompt, objective feedback is provided**

on a regular basis.

6.4 **Job incumbents who perform below minimum standards during requalification or continuing training are provided remedial training and are not permitted to work alone on associated job duties.**

Subfactors for Factor 7 (Evaluation of Training System)

7.1 **There is evidence that a method is in place to systematically evaluate the effectiveness of training programs and to revise the programs as required.**

7.2 **Examination and operating test results are evaluated so that tests are improved and feedback is provided to improve training.**

7.3 **Instructor and trainee critiques of training are used for program evaluation.**

7.4 **On-the-job experiences solicited from recent program graduates are used for program evaluation.**

7.5 **Feedback from supervisors about job performance problems is solicited for program evaluation.**

7.6 **Training program audit findings are used for program evaluation.**

7.7 **Facility experience is collected and analyzed to assess effectiveness of training and to revise the programs as required.**

To illustrate some of the indicators for this measure, the indicators are Subfactor 7.1 are as follows:

Consistency of evaluation method.

7.1a **Formal procedure is in place that specifies: what reviews will be conducted, how they will be done, who will conduct reviews, who will make changes to the training program, appropriate documentation of evaluation results, and dissemination of results to appropriate personnel. (Y/N)**

7.1b   **Percentage of initial procedures and changes in procedures that are reviewed by appropriate operations, maintenance, and training, and are recommended to management for approval. (%)**

7.1c   **Percent of evaluations that occur as scheduled. (%)**

7.1d   **Percentage of training programs that are analyzed using the formal procedure. (%)**

Feedback.

7.1e   **A procedure exists to collect, resolve, and document training issues, as identified by the evaluations from instructors, trainees, graduates, and supervisors. (Y/N)**

7.1f   **A procedure exists to expedite training needed in critical safety areas (as identified through the Management of Change system, near-miss investigations, or other program elements). (Y/N)**

7.1g   **Percentage of identified issues which have been resolved and documented (both routine and immediate issues). (%)**

7.1h   **Percentage of unresolved items that are up to date. (%)**

Measure WorkLoad Assessment

Assessment of the work required to implement and maintain a process safety management system is an important part of this measure development project. As the measure components and mathematical functions required to discriminate performance were being defined, additional information was being collected to guide the measure workload analysis. A first step is to identify the workload category of each indicator as follows:

a.   **The life cycle measure organization separates measurement components into categories that show when measure indicator data need to be collected. Data for assessment of management element design, development, and implementation factors need only be collected once, during their respective processes. After** each preparatory step is completed, data for those indicators need not be collected again. Thus, these indicators are not major contributors to the measure operation support workload.

b.　The performance rating data can be used to systematically identify indicators that can be eliminated. Performance ratings of variations of the indicators specified that the impact of some indicators is much less than variations in other indicators. Their elimination has minimal effect on the measure.

c.　Superior performance in an element might be achieved in multiple ways. Success in one of those ways is measured with a corresponding set of indicators. Consequently, data for the indicators associated with the success method preferred in the user's organization need to be collected, and data for indicators that are employed by the other methods need not be collected.

d.　Indicators may be found to be correlated or associated. Here, we define the term "correlation" to mean that there is a logical relationship among a group of indicators such that knowledge of the data for one indicator gives knowledge of the data for the others of that group. Consequently, data for all indicators of the group need not be collected. "Association" among a group of indicators means that they vary together like a correlated group, but that the relationship among the indicators is one of practice or convention rather than necessity or logic. For example, superior organizations consistently achieve superior data values for a group of associated indicators, while lower performing organizations fail to achieve superior data values for any of the indicators in the group. Both correlation and association provide a basis for simplifying the data collection process, reducing its workload.

e.　The sampling interval for data collection depends on the nature of the indicator. Many indicators require update of their data only when certain events occur. Others need to be updated on long term intervals. Only some need to be updated frequently. Thus the workload to maintain the measure as continuously useful can be reduced by updating indicator data values only when needed.

The measure maintenance workload analysis makes a distinction between the workload required to accomplish a task measured by an indicator from the work required to collect the data to determine an indicator data value. For instance, one indicator tests for the existence of a written procedure. The work required to actually develop and write the procedure is not considered as part of the measure support workload, even though the indicator may highlight the need for the written procedure and thus trigger its preparation. Indicator data collection support work is just the work to determine that the written procedure does or does not exist.

Determination of the actual workload for installing and operating the performance measures will be completed as part of the planned measure plant testing, which is described in the following section.

Plant Evaluation Plans

MOC and T&P performance measures need evaluation in actual plant environments. The plan is to solicit interest from plant management for the joint evaluation. Solicitation materials will specify what is to be accomplished and will include an estimate of the plant resources needed. Plants will be selected from those expressing an interest to provide a variety in plant size, process, and location. Three plants will be selected for evaluation of the MOC measures and another three plants for evaluating the T&P measures. A pilot test of the evaluation will be conducted with the first plant for each measure. Then, the evaluation method will be modified as necessary and used at the other two plants.

Selected plants will receive the PSM performance measure documentation, a list of information needed for the evaluation, and the measure calculator which will facilitate the measure implementation. Plant personnel will "paper install" the PSM system by identifying or estimating the installation resources, documentation, training, data collection, supplies, and any other activities or resources needed to implement and operate the system. Project personnel will not be on site during this phase but will be available to answer any questions that might arise. After the plant has had time to assimilate the performance measurement, likely several months, the members of the project team will visit the plant to obtain needed data. Results from each plant will be summarized and recommendations will be developed. These recommendations may require updating of the performance measures and documentation.

Specific data sought at the plants include: the effectiveness of documentation in assisting the plant personnel to install the measures; the clarity of the measure indicators, and of the written statements of intent and application; additional information required to support installation and operation of the measures; work resources required to obtain and process indicator data; the practicality of building the data bases specified by the measures; acceptance of the measure ratings obtained; estimates of the indicators that are likely to be associated or correlated at that plant; and, the perceived utility of the measures.

Discussion

Lessons Learned

The process of developing the MOC and T&P performance measures has been similar to the process we have used in other projects — starting with understanding of the measure development process, communicating what measurement strategies are technically and socially acceptable, identifying and subsequently refining the initial measure organization and indicators, and finally testing of the performance discrimination characteristics of the measures. Some important differences in the CCPS setting have aided the indicator development process. Subcommittee members have been involved in the methods of identifying indicators (as distinguished from just identifying indicators), engineering analogs have been used to help think creatively about management systems, and the subcommittee members have been able to concentrate on details for long periods of time. This level of involvement and depth of thought are instrumental to making good measures. The experience leaves us with the following observations:

- Committee members, especially those who worked on the task forces, are now familiar with the method and can use it to identify and review indicators and performance ratings.

- Analysis tools such as HAZOPs are useful in identifying potential indicators. Such analyses should be incorporated into the subsequent measure development work.

- Groupware systems, such as the IBM TeamFocus, are very useful in certain phases of the measure development effort. Groupware should be incorporated into the development process to improve efficiency in those phases.

- We can make effective use of subcommittee time. One means is to use tools familiar to the members such as HAZOP analysis and groupware. Another means is to call on specialists and subgroupings of subcommittee members as needed to address each element.

- **The concept of a "continuously useful" measure can be implemented in a practical manner. The data need to be up-to-date, but this does not mean that all indicators needs to be updated simultaneously and frequently. Many of the indicators are known to change infrequently or in connection with certain events. Therefore, only a subset of data need to be collected with high frequency in order for the whole measure to be "continuously useful."**

- **The "life cycle" organization that was used for both the MOC and T&P performance measures appears to be a suitable structure for building the performance measures. Early efforts on the MOC and T&P measures used several other structures which presented difficulties. The life cycle organization was not preselected — it evolved as the need for the structure was felt. Only after the measures took form did we recognize it as the life cycle organization.**

Implications for Building Performance Measures for Additional PSM Elements
As the two MOC and T&P measures were developed, it became apparent that these PSM elements are related to each other and the remaining CCPS PSM elements. For instance, training is part of each element, and it may be that training performance should be assessed in a single element rather than in each separate element. Another example of a "horizontal relationship" among elements is "test and improvement" of the PSM system. The pervasive need for that function was apparent in every factor and subfactor. For instance, indicators for test and improvement could have been included for each MOC factor. Factor 1, Design, could have included indicators testing the existence of procedures for design improvement on the next PSM design or redesign. Likewise for the Development and Installation Factors, and so forth. This unnecessary duplication of measurement coverage can be avoided by using generic measures generally applicable to multiple PSM elements.

High Performance Risk Management

P. J. Bellomo
Arthur D. Little

In response to environmental, health, and safety (EHS) regulations promulgated during the last two decades, industry has added highly specialized EHS staff and technologies to strive for regulatory compliance. The additional resource allocations have increased manufacturing costs.[1] These increased manufacturing costs, of course, run counter to industry's goal of competitiveness.

Unfortunately, current trends suggest that EHS regulations will continue to increase in number, which seems to imply that the problem will only get worse. In this business climate, an ominous transformation in perspective is taking place, as indicated in Exhibit 1.

If industry continues to respond to EHS regulations as it has in the past, then a funeral, at least for many manufacturing operations, is no doubt in store. What is needed is a different *response* .

Exhibit 1

A Sign of Crisis

Will EHS regulations spell industry's demise? (late 20th century question)

↓

When is the funeral? (early 21st century question)

---

[1] Resource allocations for EHS issues have not only increased manufacturing costs, but in the case of worker/public health and safety, have returned, at best, only marginal performance increases.

If industry continues to respond to EHS regulations as it has in the past, then a funeral, at least for many manufacturing operations, is no doubt in store. What is needed is a different *response* .

In this paper, a business management theory referred to as "High Performance Business" (HPB) is applied to industrial EHS activities. Quite simply, HPB offers an alternative response to burgeoning regulations.

Ideally, application of HPB theory will substantially reduce the EHS risk posed by manufacturing operations; and not only will this EHS risk reduction be sustainable, but it will be satisfactory to regulators, surrounding communities, workers, and owners. In addition, application of HPB theory has the potential to entirely eliminate the EHS function from some organizations; in other organizations, HPB will cause a drastic shift in resources away from EHS and toward such functions as Operations, Maintenance, and Engineering. The overall result: increased EHS performance at a lower cost.

To facilitate understanding of the concepts and practices advocated here, it is important to first establish some working definitions. Risk Management terminology is established in the following section, with HPB concepts and terminology provided in the subsequent section.

WORKING DEFINITIONS

One of the most confusing terms associated with EHS activities is *risk*. This confusion results from misuse by lay people, the media, industry representatives, regulators, and even Risk Management practitioners. Nonetheless, *risk* can be a rather precise and useful term, describing a concept which guides a tremendous amount of decision-making. For this reason, a working definition of Risk Management will be offered and then used continually throughout this work.

The ideas presented in this paper will correspond with the working definitions provided in Exhibit 2. Note that in lieu of an alphabetical listing, working definitions have been supplied in an order that progressively builds understanding of more complex concepts.

Conspicuously absent from this discussion is the concept of "Loss Prevention." Activities that might be classified under Loss Prevention involve the protection of material assets and the avoidance of business interruption (e.g., the fire protection of an automated facility). Although outside the focus of this paper, HPB concepts have direct application to Loss Prevention activities as well.

Exhibit 2

Working Definitions of Risk Management

Hazard, possible source of harm to workers, consumers, the general public, the environment, and/or material assets (e.g., both a tank containing a liquified toxic gas and a drum of contaminated waste are hazards).

Incident, an acute or chronic event in which workers, consumers, the general public, the environment, or material assets experience harm as a result of exposure to one or more hazards (e.g., a tank explosion and a tank leak are both incidents, as well as the long-term contamination of a water supply from improperly disposed waste drums).

Risk, the potential for realizing the undesirable impacts of a hazard; the notion of risk couples the components of an incident's likelihood and its severity (e.g., the storage system for a highly hazardous toxic gas might pose low risk if it is well-designed, operated, and maintained).

Management Systems, processes for planning, organizing, conducting, and controlling work within a company.

Risk Management, the aggregate of management systems designed and practiced in order to prevent and/or mitigate EHS incidents.

A MODEL OF HIGH PERFORMANCE

Arthur D. Little's formulation of the HPB model is an outgrowth of its international management consulting activities. This model, represented graphically in Exhibit 3, explicitly recognizes that successful businesses place primary importance on the satisfaction of their stakeholders.[2] These stakeholders include customers, workers, owners, as well as occasional third parties.

To summarize, in a HPB, organizations and resources are aligned with a set of business processes which offer satisfaction to all stakeholders. To take a classic example, "Just-in-Time" manufacturing comprises deliberate processes which align an organization and a set of resources to satisfy a customer's desire to be serviced quickly while also satisfying an owner's requirements for lower inventory-carrying costs.[3]

In the arena of EHS issues and regulations, there are a number of stakeholders, each with their own set of "satisfaction attributes." Regulators and surrounding communities gain satisfaction from safe and clean manufacturing operations. Workers gain satisfaction from competitive wages complemented by safe and clean working environments. While safe, clean manufacturing operations satisfy owners, competitive manufacturing costs are of primary importance.

In a manufacturing operation that addressed EHS issues within its HPB structure, resources and organizations would be aligned through a set of business processes that delivered satisfaction to regulators, surrounding communities, workers, and owners, without adversely affecting the satisfaction of customers.

---

[2] For a concise and thoroughly enlightening introduction to High Performance Business concepts, see Nayak, Drazen, and Kastner, "The High Performance Business: Accelerating Performance Improvement," *Prism*, First Quarter 1992, pp. 5-29.
   [3] *Ibid.*, pp. 19-20.

Exhibit 3
The High Performance Business

The High Performance Business

Satisfy the primary stakeholders...

...by improving critical business processes...

...and aligning resources and organizations.

But the reality of the current situation, depicted in Exhibit 4, represents a conflict. Manufacturing operations are at odds when it comes to satisfying their owners' desires for decreased costs and other stakeholders' desires for EHS performance.

Exhibit 4
Conflicting Satisfaction Attributes

Stakeholders	Select Satisfaction Attributes	Conflicting Attribute
Regulators	EHS Regulatory Compliance	Manufacturing Costs
Surrounding Communities	Clean Environment	Manufacturing Costs
	Public Safety	
Employees	Safe/Healthy Work Environment	Manufacturing Costs
Owners	Manufacturing Costs	EHS Regulatory Compliance
		Clean Environment
		Public Safety
		Safe/Healthy Work Environment

The conflict, as presented here, may be somewhat oversimplified, and perhaps even misleading. It might be argued that safe and healthy work environments can result in dedicated and productive employees, which in turn can bring about higher productivity and decreased manufacturing costs. It might even be argued that environmental regulations force manufacturing operations into waste reduction and recovery, which typically decreases long-term manufacturing costs.

But somehow, such arguments only seem to hold true in theory. Or to the degree that these arguments are true, they represent a negligible portion of EHS compliance efforts. Overall, the trend has been that industry's response to EHS regulations has increased manufacturing cost.

THE PROBLEM OF RISK MANAGEMENT

Unlike other management processes, EHS risk management activities have seldom, if ever, received the close scrutiny of management theorists. Risk management was thought to be the work of the technician.[4] As EHS regulatory requirements became increasingly stringent, technicians were assigned to meet the challenges. As the challenges grew, the technicians demanded more resources. Soon entire new organizations took root, technicians assumed responsibility for managing EHS risk, and manufacturing costs increased. Ultimately, the level of satisfaction of some stakeholders (i.e., regulators, surrounding communities, workers) began to increase in exchange for the dissatisfaction of owners.

Nowhere is this history more evident than in the chemical, pharmaceutical, and refining industries. Exhibit 5 depicts the organizational structure of a fictitious manufacturer, Low Performance Petrochemicals, Inc. (LPPI).

---

[4] The term *technician,* as used here, comprises engineers, scientists, and EHS specialists, such as risk analysts, industrial hygienists, and occupational safety specialists.

Exihibit 5

Low Performance Petrochemicals, Inc.:

Assignment of Risk Management Responsibilities

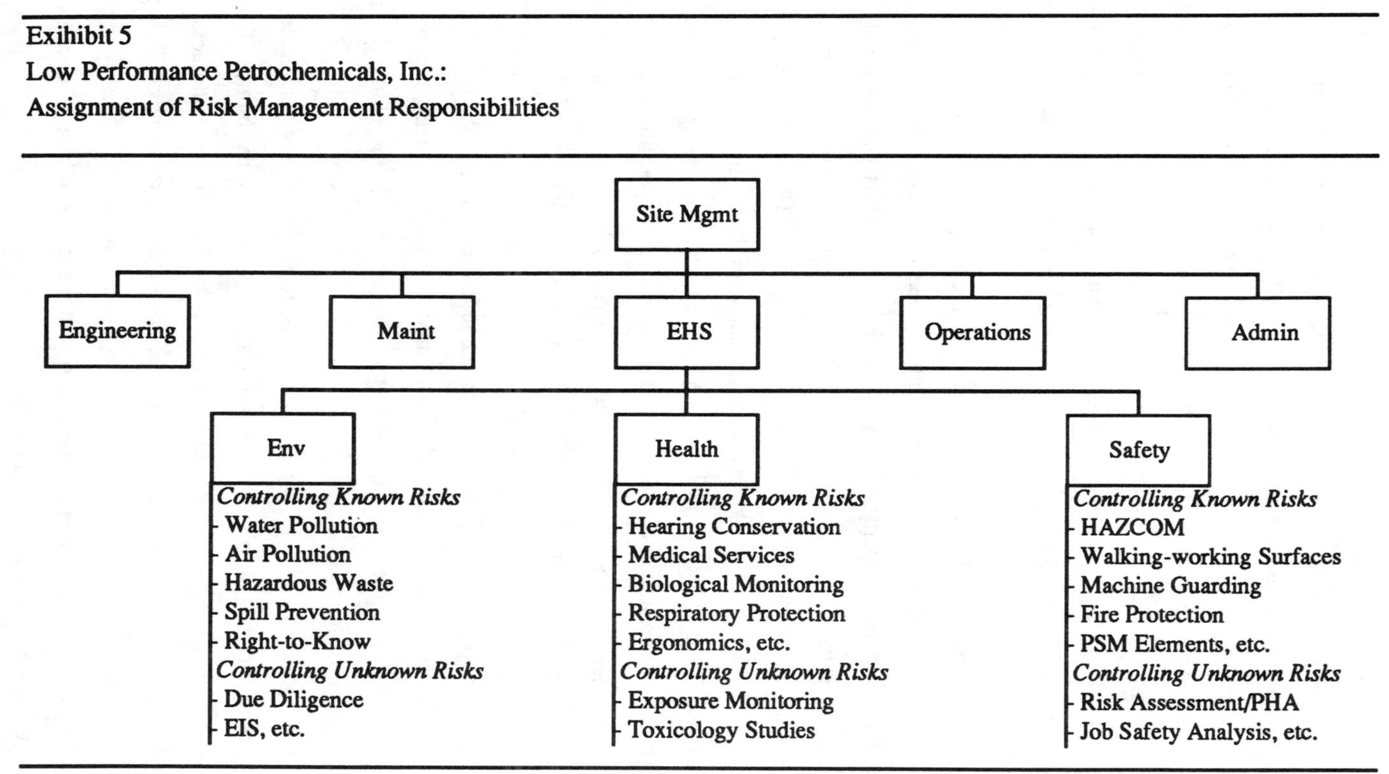

As EHS regulations grew, LPPI slowly added EHS staff to deal with compliance issues. The EHS technicians conducted audits and assessments; they demanded new hazardous material storage and handling procedures of Operations personnel. The EHS staff required new reviews which slowed new product development and capital project cycle times. Overtime increased as Operations personnel were shuffled through a barrage of expensive video training sessions conducted by the EHS technicians. In fact, as Exhibit 5 indicates, EHS technicians began to take responsibility for ensuring LPPI's compliance to a number of EHS regulations.

In short order, LPPI's EHS technicians seemed a police force. Individuals responsible for Operations, Maintenance, and Engineering continued to concentrate on manufacturing: high quality, high productivity, low cost. It was the job of the EHS police to step-in and shut things down when "laws were broken." And at no time was this dichotomy more apparent than when LPPI's EHS police conducted their audits. The analogy is strikingly accurate, and completely disturbing.

Look again at LPPI's distribution of EHS responsibility. Imagine the humor that operators on the manufacturing line found in LPPI's hypocritical adage "you are most responsible for your own safety." Operators knew the organizational structure: if they were responsible for their safety, then what did all those EHS technicians get paid to do. Operators directly experienced the contradiction; they experienced the conflict. Their job was to manufacture product and the technician's job was EHS compliance. Occasionally it appeared to make more sense to overlook EHS issues to satisfy their primary goal (i.e., production). If the police did not catch them, then no harm done.

An old joke in the Risk Management business comes from the dialogue exchanged in Exhibit 6. While owners might find momentary humor in this exchange, grave disappointment should soon follow. Owners of manufacturing operations are currently allocating vast sums of resources to create the conflict. At the root of the problem: ill-conceived business processes, unresponsive organizational designs, and misallocated resources contributing to a conflict between stakeholders who *must* be satisfied.

Exhibit 6

A Comedy of Conflict

Managing for Low EHS Performance

REGULATORY/CORPORATE EHS AUDITORS:	Hi, we're here to help.

MANUFACTURING OPERATION STAFF:	We're pleased to see you. *[big lie]*

MANUFACTURING OPERATION WORKERS:	We didn't call them. *[another big lie]*

A point here can not be overemphasized: owners of manufacturing operations actually pay to create a conflict in their organization -- a conflict that adds no value to products and results in dubious liability reductions. This does not mean that EHS technicians in general, and auditors in particular, do not have value to offer. It does mean, however, that for the most part industry does not know how to manage EHS services such that value is delivered in a meaningful, and useful way.

LOW PERFORMANCE PROBLEM, HIGH PERFORMANCE SOLUTION

The case of LPPI demonstrates the general inefficiency of industry's response to EHS regulations. It can be essentially argued that industry has failed to *manage* its risk management activities.

> Part of the problem with large, cross-organizational processes today *[e.g., managing EHS risk]* stems from their history. These processes have evolved over decades in piecemeal fashion. Typically, they involve many people doing specialized tasks. This division of labor made sense in the days when information could not be easily shared or processed. [5]

[5] Curtice, Chait, and Lynch, "Process Thinking: Today's Path to Improved Performance," *Prism*, First Quarter 1992, p 34.

Indeed, manufacturing operations have, for the most part, considered EHS regulations a series of disjointed crises. And in what amounts to a ubiquitous knee-jerk reaction:

- resources have been indiscriminately allocated;
- organizations have been passively established; and
- processes have been thoughtlessly generated.

All this in hopes of diffusing the crises. Yet surely it is clear to all that expectations of EHS performance are not only here to stay, but these expectations will grow with time.

In summary, the history of EHS risk management activities reveals a piecemeal evolution of organizations, resources, and processes. Corporate EHS policy statements abound. This, however, compares unfavorably to the scarcity of deliberate and well-conceived corporate EHS strategies. Perhaps most disturbing is the fact that there is a simple, common sense solution:

- in HPB terms, align critical organizations (e.g., Operations, Maintenance, and Engineering) and an appropriate set of new or improved resources (e.g., Management Information Systems, Just-in-time/Task Aligned Training),[6] with reconfigured business processes that satisfy both the imperative for EHS compliance and the imperative for competitiveness;
- in simpler terms, ensure that manufacturing activities represent *inherently* low EHS risk.

To begin to move from the theoretical to the applied, it is instructive to look again at LPPI's Risk Management responsibility assignments. In the following section, OSHA's recent Process Safety Management (PSM) regulation will be addressed as a case in point for moving LPPI toward high performance.

---

[6] Curtice, *et. al.*, *op. cit.*, p. 26. The main notion here is that training should not be done simply because "training is good." The training should be timed so that the knowledge and skills imparted to trainees will be demanded immediately in their regular jobs. Furthermore, the training should be highly specified to deliver the precise knowledge and skills that will be needed.

PSM: A NATURAL DEMAND FOR HIGH PERFORMANCE

Strategies for compliance to OSHA's PSM regulation are numerous.[7] Nevertheless, it is helpful to quickly review the key issues surrounding the regulation.

In what is essentially a departure from its trend toward *specification* standards, OSHA published a *performance-based* standard to reduce the risk posed by releases of "highly hazardous" chemicals.[8] There are 14 "elements" that OSHA has identified as part of this standard, listed in Exhibit 7.

Exhibit 7

OSHA's PSM Elements

Managing for Low EHS Performance

---

1. Employee Participation
2. Process Safety Information (PSI)
3. Process Hazard Analysis (PHA)
4. Operating Procedures (SOPs)
5. Training
6. Contractors
7. Pre-startup Safety Reviews (PSRs)
8. Mechanical Integrity
9. Hot Work Permits and other Safe Work Practices (SWPs)
10. Management of Change (MOC)
11. Incident Investigation
12. Emergency Planning and Response (EP&R)
13. Compliance Audits
14. Trade Secrets

---

[7] See Bellomo, P.J., "A Step-wise Approach Leads to Process Safety Management Compliance," *Oil & Gas Journal*, June 1992 for one approach, with references to related works.
[8] See 29 CFR 1910.119, "Process Safety Management of Highly Hazardous Chemicals; Final Rule," February 1992.

Of interest here, is the performance-based nature of OSHA's standard. This regulation is a natural fit for applying HPB concepts. In Exhibit 8, the Risk Management organizational assignments for LPPI are shown once again. This time, the only Risk Management component appearing is PSM. Note that in Exhibit 5, the PSM component appeared under the Safety function. In contrast, Exhibit 8 shows a redistribution of responsibility, this time broken-down by individual element.

The allocations of PSM responsibility in Exhibit 8 probably represent an extreme departure for some organizations. Nonetheless, this reorganization makes perfect sense in terms of the HPB model: primary responsibility for "compliance" to each element is given to the organizational function whose business processes would typically comprise the source of the EHS risk. And to ensure *ownership*, written job scopes would be expanded and performance measurements installed.

In such a reorganization, the EHS function would primarily act as a regulatory interpreter, guiding Operations, Maintenance, and Engineering in the improvement and/or development of their processes to ensure compliance. In its final form, this approach would result in each PSM element having a written management system to describe its compliance process. [9]

Of course, a shift in resources would be required. "Work groups" from Operations, Maintenance, and Engineering would need to be temporarily assigned to upgrading/developing the appropriate management systems; these personnel would need training to understand regulatory requirements. Then, after the work groups designed their solutions, the remainder of the respective organizations would need training to understand and implement the new processes. Apart from training, a high performance approach to PSM may require material resources. These might include "enabling technologies," such as a Management Information System (MIS) or remote communications devices.

---

[9] Elsewhere I have described the technical basis for this HPB approach (from the standpoint of a Risk Management practitioner), along with the characteristics inherent in a good management system. See Bellomo, P.J., "Risk Management for the 21st Century," National Petrochemical Conference, Mexico City, Mexico, October 1992, sponsored by Technology Training Corporation.

Exihibit 8

LPPI's Move Toward High Performance:
Rethinking PSM and Redistributing Responsibilities

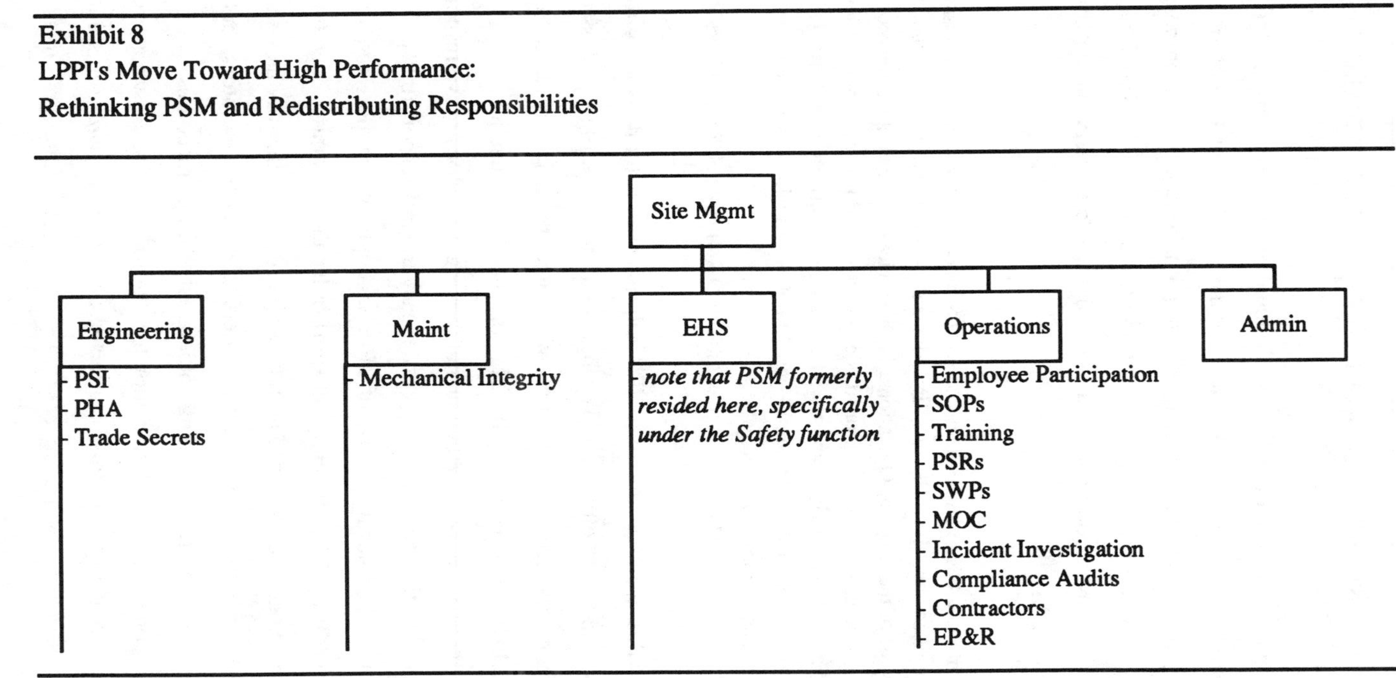

AVOIDING PITFALLS, ACHIEVING HIGH PERFORMANCE

Imagine a manufacturing operation, with numerous hazards, where the level of EHS risk satisfied all stakeholders, and imagine that this situation existed without the presence of specialized EHS staff. Would it not be ludicrous to propose adding EHS staff? What does this suggest about the goals of an EHS Risk Management program? The PSM Coordinator at a large sour crude refinery on the Gulf Coast put it something like this:

> As I see it, my job is to eliminate my job. A PSM coordinator should be transitional. All these elements --PSI, PHA, SOPs, Mechanical Integrity, etc. -- should be an integral part of the way our Operations, Tech Services, and Maintenance people do their jobs. My educational background is engineering and my refinery experience is Operations, and apart from some trivial complaints, I have to admit that OSHA's PSM regulation is consistent with the way I think we should be doing business...my task is to manage transition, then get out of the way.
> —PSM Coordinator at 200k BPD Refinery in US Gulf Coast Region

In this instance, a move toward a HPB solution was natural. And as this PSM coordinator expects to eliminate his job, should not more EHS professionals be looking to eliminate their jobs? After all, where does the EHS risk reside? Certainly not in the activities carried-out by the EHS organizational function.

Still, there are pitfalls. When manufacturing operations move responsibility for EHS risk to other functions, it is often easy to shift responsibility to a single person in that function. In the case of LPPI, a couple of operators had a knack at writing procedures. Unfortunately, in short order the remaining operators began seeing these "procedure writers" as safety people, dictating the way they should run their operations. Sure, the procedure writers *used* to be operators, but for six months they had a "cushy" job doing nothing but writing procedures.

The lesson to be learned: transition to a HPB requires *inclusive* approaches to process improvement such that the affected organization *owns* the solution and feels *accountable* for its success. A "mini EHS function" inside other organizational functions is not the answer. In the case of LPPI, a simple rotation of writing responsibilities solved the problem.

Finally, this discussion reveals a subtle but important change in organizational dynamics under high performance Risk Management. As indicated in Exhibit 9, this approach may require increased human resources allocated to certain functions, such as Operations. The percent of time any individual will spend manufacturing will go down because of the increased EHS responsibility. On the other hand, resources should be available from a down-sized EHS function.

This entire philosophy should resonate with all organizations dedicated to quality. If no other lesson was learned from the so-called "quality revolution," it certainly has become clear that the most effective manufacturing approaches have quality *built-in* to the system. By the same token, high EHS quality (i.e., low EHS risk) will result when key organizational functions and resources are aligned with new/improved processes: EHS quality will then be built-in to the system and an increased level of satisfaction will be delivered to all stakeholders.

Exhibit 9

High Performance Risk Management

Typical Trends in Shifting Resources

Function	*Direction of Shift*
Operations	
• Head count	Increase
• Hours available for manufacturing	Decrease
• Use of enabling technologies (e.g., MIS)	Increase
EHS	
• Head count	Decrease
• Hours required for compliance programs	Decrease
• Expenditures	Decrease

The Life-Cycle Approach for Nurturing Process Safety Management Systems

J. Steve Arendt
C. M. Mitchell
A. C. Remson
JBF Associates, Inc.

Concern about the potential for major accidents involving facilities that manufacture and use hazardous chemicals has led industry and government to take measures to help improve safety. One such initiative is the establishment of formal process safety management (PSM) systems. This push to institutionalize PSM programs within companies and at facilities has resulted in PSM systems with a wide range of implementation strategies and effectiveness. This paper presents an organized approach for implementing and maintaining PSM systems throughout the life of a process facility.

INTRODUCTION

Considering that tens of thousands of facilities now handle hazardous chemicals, the occasional occurrence of a significant accident is not surprising. However, to many observers, it appears that the frequency of such severe accidents is increasing. In contrast to this observation, and even though no one can guarantee a completely accident-free facility, companies should have "zero accidents" as a goal. Furthermore, the public has a right to expect industry to manage the risk of their operations in a responsible way, protecting people and the environment. Investigations into the causes of accidents have taught us that prevention efforts should address failures associated with four key areas, shown in Table 1.

For many years companies concentrated on making their equipment more reliable and correcting perceived problems with technology. This industrial process improvement usually resulted in tremendous increases in productivity and generally resulted in corresponding improvements in traditional workplace safety. However, in recent years, reviews of major accidents have shown that, collectively, human errors and the lack of adequate management systems have been the most

Table 1   Accident Prevention Focal Points

- Technology
- Management systems
- Human factors
- External events

significant contributors to such accidents. Although great progress has been made in developing methods for analyzing potential human errors and for managing chemical processes more safely, major accidents are still occurring and many sectors of society are asking for a major overhaul of the technological and institutional means of ensuring the safety of chemical operations.

Many catastrophic accidents in the chemical industry over the past 20 years can be traced to the lack of or the failure of management systems for process safety; nonetheless, industry has not devoted as much effort to the systematic nurturing and refinement of its PSM systems as it has to the development and improvement of chemical process technology. However, this emphasis is shifting. Industry has now begun to recognize the necessity and the value of formal PSM systems, means to measure PSM system performance, and approaches for continual PSM improvement. Moreover, a variety of industry organizations and government regulatory agencies have proposed structures for "complete" PSM systems.

OVERVIEW OF PSM-RELATED REGULATIONS AND INDUSTRY INITIATIVES

The recent increase in regulatory efforts dealing with prevention of major chemical plant accidents began in the aftermath of the tragedy in Bhopal, India. In December 1984, over 2,000 people were killed by a large release of methyl isocyanate from a plant operated by Union Carbide. Table 2 outlines the significant federal regulatory initiatives relating to process safety and accident prevention that have emerged since this watershed event.

The initial reaction by U.S. legislators was to pass the Superfund Amendments and Reauthorization Act (SARA Title III) and subsequent regulations created by the Environmental Protection Agency (EPA) involving Emergency Preparedness and Community Right-to-Know. This triggered an avalanche of previously unavailable technical information from industry concerning the identity and inventories of toxic chemicals at facilities and the amounts of these materials that were annually released to the environment. The end result was an increase in the regulators' and the public's awareness of the hazards of toxic chemicals and the potential for major chemical accidents.

Table 2   Federal Regulatory Efforts Related to Accident Prevention

- SARA Title III
- OSHA PSM (29 CFR 1910.119)
- 1990 Clean Air Act Amendments

At about the same time, the Department of Labor's Occupational Safety and Health Administration (OSHA) began performing wall-to-wall inspections of chemical plants. One result of these efforts was OSHA's conclusion that they could not hope to consistently regulate the safety concerns associated with potentially catastrophic accidents without a new regulatory standard. Led by this conclusion, they began a rulemaking effort that culminated on February 24, 1992, with the publishing of *Process Safety Management of Highly Hazardous Chemicals* (29 CFR 1910.119).[1] OSHA requires that companies possessing a hazardous substance in excess of a specified threshold amount develop and implement a process safety management system (Table 3). The 14-element program must be maintained over the life of the facility and is directed at protecting workers within the facility.

In November 1990, President Bush signed into law the Clean Air Act Amendments (CAAA), which contain accidental release prevention provisions.[2] This legislation mandates that EPA create regulations to require facilities possessing listed chemicals above a threshold amount to develop and implement extensive risk management programs (RMPs). These RMPs must contain three specific components: (1) a hazard assessment of potential accidents, including worst-case scenarios, (2) an accident prevention program, and (3) an emergency response program. The focus of EPA's regulations is to protect the public and the environment from the effects of catastrophic accidents. EPA's proposed rules will likely include requirements for companies to adopt a PSM system similar to OSHA's, but with some specific differences to address public safety and environmental concerns. Further, these rules will have additional features requiring companies to (1) adopt an overall management system to integrate and implement the components of the RMP and (2) provide a description of their RMP to regulators and the public.

Federal government agencies are not the only government regulators that became active in the chemical accident prevention arena. Since Bhopal, a number of states have

Table 3 Elements of OSHA's PSM Regulation

• Employee participation	• Mechanical integrity
• Process safety information	• Hot work permit
• Process hazard analysis	• Management of change
• Operating procedures	• Incident investigation
• Training	• Emergency planning and response
• Contractors	• Compliance audits
• Pre-startup safety review	• Trade secrets

also taken steps to help prevent chemical accidents by adopting extensive regulations. Many provisions of these state regulations have features similar to PSM systems. This similarity is not surprising, since the state regulations were built upon the same previous industry and federal initiatives.

As a result of these efforts, state and federal regulators alike now face the monumental task of enforcing their standards and monitoring compliance in a diverse set of industries that pose a complex spectrum of risks to workers, the public, and the environment. It is unfortunate that there is no recognized "standard" for PSM systems against which the quality of programs and implementation can be judged.

Lacking such a standard, industry has taken the initiative in advancing the state-of-the-practice in process safety and major accident prevention since the mid-eighties. Through the work of various industry groups (Table 4), significant advances have been made in the development and application of accident prevention strategies.

For example, in 1985 the American Institute of Chemical Engineers created its Center for Chemical Process Safety (CCPS). The charter of CCPS focused on advancing the state-of-the-practice in accident prevention. In 1989, CCPS published its *Guidelines for Technical Management of Chemical Process Safety*.[3] This ground-breaking document has been a leading resource for industry safety professionals involved in developing PSM systems. Of particular note is Appendix A of these Guidelines, which describes characteristics of management systems (Table 5).

In 1988 the Chemical Manufacturer's Association (CMA) began its Responsible Care initiative to declare the industry's commitment for responsible management of chemicals. Responsible Care essentially establishes a "safety value" system that the 180+ CMA-member companies embrace as a condition of membership. These member companies have adopted the *Process Safety Code of Management Practices*, a 22-point program aimed at designing, building, operating, and maintaining a plant safely throughout its operating lifetime.[4] This *Process Safety Code* is complemented by five other sets of management practices that comprise a holistic and proactive approach to

Table 4 Industry Accident Prevention Efforts

- Center for Chemical Process Safety guidelines series
- Chemical Manufacturers Association Responsible Care initiative
- American Petroleum Institute Recommended Practice 750

managing the safe use of chemicals.

In 1990, the American Petroleum Institute developed its Recommended Practice 750 — *Management of Process Hazards.*[5] This document established practices and guidelines for oil and gas industry facilities to follow in developing programs for managing process hazards.

Altogether, industry's efforts at improving the state-of-the-practice in chemical process safety have been remarkable. At the least, these efforts have improved the general level of awareness and safety. However, the design and operation of PSM systems historically have been conducted in a relatively unorganized and sometimes haphazard fashion. Many companies' initial PSM efforts have been created through ad hoc, cut-and-paste, or copy-your-neighbor methods. They have subsequently evolved through the years with feedback provided by normal operations, incidents, and lessons learned by other companies and facilities. As a result, PSM systems have not enjoyed as systematic a development and improvement process as many other aspects of process technology.

Except for CCPS, none of these industry efforts to promote the use of formal PSM systems has dealt with the fundamental issues: What are the essential features of a PSM system? What are the attributes of an effective PSM program? How can the variability

Table 5   CCPS Characteristics of a Management System

Planning
- Explicit goals and objectives
- Well-defined scope
- Clear-cut desired outputs
- Consideration of alternate achievement mechanisms
- Well-defined inputs and resource requirements
- Identification of needed tools and training

Organizing
- Strong sponsorship
- Clear lines of authority
- Explicit assignments of roles and responsibilities
- Formal procedures
- Internal coordination and communication

Implementing
- Detailed work plans
- Specific milestones for accomplishments
- Initiating mechanisms

Controlling
- Performance standards and measurement methods
- Checks and balances
- Performance measurement and reporting
- Internal reviews
- Variance procedure
- Audit mechanisms
- Corrective action mechanisms
- Procedure renewal and reauthorization

in PSM system application in a large company and in industry be understood on the basis of differing needs? How can the difference in these programs be reconciled in a context of regulatory compliance? Obviously, PSM development needs to be approached on a more systematic basis. The remainder of this paper deals with the issue of developing a systematic approach to establishing and nurturing formal PSM programs.

THE LIFE-CYCLE APPROACH

Most major chemical companies use a systematic approach for identifying viable commercial products, developing the necessary process technology to manufacture such products, designing and constructing full-scale production facilities, and operating the plants to manufacture products of the desired quality. Although the specific project management and plant operating phases may have different names from company to company, they all involve the same basic activities listed in Table 6.

Because the approach that a company uses to develop, construct, and operate new manufacturing processes can have a major impact on the bottom line performance, companies have devoted a great deal of energy and management attention to fine-tuning these approaches. However, not as much attention has been paid to the development and operation of systems for managing process safety. Since the importance of having effective PSM systems has become well-known, companies are now looking for better ways to establish consistent PSM systems across their diverse operations. One approach for companies to consider is to apply existing process development and plant operation paradigms to similar efforts to create, adopt, and improve PSM programs.

Figure 1 illustrates a proposed "life-cycle" approach for PSM systems. In the design phase, management describes their requirements and performance expectations for the PSM system. Then, during the development stage, a team creates a written description of the how the PSM system is to be built and operated. Once the design is finished and the PSM system is "fabricated" (i.e., developed), it can be installed in a facility (e.g., training of employees to participate in a management of change [MOC] system). Often, the installation of PSM systems will involve pilot testing the system, much like the startup testing that is done in a chemical plant. Finally, the PSM system is ready to operate. As it manufactures its "products" (e.g., properly

Table 6 Process Technology System Life Cycle

- Research
- Design
- Construction
- Startup
- Operation
- Modification
- Decommissioning

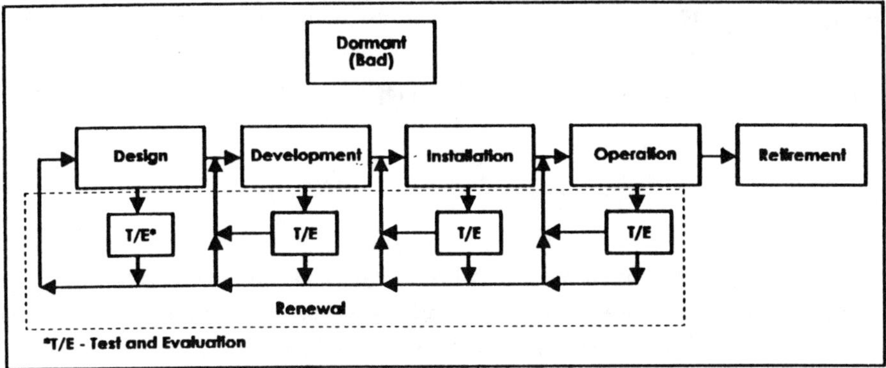

Figure 1   PSM Life Cycle

reviewed and authorized changes for an MOC system) through the years, periodic improvements (e.g., normal maintenance, debottlenecking, and expansions) are made to the PSM system to increase its capacity, improve its reliability, and improve product quality. In fact, at each phase in the life cycle of a PSM system, the opportunity exists to derive lessons from the completed activity (e.g., design, development, installation) and to recycle this learning back into the company's institutional memory. Thus, the next time a similar PSM activity is performed (e.g., installation of an MOC system at another plant), it can be done more effectively. In Figure 1, this renewal process is represented by the test/evaluation elements in the dashed box. This activity represents an opportunity to factor total-quality-based continuous improvement concepts into the PSM life cycle.[6]

Another aspect of the process technology paradigm is noted in Figure 1. Just like safety systems (e.g., scrubbers, flares) constructed of hardware, it would be undesirable to unknowingly allow a PSM system to become dormant (i.e., become out-of-service). In this situation, plant management might believe they have an active, effective PSM system in place, but in reality, the system has fallen into disrepair, or is ignored, thus leaving gaps in the risk management strategy originally put in place by management. If a component of a PSM system is no longer needed, then plant management should formally decommission and dismantle the PSM component. Such definitive action will help preserve the integrity of the plant workers' perception of management's commitment to process safety.

Applying the development methods used for chemical processes to the implementation of PSM systems could lead to significant improvements in the overall effectiveness of industry PSM systems. Using a life-cycle approach such as the one described above would allow companies and industry to develop consistency in their overall approach to implementing PSM systems. One of the most fruitful areas where management should focus their attention is in the design of high-performance PSM systems. The following section presents some ideas on how PSM design could be performed in a fashion similar to that of designing chemical process technology.

DESIGNING PSM SYSTEMS

Have you ever considered the following types of questions concerning PSM systems: What is the capacity of my system? What sort of throughput is my system rated for? Do I have an emergency shutdown switch for my PSM system? Are their adequate relief devices in my system? What kind of alarms and detectors should I install to warn of an imminent PSM system failure? These and other analogies should be pursued to take full advantage of the opportunities for applying process design approaches to PSM system design.

One of the most neglected areas in "PSM science" is the area of design. Why do companies not make the effort to design their PSM systems for a plant with the same concern as they design the equipment for the plant? A PSM system's performance can have just as dramatic an impact on the bottom line as the performance of a production unit. To design more effective, reliable, and capable PSM systems, consider applying the process design strategies to the design of PSM systems. As in the previous section, comparison of examples of hardware design to PSM system design will help us see if such analogies make sense and are useful.

Some of the most basic process design documents used in industry are process flow diagrams and P&IDs. Similar types of diagrams could be extremely useful in organizing a PSM system. Figure 2 is an example of a generic simplified flow diagram for a PSM system. On this diagram, three generic flowpaths are portrayed: work flow, information flow, and document flow. The work flowpath is the main "process flow" in a PSM system. It represents the essential manufacturing process that produces the end results of the PSM system. For example, in an MOC system, the work flow can be characterized by the activities involved in the origination of a request for change, determination of whether it is a replacement-in-kind or a change, review of the safety implications associated with the proposed change, and authorization for the change to be implemented. The main product of an MOC system is a properly authorized change request.

Auxiliary PSM "process flows" exist on the diagram and are essential to the smooth operation of a PSM system throughout the life of a plant. Created by certain PSM

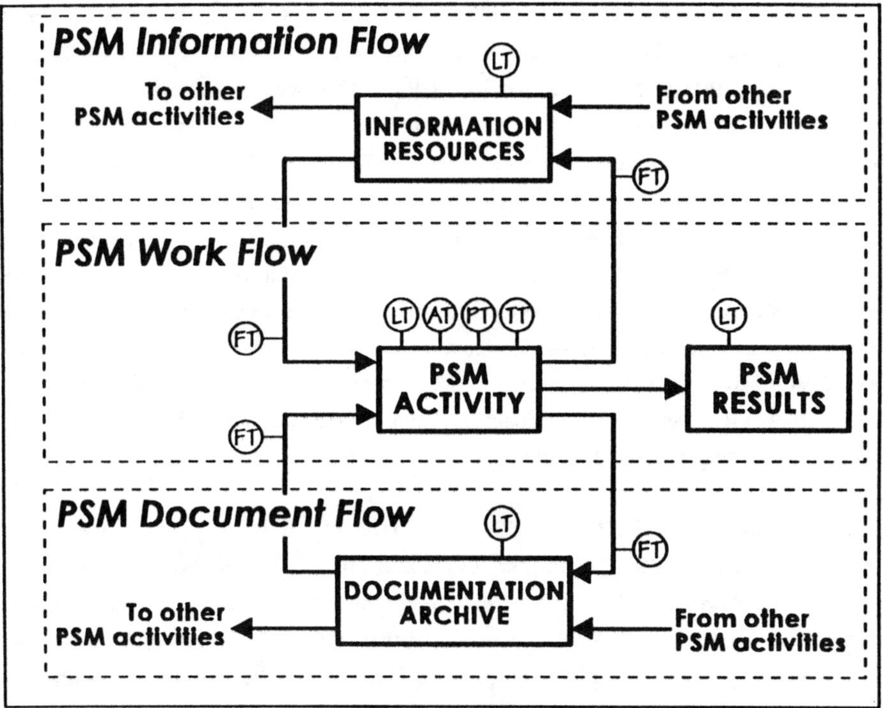

Figure 2   PSM System Flow Diagram

activities, information flows — normally through people — as inputs to other essential PSM activities. An example of information flow into an MOC system is the periodic communication of the implementation of process changes to people that could be affected by the change.   Thus, operators, mechanics, and supervisors who receive such information are in the flowpath and use it to correctly perform their process functions, as well as to function within the overall PSM system.

The flow of written or electronic documents within a PSM system is also an important path. These records are used by participants in the PSM activity to perform their PSM work, and the results of this work are often placed in documents that must be provided as inputs to other PSM systems and retained in the plant's institutional memory. For example, an MOC system uses process drawings and procedures as inputs to the review of proposed changes. If the proposed change is authorized, then documents flow as MOC work products back to the documentation archive. Many other specific process flow analogies can be used by PSM system designers to ensure that they have a fully integrated system supported by all the "utilities" needed to "process" the PSM "feed" material into high-quality finished products.

Figure 2 illustrates another type of process technology analogy that may be useful to PSM designers. The concept of measuring the performance of PSM systems is emerging through research sponsored by CCPS.[7] One way of gathering data on the performance of PSM systems is to install "sensors" in the PSM process system to allow PSM operators to constantly monitor the operation of the system and to quickly respond to upset conditions. Noted in Figure 2 are several transmitters for critical "PSM process parameters." The flow of feed materials into the PSM work activity (i.e., information and document flow) can be monitored so that if the flow deviates outside acceptable limits, then corrective action can be taken before a problem occurs. For example, what would happen if the flow of updated procedures to MOC reviewers stopped? Either they would stop MOC production, or they might unintentionally create an off-specification MOC product (i.e., an authorized change that contained a latent procedural safety defect) because the procedure was unavailable to the reviewers.

Consider a component in the PSM activity block. A PSM "level device" could transmit a signal to a PSM "controller" in the event of an accumulation of unprocessed material in a PSM "vessel" (e.g., an MOC reviewer is on vacation for a week and the in-basket is filling up with MOC requests). In this case the information could be used to prevent an overflow condition and possible "rupture" of the vessel (i.e., mechanics circumvent the missing reviewer and install the change anyway). Sensors could be devised to analyze the composition of the PSM work product to determine its quality. PSM "temperature and pressure" devices could determine when process conditions (e.g., high demand, immediate deadlines) exist that could lead to the production of off-specification PSM products.

For critical parameters, PSM system designers can install feedback control loops, alarms, interlocks, and emergency shutdown devices to ensure the proper operation of the PSM system. Moreover, the use of these analogies may assist in the real-time monitoring of the performance and efficiency of the PSM system. Many other process design analogies could be used to aid in the design of PSM systems. Research is needed to fully develop all of the reasonable PSM analogs to the basic elements in a typical design specification package for new chemical process equipment.

CONCLUSIONS

Companies that operate highly hazardous processes have paid insufficient attention to the systematic design, development, installation, operation, and improvement of PSM systems. One solution may be for companies to apply their experience in designing and operating their chemical process systems to PSM systems. But, the ideas presented in this paper may not be relevant for every PSM implementation situation. It is not our intention to force-fit or overstate the application of process technology analogies to PSM systems. Rather, we hope that these ideas spark others to thoughtfully consider creative new strategies to design, build, and operate PSM systems in chemical process facilities. The results of such efforts will most certainly lead to more consistent, higher performance, and more effective PSM systems. Better PSM systems ultimately will reduce the risk associated with major chemical accidents and improve the competitiveness of the chemical industry.

REFERENCES

1. 29 CFR 1910.119, *Process Safety Management of Highly Hazardous Chemicals*, Occupational Safety and Health Administration, Washington, DC, 1992.

2. 1990 Clean Air Act Amendments, Section 301(r), Public Law No. 101-549, U.S. Congress, signed by President Bush on November 15, 1990.

3. *Guidelines for Technical Management of Chemical Process Safety*, Center for Chemical Process Safety, AIChE, New York, 1989.

4. *Resource Guide for the Process Safety Code of Management Practices*, Chemical Manufacturers Association, Washington, DC, 1990.

5. American Petroleum Institute Recommended Practice 750, *Management of Process Hazards*, 1st. ed., Washington, DC, 1990.

6. "A Practical Approach for Integrating PSM and TQM Programs," *Proceedings of the International Process Safety Management Conference and Workshop*, September 22-24, 1993, San Francisco, CA, Center for Chemical Process Safety, AIChE, New York, 1993.

7. "The Measurement of the Performance and Effectiveness of PSM Systems," Project 13 progress report to the CCPS Technical Steering Committee, October 13, 1992.

Issues and Strategies in Risk Decision Making

Gregory L. Hamm
Rick G. Schwartz
Applied Decision Analysis, Inc., 2710 Sand Hill Road, Menlo Park, CA 94025

Management has a responsibility to the public, its shareholders, and its workers to effectively deal with the risks from storing, handling, processing, and distributing hazardous materials. Risk decision making refers to the choices an organization makes to deal with these risks. In order to manage risks effectively, management must strive to achieve quality risk decision making throughout the organization.

This paper has three sections. The first section defines quality risk decision making, key elements of which are (1) an organizational risk policy that expresses the values and philosophy of the organization toward risk management problems, and (2) the guidance provided on how decisions should be made in support of the this policy. The second section describes the types of risk policies that an organization may have, and the how a risk policy may be created. The final section addresses implementation issues, including the steps that an organization can take to improve its risk decisions, and the tools available to assist in risk decision making.

QUALITY DECISIONS

Risks to safety and health are intrinsic to virtually all human activities. The chemical processing industry has been particularly sensitive to and concerned about these risks. There are recognized hazards associated with chemical products and byproducts, and with the materials and processes used in their manufacture. Society generally recognizes the inherent hazards in this industry, but also recognizes the significant benefits to society that the industry provides. There is fairly broad, though not universal acceptance of the concept that to manage the risks of the industry effectively does not mean that all hazards must be completely eliminated. Instead, effective risk management is viewed as resulting from striking a proper balance between the industry's benefits and risks.

Effective risk management will:

- recognize the responsibility of the organization toward parties both inside and outside the company;
- determine a rationale for the amount of resources allocated toward risk reduction;
- maximize the degree of risk reduction achieved with the amount of resources allocated; and
- make decisions that are defensible to those who may review them. both internal and external to the organization.

A necessity for effective risk management is quality decision making. A quality decision is one that is consistent with the values of the decision maker, has a logical basis, can be clearly understood by outside reviewers, and makes the best use of available information and resources. Because there is uncertainty about the consequences of a decision, a quality decision does not necessarily equate to a good outcome. Again, risk is never entirely eliminated, and even the best decision may not prevent economic harm, injury, or even loss of life. However, in the long run quality decision making will achieve the greatest benefits achievable given the values of the decision maker.

An organization's risk policy expresses the values and philosophy underlying its risk decisions. To implement this policy, guidance and tools are required. Guidance translates corporate policy into specific directions for action. Decision aids are tools that support risk decision making and may help determine how policies will be sustained in practice. A decision aid provides a process for analyzing a problem, and rules for determining the preferred course of action. Decision aids can range from simple techniques such as Voting, to more sophisticated techniques such as Cost-Benefit Analysis and Decision Analysis. The selection of a decision aid and specifications for its use are key parts of the guidance an organization provides for risk decision making.

Management backing is crucial to both developing and implementing quality decision making. Decision quality is fostered by organizations that have:

- Commitment: The organization commits itself philosophically to making decisions that are logical, rigorous, and defensible, establishes clear guidance for these decisions, and consistently allocates sufficient resources to support quality decisions.
- Responsibility: Management and staff members understand the responsibilities they have for effective decision making, whether their role is to make the decision, analyze the decision, or provide required input to the decision.

- Training: Staff members understand the corporate philosophy and have adequate expertise in the decision aids that can support quality decision making.
- Experience: Quality decision making is made routine throughout the organization, and experience has been accumulated in a wide range of decision contexts.

To establish clear guidance, the organization must translate its values and philosophy into specific directions for action. A.M. Dowell [1] has suggested a "Pyramid Model" that describes how an organization builds a connection between corporate values at the top, and worker practices at the bottom. This model is depicted in Figure 1.

Following Dowell's terminology, *values* express what corporate management wishes to stand for, whether stated explicitly or not. Values might reflect concern for one or more of the aspects listed in Table 1. *Policies* express the principles that management will use in pursuit of these values, such as "avoid activities that pose an undue risk to the public." *Criteria* or *guidelines* provide a target for implementing a policy, such as requirements for the range of alternatives to consider, the decision aids to be used, or the weights to be placed on competing objectives (e.g., cost vs. safety). *Standards* and *procedures* instruct design and operation in order to fulfill criteria/guidelines, e.g., the maximum allowable quantity of toxic material stored at one location, or the required number and type of workers present during a hazardous operation. *Practices* and *behaviors* are what workers actually do, which may or may not coincide with the expectations expressed by the higher levels of the pyramid. In this paper, we will refer to all these elements as the organization's risk decision structure.

FIGURE 1. The Pyramid Model of Decision Guidance (from Dowell [1])

External Impacts	Internal Indirect Impacts
External Impacts	**Internal Indirect Impacts**
Employee fatalities	Public image
Employee injuries	Employee image
Public fatalities	Stockholder image
Public injuries	Capital availability
Environmental damage	Customer credibility
Public concern	Regulatory constraints
Internal Direct Impacts	Personal legal liability
Costs of safety measures	Personal image
Damage from accidents	(bonuses, promotions)
Lost business from accidents	
Liability costs	
Insurance costs	
Regulatory fines	

TABLE 1. Possible Values Considered in Risk Decision Making

This paper focuses on the upper levels of the pyramid, the translation of values and philosophy into policies and further into criteria and guidelines and the implementation of these criteria and guidelines. While we cannot give detailed instructions on implementation, we will discuss some critical success factors, and hopefully will sensitize the reader to relevant issues.

DEFINING AN ORGANIZATIONAL RISK DECISION STRUCTURE

An organizational risk decision structure can come into existence in many ways. In some cases, it may be developed through a specific effort and be stated explicitly at the highest level of the organization. More often, it will develop informally over time. The structure will be set by precedent and evolve as new decisions are faced and executed. In this case, the structure will not be specified entirely by management, but result instead from an interplay between management and staff, incorporating cumulative experience. Staff will pick decision aids, analyze problems, and recommend actions. Management will approve some recommendations and reward some analyses. Over time staff and management will understand the organizational risk decision structure without it having been expressed formally.

There are certain advantages to the less formal approach to developing an organizational risk decision structure. This approach allows for broad participation in the formulation of a structure, a gradual development of formal approaches to risk decision making, a high degree of flexibility depending on the decision context, and a reflection of organizational experience in the structure that evolves. As drawbacks, the informal approach may result in structures that are relatively slow to change organizational thinking, less effective in codifying organizational experience, and more vulnerable to misinterpretation.

An alternative approach, at the other extreme, is for management to develop and express an explicit structure at the highest levels. This can more rapidly change decision making throughout the organization, and ensure greater consistency between the decisions and the values of the organization. An approach is also possible that lies between these two extremes, such as a formal structure that is gradually enunciated and clarified through a more informal, evolutionary process.

The risk decision structure may be implemented very differently in different organizations. Even within one organization different elements of the structure will guide decision making at different levels. In many organizations, policy is set by a central body, criteria and guidelines are set at a wider and lower level, and standards and procedures are developed by still broader and lower levels. In other organizations, every step of the structure is determined centrally. In such an organization, policies, criteria/guidelines, and standards/procedures may be almost indistinguishable. We can not provide an exhaustive list of all implementations of the structure. Rather, we will discuss a number of distinct, formal, and hypothetical possibilities at the policy level. We say hypothetical because it is unlikely that any of these could be implemented in the pure forms described below. The following structures will be discussed:

- tip-of-the-pyramid,
- bottom-of-the-pyramid,
- all reasonable actions,
- risk limit criteria,
- budgetary control with cost-benefit criteria for individual projects, and
- formal multiattribute criteria.

Tip-of-the-Pyramid

This structure attempts to avoid criteria/guidelines and standards/procedures. It consists only of a policy statement by the organization about its regard for the public and its intent to take reasonable actions to reduce risk. Because it is incomplete, this structure does not guide independent, decentralized decision making. Usually, this form of structure suggests an underlying informal framework of criteria/guidelines, and standards/procedures, established by precedent.

The benefits of a tip-of-the-pyramid structure are that it indicates management's concern for risks, while allowing maximum flexibility in individual decision making. Although this flexibility is valuable, a key problem with such a structure is that the lack of direction can lead to inconsistent application of the values expressed in the policy, resulting in inefficient allocation of organizational resources. Another result is that decision making itself may be cumbersome and time consuming. A tip-of-the-pyramid structure can imply a lack of support by top management for formal, quantitative, or logically rigorous decision making approaches.

Bottom-of-the-Pyramid

Standards for equipment and facilities and procedures for operations are a part of most organizations' control systems. These standards are often determined by experts in the organization, by industry associations, or by federal or state regulations. A formal bottom-of-the-pyramid structure is one in which the organization seeks to control decision making solely through standards and procedures determined centrally. A formal bottom-of-the-pyramid structure is also incomplete and must be supplemented by informal policies and criteria that support the centrally set standards.

Developing these standards at the corporate level can provide central control of risk decision making and ensure that the knowledge of the best technical experts in the organization is utilized. Standards can be extremely effective in reducing specific risks, and their clarity minimizes the chance of misinterpretation. Standards can greatly reduce the cost of risk decision making throughout the organization, since any individual needs only to determine if their decision fulfills the applicable standards. For these reasons, almost all organizations use standards. Because the informal policies, criteria, and guidelines behind the standards are not announced, the organization has great flexibility in setting the standards. Standards also avoid the need to openly express difficult tradeoffs between costs and risks, and reduce the legal/political liabilities that management may believe are inherent in such risk decisions.

A number of problems are associated with this approach. First, it is impossible to define standards that cover every combination of technologies and procedures that a chemical manufacturer may encounter. In many situations, decision makers will be left without guidance. Second, standards are hard to adapt to specific situations. This can result in spending a great deal for a small reduction of risk, or accepting a significant risk that could be reduced at relatively little cost. Third, if standards are not kept up to date, they will not reflect the latest technological opportunities to reduce risks or costs. Finally, if standards are set arbitrarily, they may result in too high or too low expenditure on risk reducing activities.

All Reasonable Actions

Under this approach, the policy directs the organization to take all economically reasonable actions to reduce risks. Without criteria and guidelines, such a policy is a variation of the tip-of-the-pyramid structure. To specify the structure fully, criteria defining economically feasible actions must be stated. For example, the policy might direct that the organization pursue all actions that are compatible with the firm or a unit achieving a specified profit goal.

A recent paper on the corporate use of risk analysis provides an example of a statement that might be part of a such a policy: "harms to persons, property and the environment, and which arise from any given activity, are either eliminated, or, where this is not possible, reduced to a level which is as low as is reasonably practicable, or as may be required by legislation."[2]

The criteria for economic feasibility will determine the practical implications of such a policy. At one extreme, an organization might characterize "economically feasible actions" as being any for which funding can be obtained, even including actions that could threaten the existence of the organization. This might be referred to as an "at all costs" definition. At the other extreme, "economically feasible actions" may include only those actions that do not affect organizational profits. This might be referred to as a "preserve profits" definition.

An "at all costs" definition implies a high level of expenditure toward reducing risk, and provides a strong message both inside and outside the organization that there is a commitment to safety. On the other hand, a "preserve profits" definition safeguards the economic viability of the organization but may result in few actions to reduce risks. These extreme criteria will be fairly easy to interpret throughout an organization, and thus should help reduce the cost of decision making.

While an "at all costs" interpretation of all reasonable actions may be appropriate in specific contexts, it is unlikely that such an approach will be economically feasible over a significant length of time. Opportunities to reduce risks always exist, but they may not be significant. Many projects with only small contributions to risk reduction may be undertaken; but, in other cases, organizational profitability may be preserved even at a significant risk to the public. Under versions of the policy that preserve organizational profitability, the firm may be accused of placing money before lives when accidents do occur.

This policy approach avoids having decisions embrace explicit tradeoffs among financial costs, risk reduction, and other values.

Risk Limit Criteria

None of the three previous structures specify criteria and guidelines that would support decentralized decision making. The next three structures provide such guidance. In realistic applications of these structures, the resulting criteria and guidelines would guide decision making in unique situations and help establish standards and procedures that apply to more common situations.

Risk limits are probably the most popular quantitative policy approach. They have been adopted by a number of government agencies as a basis to regulate risks. Such a policy typically expresses limits that apply separately to the risk placed on individuals and the risks placed on society. Individual limits may be stated as the yearly risk of death or injury to an individual. Societal limits are generally expressed as a line or lines in a plot of fatalities or injuries versus probability (i.e., a maximum acceptable F-N curve). Examples of the criteria associated with such policies are shown in Figure 2, from Van Kuijen [3].

The Netherlands policy has three regions. Projects with calculated risks in the acceptable region may be implemented under all circumstances. Projects with calculated risks in the unacceptable region are not allowed. Projects in the reduction-desired region may be accepted under some circumstances.

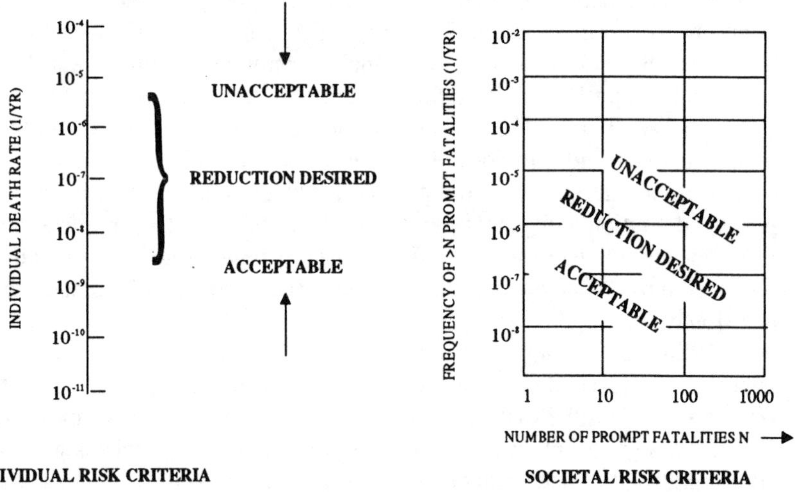

FIGURE 2. Risk Criteria for the Policy on Chemical Hazard Prevention in The Netherlands (from Van Kuijen [3])

Such a policy and consequent criteria have a number of benefits. A policy that limits risks will generally:

- promote social equity by limiting the risks to individuals and groups,
- be explicit and clear,
- promote consistent evaluations, and
- allow decentralized evaluations.

If the policy and criteria strictly define only two regions, acceptable and unacceptable, tradeoffs among costs and risks are not required. This simplifies the analysis in comparison to more complex quantitative approaches.

The major drawback of a strict risk limiting approach is its inflexibility. If projects already meet the acceptability standards, further reductions in risk may be ignored even if such reductions are costless or inexpensive. Similarly, projects with high social value that only slightly violate the standards may be rejected. Because of this discontinuity, such a policy is unlikely to lead to the maximum reduction in risk that can be achieved for a given level of expenditure. The policy may tend to encourage the division of projects or operations into many of a smaller scale. This avoids operations that individually violate risk limits, but may result in an increase in the overall risk. Another drawback is that such risk limits do not distinguish among acceptable projects. If several projects are acceptable, the policy does not indicate which to undertake. For all these reasons, most policies and criteria of this type, as in The Netherlands, define a region in which projects may be acceptable under some circumstances. However, when such a region exists, the policy and criteria are incomplete and provide little direction for how to act.

The Netherlands provides one example of a policy based on risk limits. Baybutt [4] provides an example of the use of risk limits in designing a tank farm.

Budgetary Control with Cost-Benefit Criterion for Individual Projects

Under this structure, a total budget for risk reduction is set, and within this budget those projects that together provide the highest level of benefit are undertaken. If the projects are relatively small and the benefits independent of one another, the best projects to undertake are those with the highest ratio of benefits to costs. If there are interactions among the projects, such as tradeoffs between undertaking many small or one large project, then a more sophisticated analysis is required.

The major advantage of this approach is that it maximizes the benefits from the projects given the level of investment. The approach allows decentralized decision making while preserving central control over expenditures. If the structure includes guidelines on the tradeoffs to be made among injuries, fatalities, environmental effects, and other effects, the policy can support consistent decision making that addresses a number of criteria. Finally, the approach avoids the explicit tradeoffs between individual health and safety risks and expenditures, that many organizations find unacceptable.

The key disadvantage of this approach is that it does not provide guidance on the overall level of expenditures. A second disadvantage is that it may be hard to separate the costs of risk reduction from other costs of a project. For example, when constructing a new plant, what portion of the costs should be considered standard plant expenditures and what portion should fall under the risk reduction budget? Finally, this approach requires a quantitative analysis of each project, and the analysis may become particularly difficult if a large number of projects that affect each other must be considered.

The U.S. government provides several example of the use of budgetary control with cost-benefit criteria for individual projects. The Fiscal Year 1992 Budget of the United States Government, Part Two, IX.C, Reforming Regulation and Managing Risk-Reduction Sensibly [5], identifies a number of government agencies that since 1991 are using formal cost-benefit analyses to prioritize the projects they undertake. The approach has also been used to support risk decisions at major government laboratories. [6]

Formal Multiattribute Criteria

This policy involves explicit tradeoffs among quantitative criteria associated with specific corporate values. Included among the criteria are explicit tradeoffs to be made among fatalities, injuries, and costs. Two simple examples of criteria are:

1. measures to reduce fatalities should have an efficiency of at least two expected lives saved per $10 Million in expenditures, and
2. equal weight should be given to a 10^{-6} chance of a fatality and 0.1 chance of a severe injury.

Criteria used in actual decision making would be defined in more detail than these simple examples, and might cover a broad range of values such as those listed earlier in Table 1.

A multiattribute policy allows for decentralized decision making and aims to optimize organizational welfare, social welfare, or a combination of the two. It provides a clear method for selecting among competing alternatives even when a number of values and consequent criteria are of concern.

The approach requires clear criteria that state the tradeoffs to be made, and management is often reluctant to express such judgments due to the perceived risk of unfavorable publicity or organizational liability in the event of an accident that might have been prevented under a different set of criteria. The approach also requires familiarity with quantitative decision aids, and a commitment to develop comprehensive criteria.

The U.S. government has conducted a number of studies utilizing multiattribute decision analysis. These include decisions by the Environmental Protection Agency on the allocation of Superfunds to hazardous sites, decisions by the Department of Energy on environmental restoration at its laboratories and facilities, and decisions by the Department of Energy on the siting of nuclear waste repositories.

Within the chemical industry at least one firm is considering the adoption of an explicit and consistent multiattribute approach. A team at BP International has proposed a two-part policy for the evaluation of risks [7]. The two-part policy is based on the idea that explicit tradeoffs are useful for achieving the highest level of social benefits, but that individual limits on risk are necessary so that risks do not fall unfairly on a few individuals. The policy proposes individual risk criteria that no operations create a risk of death greater than 10^{-3} per year for a worker and 10^{-4} per year for a member of the public. The policy also proposes that for decisions regarding societal risk, a value in the range of $\$10^7$ per statistical fatality be employed.

IMPLEMENTING AN ORGANIZATIONAL RISK POLICY

Various strategies can be used to implement a risk policy. Some organizations simply hope it happens as a natural result of good management. Some organizations commit themselves to rapid and massive change, organized from the top, and backed by significant resources. Others empower specific groups to experiment, and then foster those approaches that work best. Because strong cooperation is needed among different parts of the organization involved in risk decision making, we have found that absent management's active support for change, change occurs slowly. Active commitment by organizational leaders can spearhead industry leadership in the improvement of risk decision making. The remainder of this section discusses the steps an organization can take to improve its risk decision making, the decision aids to consider using in these decisions, and practical issues when undertaking projects for risk decision making.

Steps to Improved Risk Decision Making

Earlier we identified four key characteristics of firms that foster quality decision making: commitment, responsibility, training, and experience. In this section, we consider the steps required to develop these traits.

Commitment. The first step towards commitment is to adopt a clear policy statement and to give that statement substance through criteria/guidelines and standards/procedures. In doing this, there is a tradeoff between flexibility and control. If a very general policy is chosen and not specified through criteria/guidelines and standards/procedures, management must be prepared to accept a variety of interpretations and a level of inconsistency in decision making.

The second step towards commitment is to provide sufficient resources to support quality decision making. This includes training, access to expertise, provision of resources for decision making, and rewards for quality decision making. Training is discussed further in the next section. Access to experts may mean hiring consultants when a new decision aid is first being introduced in a firm, or when a more complex decision is being addressed. There are also benefits to creating one or more internal centers for decision support. Staffs of these centers participate in many projects and thus can rapidly build expertise. These individuals can then serve as a resource for the rest of the firm. A decision support center will usually include one or more persons who identify professionally with the field of management science, and will keep abreast of current theory and practice in this field.

To support individual projects, a rule of thumb is that 1% of expected costs or revenues is a reasonable expenditure for analysis. With most quantitative decision aids, it is possible to execute a first cut at analyzing the decision, and then use sensitivity analysis or other techniques to assess whether further analysis and expenditures are appropriate.

The third step is to demonstrate firm support for quality decision making. Management should emphasize the importance of understanding and documenting the rationale behind its decisions. It should expect that a significant decision will be underpinned by a well-prepared report, and reward those who provide them. Management should also ensure that its reward structure does not place undue emphasis on outcomes rather than decisions. This is one of the most difficult steps because identifying quality decision making is more difficult than identifying good results. In examining both successes and failures, management must use its best judgment to identify the effects of good or poor decision making and of good or poor luck.

Responsibility.   Management must understand the responsibilities of different groups, including management, in achieving quality decisions. Different groups or individuals have different responsibilities in creating quality decisions, as depicted in Figure 3. First, management is responsible for clearly stating the organizational goals and the values that are to be maximized through the decision. Second, experts throughout the firm are responsible for supplying honest information and evaluations. By honest, we mean without bias and without exaggerating the accuracy or inaccuracy of their information. Finally, individuals or leaders of a team making an analysis are responsible for using the inputs in a logical and unbiased fashion and for gaining buy-in to proposed solutions to ensure that they are executed.

Training. Training is essential to improving decision making. It is required for both analysts and decision makers. Decision makers need to understand the impacts of different policies and the limitations of different decision aids, the support needed to implement these aids, and the required staff, resources, time, and other inputs to the decision process. A similar level of familiarity is useful for analysts who provide input to decision making but will not lead decision teams. Finally, analysts who will direct or facilitate decision making teams need to have a detailed understanding of the decision aids to be used.

Training in the use of a decision aid can be given organization-wide or it can be focused on in-house specialists. Focused training builds internal expertise more quickly. Training can be both formal (e.g., through internal or outside courses) and informal (e.g., though mentoring by experienced in-house practitioners). However it is done, training must be accomplished before the immediate need for the analysis arises.

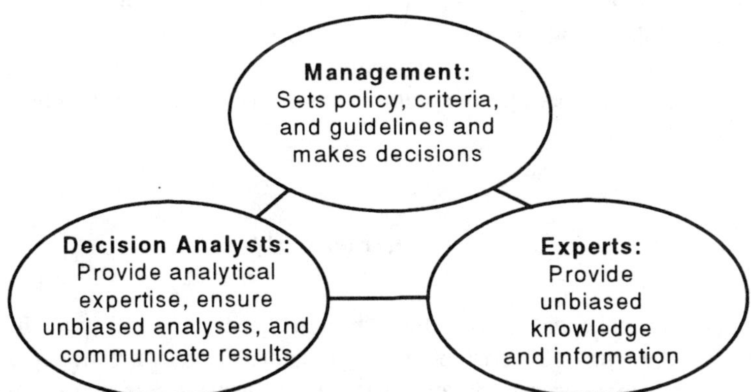

FIGURE 3. Decision Making Responsibilities

Experience. The first step in developing experience in quality decision making is to demonstrate that clearly stated policies, and criteria/guidelines that specify the use of formal decision aids, do improve decision making. One approach is to work with in-house or outside experts to create some winning projects that exemplify quality decision making, and then share these success stories, emphasizing the analysis that led to the right decision.

A second step is to preserve organizational memory. One author states that "organizations have no memory; only people have memories and they move." [8] This emphasizes the need to actively work to create an organizational memory. As noted earlier, an organizational risk decision structure usually develops over time rather being fully formulated at one point in time. By preserving organizational memory, management promotes the refinement of organizational policy while also ensuring that its policy is stable and consistent. Organizational memory fosters internal expertise, contributes to training, and allows successes and failures to be recognized.

Organizational memory is developed and preserved in three ways: by fostering sources of expertise, by requiring formal documentation, and by sponsoring internal learning, both formal and informal. One of the advantages of developing a core team of internal decision experts is that this group becomes a focus and a repository for organizational memory regarding decision making. Because they will lead or participate in many studies, they will accumulate much of the organizational history. Further, their files will become a documentation center for the organizational memory. Second, management can encourage that analyses of decisions be documented and circulated both internally and externally. Some firms have regular newsletters that are published by their centers for decision support. Third, an organization should have internal experts help teach others. While outside experts or courses may have broad experience in both projects, techniques, and teaching, only internal experts can communicate the organizational memory and the unique constraints on decision making within the organization.

Decision aids

A decision policy is specified through the criteria and guidelines derived from it. An important part of these guidelines is the direction given on the decision aids to be used in the organization. However, there is no simple answer to the question of "which aid is the best one to use for my decision?" The answer to this question will depend on the decision structure (values, policies, criteria/guidelines, and standards/procedures) adopted by the organization, and by the characteristics of the problem. This section describes important problem characteristics, the range and types of decision aids that are available for risk decision making, and suggests issues to consider in selecting a decision aid for a given problem.

Problem characteristics. Some key dimensions on which to characterize a risk decision problem include:

- Problem Scale, e.g., financial stakes, degree of uncertainty, problem complexity, severity of risk, time and budget availability, and information and people availability.
- Problem Importance, e.g., population at risk, severity of risk, need for outside review, and concern for multiple stakeholders or objectives.
- Group Involvement, e.g., impact on different parts of an organization, desire to involve multiple decision makers, desire to involve outside stakeholders.
- Need for Quantification, e.g., necessity to allocate limited resources, desire to use a quantitative basis for decision making.
- Need for Rapid Decision Making, e.g., a necessity to act quickly on a problem.
- Other Constraints on Decision Making, e.g., attitude toward use of subjective information, concern for confidentiality, legal constraints.

Decision Aid Classification. Decision aid classifications have been suggested by several authors [9, 10, 11, 12, 13]. We have found it useful to develop a classification of decision aids oriented toward the problem addressed and the organizational risk decision making structure. The classification scheme is depicted in Figure 4: Classification of Decision Aids.

The first dimension, "Treatment of competitive response," separates those aids that deal explicitly with multiple parties with conflicting goals, and those aids that support decisions of one decision maker or a group of decision makers with consistent goals. In this discussion we assume that participants share consistent goals, and therefore address this latter class of decision aids.

The second classification dimension, "Number of alternatives," considers how many alternatives can be addressed. Decision aids that consider many (hundreds, thousands, or infinite) alternatives require quantitative modeling of the constraints on the alternatives, the relationships among alternatives, and the relationships among alternatives and outcomes. This structure allows a systematic, efficient search for the best alternative. Mathematical Programming techniques, including Linear Programming, Non-Linear Optimization, and Compromise Programming, are examples of these methods. These aids are usually used to refine the design of a single type of alternative rather than to choose among a group of alternatives with diverse characteristics. In this discussion we assume that the options are diverse but finite, and we do not discuss mathematical programming further.

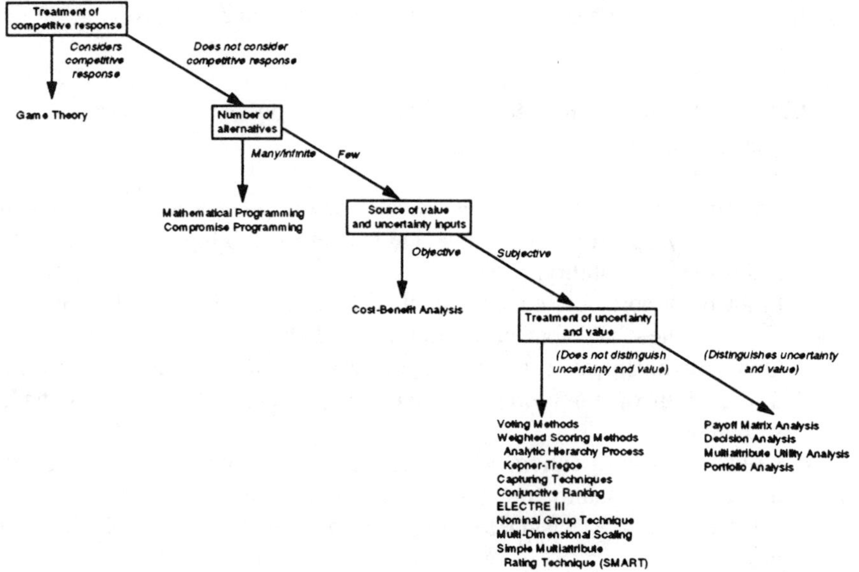

FIGURE 4. Classification of Decision Aids

The third classification dimension, "Source of value and uncertainty inputs," makes an important philosophical distinction. One branch emphasizes objectivity in the analysis while the other considers subjective values and information. Objectivity favors the use of empirical data such as market prices, observed failure rates, and historical experience, rather than expert opinion and personal values. The objective approach, of which Cost-Benefit Analysis is a well-known example, strives for decisions that are independent of the preferences or judgments of the decision maker.

The alternative, subjective approach recognizes that hard data may not be available, that experts can provide valuable, non-empirical data, that the best judgmental information (combining expert belief with experience and hard data) should be used, and that the decision maker's own information and preferences should drive the analysis. As the figure shows, a large number of decision aids support the subjective approach.

The fourth and final dimension, "Treatment of uncertainty and value," concerns whether the method distinguishes between (1) the uncertainty about the consequences of a decision and (2) the value for these consequences. Methods that simplify the problem by ignoring the distinction between uncertainties and values include: Voting Methods and Weighted Scoring Methods. Weighted scoring methods include the Analytic Hierarchy Process (AHP) and the Kepner-Tregoe Process, both of which provide a structured approach for identifying objectives on which to evaluate decision alternatives, scoring the alternatives on these objectives, assessing relative weights on the objectives, and integrating these scores into an overall score for each alternative.

Proponents of methods that distinguish uncertainties from values feel that doing this clarifies the analysis. Moreover, expected utility theory (as developed by von Neumann and Morgenstern and others) and the mathematics of probability theory provide a firm theoretical foundation for this approach. Examples include Payoff Matrix Analysis, Decision Analysis, and Multiattribute Utility Analysis. These methods involve modeling and assessing the uncertainties affecting the decision, assessing the decision maker's values for the outcomes of the decision, and combining the two to determine the preferred alternative.

Evaluations of Decision Aids. Choosing the appropriate decision aid for a particular problem requires an understanding of: the organizational context including acceptance and experience with specific decision aids, the characteristics of the problem that determine what decision aid strengths and weaknesses are of most concern, and the range of available decision aids and their strengths and weaknesses. A detailed evaluation of the various decision aids is beyond the scope of this paper; however, a brief and necessarily subjective review of the strengths and weaknesses of a few aids can illustrate the issues to be considered.

An evaluation of six decision aids is summarized in Table 2, which addresses how each aid performs, under different risk decision making structures and in different problem situations. The aids are evaluated in the context of applying a policy and any consequent criteria, not in the context of designing policies or criteria. It is important to note that most decision aids can accommodate any decision situation, and refinements or extensions to these aids

Evaluation Categories	Voting	Cost-Benefit Analysis	Weighted Scoring	Payoff Matrix	Decision Analysis	Multiattribute Utility Analysis
Policy						
Tip-of-the-Pyramid	✔		✔	✔		
Bottom-of-the-Pyramid	NA	NA	NA	NA	NA	NA
All reasonable actions			✔			
Risk Limit					✔	
Budgetary Control		✔				✔
Multiattribute						✔
Problem Characteristics						
Large scale		✔				✔
Close scrutiny		✔				✔
Need for quantification		✔			✔	✔
Group involvement	✔		✔	✔		
Need for rapid action	✔		✔	✔		

TABLE 2. Decision Aid Evaluation

have been developed specifically to address broader contexts. A check mark indicates that the decision aid, as routinely applied, is relatively strong on that as compared with the other aids. An "NA" means that the evaluating the aid dimension is not applicable.

The upper portion of the table considers how well the decision aids support various decision making structures.

The first row examines the strength of the aids in supporting a tip-of-the-pyramid policy. A tip-of-the-pyramid structure does not provide the detailed criteria needed to apply the more quantitative aids such as cost-benefit analysis, decision analysis, and multiattribute analysis. Further, it suggests a disinterest or distrust in detailed quantification, and therefore favors methods such as voting, weighted scoring, and payoff matrix analyses.

A bottom-of-the-pyramid structure assumes thorough, central control of decisions. Such a policy attempts to do away with risk decision making in the organization as a whole, thus evaluation of decision aids with respect to this policy is not applicable.

A policy of undertaking all reasonable actions may best be accommodated by weighted scoring methods. Such a policy, without more detailed tradeoff criteria, suggests a move toward more quantitative analysis, but still does not provide enough direction for the more quantitative aids such as cost-benefit analysis or multiattribute utility analysis.

A strict policy to limit risks assumes there are equal benefits in all alternatives that satisfy the risk limit. This creates a one dimensional problem of choosing the acceptable alternative with the least cost. Decision analysis performs well on such a problem, particularly if costs are uncertain. If it is desired to consider both non-risk benefits and costs in choosing among acceptable alternatives, then the multiattribute aids such as cost-benefit analysis, weighted scoring, and multiattribute utility analysis can be useful.

The policy of budgetary control with a cost-benefit analysis of alternatives tends to favor the more quantitative multiattribute approaches such as cost-benefit analysis and multiattribute utility analysis.

A multiattribute policy and consequent criteria strongly supports the use of formal multiattribute utility analysis.

The lower portion of the table considers problem characteristics that point toward the use of particular decision aids.

Large, complex problems and those that will receive close scrutiny favor the use of the more sophisticated, detailed, and logically rigorous methods of analysis such as cost-benefit analysis or multiattribute utility analysis. These approaches are recommended for such problems unless budgetary limitations prevent their use.

For problems on which quantitative methods are preferred, cost-benefit analysis, decision analysis, and multiattribute analysis are all quantitative and logically rigorous.

When group decision making is desired, aids that are somewhat simpler and designed to accommodate group situations are desirable. Such aids include voting, weighted scoring methods, and payoff matrix analysis. Similarly if decisions require very quick analysis, simpler methods such as weighted scoring and payoff matrix analysis are helpful.

Project Management

Though management of a decision study is similar to management of other projects, there are some unique, critical factors that must be considered.

The first step in a successful decision study is to develop a clear problem statement and a specification of roles. It is essential that the leader of a decision study first determine if a decision is to be made or if management is simply seeking background information or a planning study. If a decision is to be made, management and the project leader will need to work together to define practical alternatives, agree on management's values, and select experts that management believes in. This close cooperation may not be as necessary in background studies. In a decision oriented study, project leaders should be reluctant to employ decisions aids for which management is unwilling to provide the proper guidance. For example, multiattribute analysis can be done without management inputs, but this greatly weakens the chance for a successful and accepted result.

The line between planning and decision making is not always clear. Frequently, a decision must be made quickly after some future uncertainties are resolved. In this case, the analysis can determine a number of different possible future outcomes and for each of them specify a course of action.

A decision study team will typically consist of a decision maker or makers, a core group of analysts, and a larger group of experts. The core group of analysts is likely to include one or two experts in the use of the chosen decision aid and one or two experts with a broad knowledge of the problem. It is extremely important that no one on the core team have a strong personal bias towards any of the proposed alternatives. Such biases frequently cause team conflicts and bring the results of the study into doubt.

The experts should cover all of the areas important to the project. Part of the core teams' responsibilities will often include identifying experts or quickly developing internal expertise relevant to the problem. An important part of dealing with experts is to educate them about the importance of their contribution to the study and the importance of their providing unbiased information. Expert selection and education are the most important ways of eliminating or reducing bias from experts.

Outside experts can be a valuable part of a decision team. Outside experts can provide both knowledge specific to a problem, and knowledge about the proper use of a decision aid. Whatever the level of education or academic training, it is difficult to successfully lead a study using aids such as cost-benefit analysis, decision analysis, and multiattribute decision analysis without prior experience on several projects and access to an advisor who has depth of experience. When a firm uses an outside expert to lead or facilitate a decision study, this can provide an excellent opportunity to educate the internal staff on a decision aid. The use of outside experts should also be considered when there is a possibility that the team will be seen as biased or when a fresh viewpoint is desired.

Two common pitfalls make many technically excellent decision studies unsatisfactory in practice. First, the emphasis on technical excellence in the analysis of the alternatives identified during the formulation of the decision problem can blind the analysts to new and superior alternatives that can surface in the course of the study. Instead, the process of examining alternatives should be used to help identify new combinations of actions or an alternative not even initially considered. Even if consideration of new alternatives means throwing out the analysis of current alternatives, new alternatives can't be ignored

A second potential pitfall is the failure to achieve buy-in. Solving a large problem almost always requires serious commitment and effort by a number of key individuals, and even the best decisions can fail to be executed without this. One of the best ways to establish buy-in is through good analyses that communicate the assumptions and results clearly, and makes evident the superiority of the chosen alternative. It is also helpful to involve all parties that will help implement the decision, in the early formulation of the problem. Finally, the need of organizational backing for the success of each alternative can be included as a formal or informal part of the comparison of alternatives.

Acknowledgment. The authors would like to acknowledge the comments and insights provided by members of the Risk Assessment Subcommittee of The Center for Chemical Process Safety (AIChE). The views are the authors' and do not necessarily represent those of CCPS or its members.

REFERENCES

1. A.M. Dowell, III, Getting from policy to practices: the pyramid model (or, what is this standard really trying to do?), South Texas Section AIChE Process and Safety Symposium, Houston, TX., February 18, 1992, 819-840.
2. V.C. Marshall, The social acceptability of the chemical and process industries: a proposal for an integrated approach," TransIChemE. Vol68, Part B, May 1990, pp. 83-93.
3. C.J. Van Kuijen, Prevention of industrial accidents in The Netherlands, UNEP Industry and Environment, July/August/September, 1988, 2-12.
4. P. Baybutt, Procedures for the use of risk analysis in decision making, presented at the Second Symposium on Heavy Gases and Risk Assessment, Frankfurt, Germany, 1991.
5. *Fiscal Year 1992 Budget of the United States Government*, Part Two, IX.C. Reforming Regulation and Managing Risk-Reduction, pp. 367 - 376.

6. D.G. Brooks, G.L. Hamm, M.W. Merkhofer, and A.C. Miller, An approach to prioritizing risk-reducing projects. In this volume.

7. M.S. Hogh and A.B. Fleishman, The development of risk acceptability criteria for worldwide application: an international energy company view, presented at the Third Conference of the Society of Risk Analysis, Paris, December 1991.

8. T.A. Kletz, Managing risk in chemical manufacture, Special Publication of the Royal Society of Chemistry, (Risk Manage.Chem.), 1991, 92-105.

9. R.V. Brown and J.W. Ulvila,. Setting Analytic Approaches for Decision Situations; A Matching of Taxonomies. (Technical Report 76-10, sponsored by Office of Naval Research). McLean, VA: Decisions and Designs, Inc.; 1976.

10. B.S. Fischhoff S. Lichtenstein; P. Slovic; S.L. Derby; and R.L. Keeney, Acceptable Risk. New York: Cambridge University Press, 1981.

11. C.L. Hwang, and M.J. Lin, Group Decision Making Under Multiple Criteria; Methods and Applications. New York: Springer-Verlag, 1987.

12. C.L. Hwang and K. Yoon, Multiattribute Decision Making Methods and Applications. New York: Springer-Verlag; 1981.

13. M.W. Merkhofer, Decision Science and Social Risk Management. Boston: D. Reidel Publishing Company, 1987.

WORKSHOPS

Regulation of Chemical Process Safety in Japan

Eiji O'Shima

Professor Emeritus, Tokyo Institute of Technology
Professor, Fukui Institute of Technology
Counselor, The High Pressure Gas Safety Institute of Japan

1. INTRODUCTION

It was in 1958 when the first ethylene plant was put into operation in Japan, for the purpose of utilizing naphtha which had no particular usage at that time. This was, of course, the start of petrochemical industry in Japan. It was an essential problem for the Japanese petrochemical industry which totally depended on the imported oil as the raw material how the international competitiveness can be achieved.

The following four strategies were the typical reactions taken by the Japanese petrochemical companies in order to strengthen their competitiveness;

    (1) Large scale production

    (2) Continuous operation

    (3) Process optimization

    (4) Structuring of complex

These strategies were particularly effective for the petrochemical industry in establishing their leading position in the whole Japanese industry by synchronizing with the high rate economical growth policy of Japanese government. The rapid economical growth took place from 1960 and lasted for about 13 years until the first oil crisis of 1973.

Fig.1 shows the trend of accidents in petrochemical complexes. It is obviously not contingent that the rapid increase in the number of accidents in that period of time showed exactly the same trend with the economical growth. The year of 1974 was

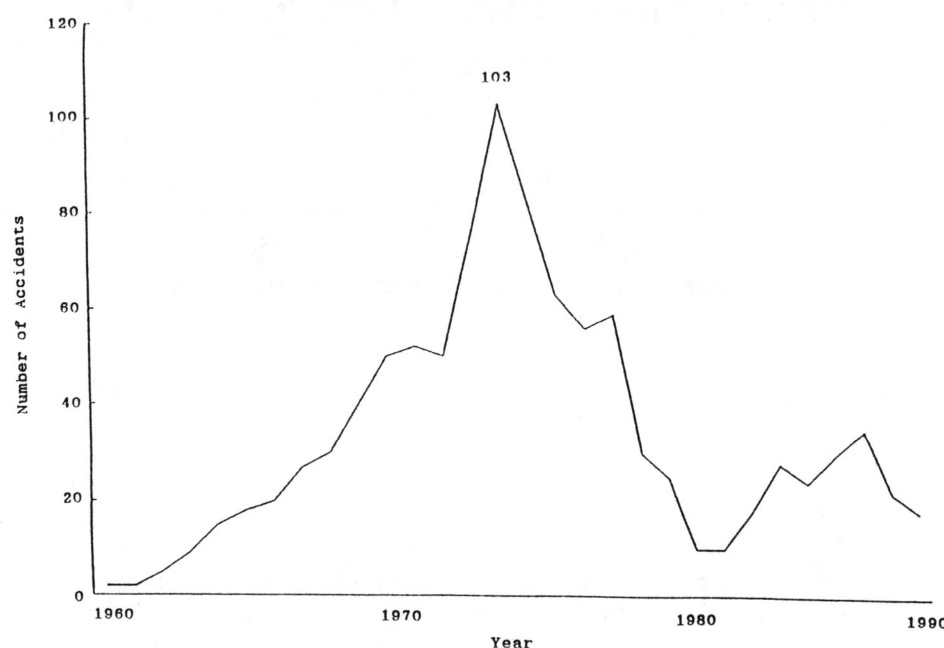

Fig. 1 Trend of Accidents in Petrochemical Complexes in Japan

the worst in the record where 103 accidents were reported. Ministry of International Trade and Industry which is the regulatory body in charge of safety control of chemical plants organized an investigating committee on the accidents and identified the causes of the accidents as indicated in Table 1.

The rapid increase of the accidents in the years of early 1970s gave, at the same time, a strong impact both to the industry and the government to force them to review their safety approaches. Taking that occasion, the government comprehensively revised the legal and political regulations. Each chemical company has changed the attitude toward safety management to remarkably increase the safety budget and to reinforce the safety education. These efforts were undoubtedly effective in reducing the number of accidents for the following period as indicated in the record.

Table 1   Causes of Accidents

		Number of Accidents
(i) Equipment-associated		
	(a) Improper design	226
	(b) Improper manufacturing, repair and inspection	85
	(c) Corrosion and deterioration	103
	(d) Defects in instrumentation and control system	26
	(e) External load, external impact and natural disaster	21
	(f) Others	13
		(474)
(ii) Operational management-associated		
	(a) Wrong operation, wrong decision and mistake in confirmation	235
	(b) Deficient information furnished and leadership improperly taken and/or unsatisfactory instruction	92
	(c) Unsatisfactory operation standards	79
	(d) Improper check	79
	(e) Insufficient technological knowledge	30
	(f) Imperfect training and unskillfulness	15
	(g) Others	15
		(545)
	Grand total	1019

2. SAFETY REGULATIONS

There are several different laws in Japan which are relevant to the problems of safety control in the chemical plants. For examples;

(1) The High Pressure Gas Control Law

(2) The Fire Service Law

(3) The Labor Standards Law

(4) The Law on Disasters in Petroleum Complexes and Other Petroleum Facilities

(5) The Toxic and Poisonous Substance Control Law

Each of these laws has, of course, its own objectives or different coverage of the problems and managed by different ministries. The Fire Service Law and the Labor Standards Law are managed by the Fire Defense Board and the Ministry of Labor, respectively. And the Ministry of International Trade and Industry is responsible for the other laws.

The High Pressure Gas Control Law aims to prevent high pressure gas related accidents by regulating the production, sale, transportation and use of high pressure gases and pressurized container. The following technical standards and procedures are the relevant regulations to the High Pressure Gas Control Law defined by the Ministry of International Trade and Industry.

(1) General Gas Safety Regulations
(2) Liquefied Petroleum Gas Safety Regulations
(3) Chemical Complex Safety Regulations
(4) Designated Equipment Inspection Regulations
(5) Refrigeration Safety Regulations
(6) Containing Cylinder Safety Regulations

The Chemical Complex Safety Regulations apply to the large scale plants where a large amount of high pressure gases are processed. There are ten sites that are designated as the petrochemical complexes where many plants dealing with vast amount of high pressure gases are densely located. About four hundred plants are subject to the Chemical Complex Safety Regulations.

The main items of the law:
(1)   Permission of the Prefectural Government
      It is necessary to obtain the approval for installation from the prefectural government prior to the construction of the high pressure gas plant. As the auditing procedures, the prefectural government carries out a comprehensive investigation whether the plant layout and the designing methods of pressurized containers and other safety devices meet the technical standards defined in the regulations.
(2)   Inspection of Designated Pressure Vessels
      Some types of the pressure vessels are designated as to have potential hazard in the case of accident by the Desig-

nated Equipment Inspection Regulations. The designated equipment is mandated to be inspected by the specialized body authorized by MITI, with respect to designing, construction material, and fabricating process.

(3) Completion Inspection

At the time of completion of the construction of the plant, the final inspection is carried out in the presence of the inspector sent by the prefectural government to confirm that the plant is constructed in accordance with the technical standard. The major items of the inspection are:

      (1) Safety distance

      (2) Safety valves and alarms

      (3) Controller

      (4) Leak test

(4) Periodical Inspection

The plant which has been under operation is mandated to be inspected annually by the prefectural government inspector in order to confirm that the plant still meet the technical standard. Such inspections, for example, as deteriorating condition of vessels and pipes, pressurizing test, leak test, test of emergency valves and control units are carried out, which usually require shut-down of the plant.

(5) Personnel in Charge of Safety

In order to assure the safety operation of the plant, every plant is mandated to designate a qualified personnel to be in charge of supervising and managing safety of the plant.

(6) Safety Code of Practice

Each plant is required to prepare its own Safety Code of Practice, which defines the basic rules of safety activities as well as the basic principles of operation and maintenance of the high pressure gas equipment. This Safety Code is to be turned in to the prefectural government for approval.

(7) Safety Education

Each plant has to prepare the safety education program for the personnels and to be submitted to the prefectural government.

(8) Technical Standard

Any of the high pressure equipment has to satisfy the technical standard of safety defined under the law of high pressure gas safety control. The standards include;

i)   standards of pressurized equipment
ii) safety distance, for example from residential areas,
     schools, hospitals etc.
iii) safety devices to be installed at the equipment under
     certain conditions, such as emergency shut-down
     valves, sensors and alarms for gas leakage, water
     sprinklers for tanks etc.

3. APPROVED PLANT SYSTEM

Under the High Pressure Gas Control Law, the periodic safety inspection is regulated, which is to be carried out annually by the governor of each prefecture with purpose of auditing the safety conditions of such plant where high pressure gas is handled. The said inspection was customarily conducted by shutting the plant down, though it is not explicitly mandated in the law.

In the meantime, it was recognized that too frequent shut-down inspection may not be appropriate or could be even unfavorable from the technical point of view. From such recognition, the regulation was changed, so that the inspection at the first year can be replaced by the on-stream inspection which is carried out while the plant being under operation, whereby the shut-down inspection is consequently to be carried out once every two years.

It is, however, necessary for the plant to be applied such policy to get the approval from the Minister of International Trade and Industry through an evaluation procedure. This policy, so-called the Approved Plant System for Chemical Complex, was started in 1987 and currently 72 plants have been approved.

The approval is specifically given to such plants where sufficient safety conditions are maintained as well as the high level of safety management is assured which are evaluated by the investigating committee organized in the High Pressure Gas Safety Institute of Japan as the trusted board of the Minister of International Trade and Industry.

The technical feasibility of the continuous operation for three or four years without shut-down inspections is now under study.

International Activities: International Process Safety Group Session

J. L. Hawksley

OBJECTIVE

To outline what the International Process Safety Group is, what it does and to discuss its contribution to improving process safety and how that could be enhanced.

The programme of the Workshop will comprise a number of presentations by members of the Group followed by a discussion period.

WHAT IS IPSG?

The International Process Safety Group (IPSG) was formed after the Flixborough explosion in the UK in 1974 (it was initially called the International Study Group for Hydrocarbon Oxidation). Membership is restricted to Companies which operate or develop chemical processes that present significant hazards. Currently there are 37 member Companies, predominantly from Western Europe and North America, but including Japan, Poland and India A condition of membership is that the Company is demonstrably active in promoting process safety. Meetings are held about every 8 months, sometimes in association with other related events.

The Chairman and Technical Secretary of the Group are elected from its membership and serve for a term of 5 meetings; currently they are J Gremmen of Dow Polyol and J L Hawksley of ICI. The Administrative Secretariat for the Group is provided by the Institution of Chemical Engineers via B Hancock of the European Process Safety Centre (EPSC).

WHAT DOES IPSG DO?

The aims of IPSG are to consider major process hazards and to exchange and share experience of all aspects of the prevention and mitigation of process accidents. Where a need is identified, the Group sets up Working Parties to study particular topics in depth. The output may be, for example, guidance notes or training packages, which are mostly published and made generally available via the UK Institution of Chemical Engineers. A strength of the Group is that it brings together a very large amount of direct operating experience of process safety (over 600 man years at a recent meeting) and provides a forum in which, where necessary, incidents can be discussed in confidence to facilitate information exchange.

Typical of the output from IPSG is a video and 35mm slide training package "Managing for Safety". Some of the material from this will be outlined in the Workshop. A similar package "Safer Piping" has also been produced. Currently nearing completion is a report on acoustic emission testing of equipment. Some guidance on programmable electronic systems for process control is being developed. A training package on Inherent Safety is planned.

THE "MANAGING FOR SAFETY" TRAINING PACKAGE

This package is built around a 23 minute video which highlights key aspects of managing for safety by following the experience of a manager taking over the operation of a production facility. The underlying theme is that the majority of accidents are due to human failings of one sort or another and good management systems are a prerequisite for minimising the likelihood and consequence of such failures. Initially the importance of training and an organised handover is stressed. It then illustrates management systems in action through a number of typical dilemmas and situations that may face a manager of any process plant. These include managing a minor leak, working with contractors, issuing work permits, coping with production pressures, preventive maintenance, near-miss reporting, start-up procedures, safety critical procedures, emergency preparedness and auditing.

Complementary written material in the package gives suggestions for using the video interactively in training sessions. Also, there are examples of typical management systems contributed by the Companies who took part in assembling the package. These are, in effect, practical "how to" guidelines from which actual systems could be developed for a particular situation.

Lastly, the package includes descriptions and presentation slides (35mm) for 25 case studies which can be used to illustrate various aspects of management failures. The case studies are grouped under the following aspects:

- understanding and managing process hazards
- the dilemma of business pressure
- work permit systems
- procedures
- management of change
- learning from previous incidents
- safety critical systems
- plant inspection

A selection of the case studies will be presented briefly during the Workshop.

DISCUSSION

The discussion period will provide an opportunity for Workshop attendees to interact with a number of members of IPSG and explore further the contribution it makes to improving the management of process safety. Some possible discussion points are:

- how is experience best shared?
- how does IPSG interact with other International/National activities?
- would more co-ordination help or hinder?
- what are the advantages and disadvantages of IPSG's restricted membership?
- should IPSG seek to be more influential?

POSTER SESSIONS

Integration of a German Parent Company's Process Safety Policies at a U.S. Subsidiary: An Example of Bayer AG & Miles Inc.

Nita M. Tosic
Miles Inc., Mobay Road, Pittsburgh, PA 15205

1.   BACKGROUND

Miles Inc. is a major U.S. manufacturer with businesses in chemicals, health care and imaging technologies, and with more than 23,000 employees. Miles Inc. is a wholly-owned subsidiary of Bayer AG, headquartered in Leverkusen, Germany. Miles Inc. has annual sales of nearly $6.5 billion, which is approximately one-fourth of the worldwide sales of Bayer AG. The operating divisions include: Agfa, Agriculture, Diagnostics, Industrial Chemicals, Organic Products, Pharmaceuticals, Polymers, and Polysar Rubber.

2.   INTRODUCTION

The purpose of this paper is to describe how the process safety policies and directives of the German parent company, Bayer AG are implemented at the U.S. subsidiary, Miles Inc. The underlying requirement is that throughout the world, subsidiaries are obligated to apply the same standards of environmental protection and safety as the parent. One complicating factor is the variations in the regulations governing process safety. It goes without saying that the local, state, and federal regulations must be observed.

3.   BAYER AG POLICY

The written environmental protection and safety policy of Bayer AG includes the following four basic principles:

- Comprehensive environmental protection and maximum safety that is reasonably achievable are given the same priority as high product quality and optimum commercial efficiency.

- Environmental protection and safety, the quality of our products and the economic interests of our company are equally important goals. Also, for economic reasons, safety will not be reduced. The measures deemed to be necessary on the basis of scientific knowledge are implemented in agreement with the authorities and the "Berufsgenossenschaft" (employers' liability insurance association) in Germany.

- Research into environmental protection serves not only the company and the chemical industry, it also helps deal with issues of general public interest.

- Subsidiaries throughout the world are obliged to apply the same standards of environmental protection and safety as Bayer AG.

This last basic principle is the most challenging one to apply in the process safety field. Typically, it requires Miles Inc. to investigate various alternative designs to ensure the same "level of safety" as that of Bayer AG is incorporated into the process design. At first glance one may think that this is easy to accomplish by copying Bayer AG's designs, but typically this cannot and should not be done due to a number of factors:

- Differences in the equipment and process controls available in the U.S.

- Variations in the regulations (e.g. U.S. Electrical Classification vs. German "Zones").

- Need to always consider best available technology, not just copying a previous design.

- Differing culture and the way it affects operation of a facility.

Due to these factors a concerted effort is made by Miles Inc. to review the process and ensure that the same high "level of safety" is incorporated in our designs.

More specifically, in the area of process safety Bayer AG's policy includes the following requirements:

- New processes are selected on the basis of maxized process safety as much as reasonably achievable; older processes are modified as appropriate.

- Technical installations are designed in such a way as to avoid hazards and reduce risks to an acceptable level.

- A safety concept which not only establishes the safe operating conditions but also provides for technical measures to overcome malfunctions and limit their consequences is developed for every process and every plant.

- Operational safety is maintained through careful servicing and regular inspection of the installations.

In the area of technology transfer from Bayer AG to the subsidiaries, the following policy statements set requirements on the exchange of know-how on production processes and units:

- Only well tried-and-tested technology is passed on to subsidiaries or third parties.

- The subsidiaries ensure that the measures for environmental protection as well as for employee and process safety are developed and applied according to the same principles as those of Bayer AG.

- Know-how from third parties relating to processes or technical installations is only adopted after thorough safety tests.

4.   APPLICATION

The application of the process safety policy at Bayer AG and Miles Inc. is detailed in handbooks for each.

The Bayer AG handbook is called the <u>Handbook, Process and Plant Safety</u> and the U.S. handbook is titled the <u>Miles Inc. Process Safety Management Program Manual</u>. At first glance, the two documents appear similar, yet different. The Miles Inc. manual follows the format and terminology of the OSHA 1910.119 Process Safety Management of Highly Hazardous Chemicals. The key requirements of Bayer AG are incorporated in the appropriate sections of the Miles Inc. manual. Obviously, some details from the Bayer AG program are deleted when transferring the requirements to Miles Inc. because they are not applicable in the U.S. For example, a safety analysis based upon the Stoerfall Verordnung (German Hazardous Incidents Regulation) is not performed in the United States. Secondly, the OSHA 1910.119

requirements are very specific in certain areas of process safety management and the incorporation of these details into the Miles Inc. Process Safety Management Program adds additional items which must be spelled out in the program procedures.

Looking specifically at the requirements for safety reviews of new capital projects, a four step review process is used at both Miles Inc. and Bayer AG (see Figure 1). Although the names of the various reviews are different, the overall methodology and timing is quite similar.

During the preliminary stages of the capital project, when the process design is being formulated, the hazards involved in the process and inherent in the chemicals and design are investigated and described. Where possible, inherently safer technology is employed through the substitution of less hazardous chemicals, lowering temperatures, pressures, inventories, etc. The process safety concepts for controlling the hazards that cannot be eliminated are formulated. As the life of the project continues, a process hazards analysis typically with the HAZOP method is performed, followed by a layout safety review and finally a pre-startup review prior to operation of the facility.

Process hazards analyses of existing process units are performed on a regular frequency at both Miles Inc. and Bayer AG. At Miles Inc. the frequency utilized is every five years and, typically, the HAZOP method is employed.

5. **VARIATIONS IN HAZOP METHOD**

At Miles Inc., we use the "classical" method for conducting HAZOP studies. The format used for note-recording is shown in Figure 2. This format is in common use within the U.S. and is one of the accepted methods listed in the OSHA 1910.119 Process Safety Management of Highly Hazardous Chemicals. When applicable, other methods of process hazards analysis are used at Miles Inc. For example, the what-if/checklist method is used for a process hazards analysis of a non-chemical process.

Bayer AG uses a more varied approach of process hazards analysis employing various methods for differing processes. Although the classical HAZOP is utilized, a more common method used is the "modified" HAZOP method. An example format for the modified HAZOP method is shown in Figure 3. The approach differs in that it is more deductive and adapts the scope of the guidewords based upon previously

FIGURE 1

COMPARISON OF PROCESS SAFETY REVIEW METHODOLOGY FOR CAPITAL EXPENDITURES

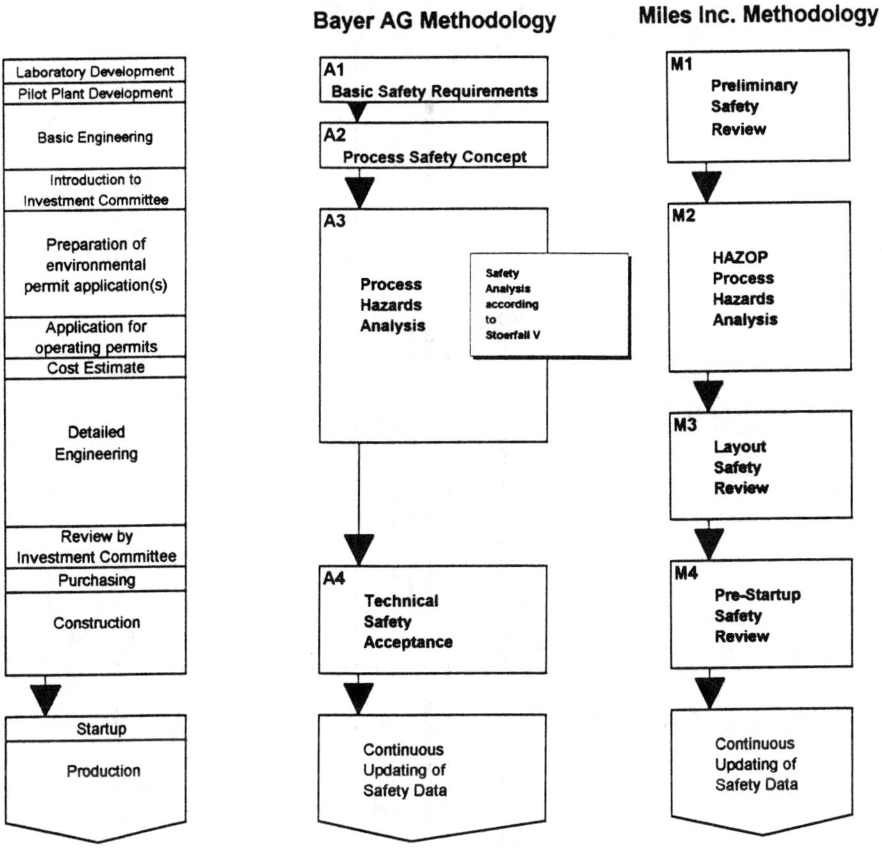

defined hazards to the details of the particular situation. With the "modified" HAZOP method, the major hazards and effects are discussed for each process stage. For each major hazard, the protection tasks are identified and the protection measures employed to solve the appropriate protection tasks are detailed.

The advantage of the "modified" HAZOP method is that less time is required for HAZOP meetings. In comparison, at Miles Inc. the HAZOP notes are recorded using a computer program during the HAZOP meetings, which requires less follow-up time.

Both Miles Inc. and Bayer AG have experienced a number of benefits from conducting these process hazards analyses:

FIGURE 2

Preliminary HAZOP Notes: Miles classical notes format   (03/24/93)

System	Section	Deviation	Possible Causes	Results of Malfunction	Exist. Protective Measures	Action Items	Responsibility
		[1]					

Hazard	Protection Task	Protection Measures	Design Details
1. Considerable damage of the combustion furnace due to explosion of flammable gas mixtures.	1. Avoid incorrect operation of the combustion process.	1.1 Flame monitoring	1.1.1 Flame monitoring (pilot and main measurement) (UV) 1.1.2 Alarm 1.1.3 Interlock turn off burning 1.1.4 Close natural gas valve and open vent valve 1.1.5 Close valves in off gas feed lines 1.1.6 Open feed lines to other destruction system
		1.2 Natural gas pressure monitoring	1.2.1 ... etc.
		1.3 Natural gas flow monitoring	1.3.1 ... etc.
		1.4 Combustion air flow monitoring	1.4.1 ... etc.
2.	2.	2.1	2.1.1 ... etc.
		2.2	2.2.1 ... etc.
		2.3	2.3.1 ... etc.

FIGURE 3: Modified HAZOP method. Process stage: Combustion.

- Rather than controlling high risks, high risks are reduced and when possible hazards are eliminated.

- Problems are most easily and inexpensively solved at the design stage rather than after startup.

- A deeper understanding of the process.

- Timely startup and faster attainment of design rates.

- Fewer engineering change orders at commissioning and startup.

6.   <u>TECHNOLOGY TRANSFER FROM BAYER AG TO MILES INC.</u>

Bayer AG attaches the utmost importance to the transfer of safe technology to Miles Inc. The transfer of safe technology takes place when new production processes are constructed with a detailed exchange of know-how between Bayer AG and Miles Inc. Miles Inc. is ultimately responsible for all safety matters, including the establishment of a mandatory process safety concept and design for the facility. Bayer AG provides Miles Inc. with the necessary assistance in all matters relating to safety.   The following basic principles apply:

- As a rule, the technology transferred to Miles Inc. are proven processes which have been tested in production. Transfer takes place only if the necessary safety infrastructure can be assured at the site.

- The extent of organizational and technical support given to Miles Inc. depends on the nature of the process and on local conditions and capabilities.

- The transfer of technology from Germany to the U.S. always takes place in accordance with U.S. legislation on safety and environmental protection. Federal, state and local laws and regulations relating to safety and environmental protection must always be observed.

- Subsidiaries operating chemical production facilities must have their own safety policy and organization. Technology transfer to subsidiaries is accompanied by concerns at the subsidiary for the safety of the process. Therefore, depending on the nature of the process transferred, Miles Inc. must determine whether the existing process safety design concepts are adequate or require modification.

- Miles Inc. is responsible for the ongoing development and consistent implementation of its process safety concept. Miles Inc. is responsible for the continuous improvement of its process safety concepts.

Considering all these responsibilities and requirements, there is ongoing communication and exchange of information between Miles Inc. and Bayer AG. The services and expertise of the specialists at Bayer AG are employed regularly and viewed as a valuable input to Miles Inc.

Applications of Risk Contours in Quantitative Risk Analyses

Marc Rothschild
Halliburton NUS

ABSTRACT

Quantitative risk analysis is an important tool in evaluating the acceptability of potentially hazardous facilities. Of the various ways to present risk, risk contours are perhaps the most misunderstood. As a result, risk contours are not utilized as completely as they could be. Risk contours, commonly used to depict the risk to individuals, are overlaid onto an area map, visually depicting the risk. While the risks given by these contours seem to be intuitive, risk contours actually present several possible interpretations. For example, even though risk is usually presented on an annualized basis, the risk contours may, in fact, be depicting the risk over a narrow time period, or even a peak risk. In addition, risk contours can present the risk to the average individual in the population, to a particular group of people, or to a particular individual. The correct interpretation of the risk contours depends on the selection of the study parameters.

It is important to clearly define the needs of the study and then establish the relevant project parameters to arrive at meaningful individual risk results. This paper presents discussion of the various potential interpretations of individual risk and how each can be used to better understand the overall risk. Discussion follows on how the parameters of a study can be selected to provide these various individual risk outcomes.

Introduction

Quantitative Risk Analysis (QRA) is slowly gaining importance worldwide as the preferred method to evaluate the safety of process facilities. In 1985, the Dutch Government was the first to introduce risk assessment into environmental policy, which was further clarified in "Premises for Risk Management, Dutch National Environmental Policy Plan, 1989" (Ref. 1). Other countries that are including

QRA's as part of their policy include the United Kingdom (Refs. 2-4) and Norway (Ref. 5). In the United States, the Nuclear Regulatory Commission has established qualitative safety goals supported by quantitative objectives for populations near nuclear power generating stations (Ref. 6). QRA combines the likelihood and the potential consequences of all conceived hazardous events into numeric terms, which can then be compared to societal (or governmental) standards. In addition, the results of a QRA can be used as a base case with which the relative effectiveness of several potential mitigation measures can be compared.

There are two classifications of risk, societal and individual risk. These two types of risk are complementary to each other. Societal risk evaluates the risk to society as a whole by taking into account the number of people who would be potentially impacted by a hazardous release. A process facility in a city would present a much greater societal risk than it would if it were in a more remote location. In contrast, individual risk does not take actual population data into account and, instead, evaluates the risk to a hypothetical individual. As a result, the above process facility would project the same individual risks regardless of its location. While societal risk is used to project the overall risk upon society, individual risk is used to assure that, in the process, no one individual would be burdened with an unacceptable risk. Although societal risk is very important in QRA, the focus of this paper is on the application of individual risk.

Individual risk is the probability at which an individual may be expected to sustain a given level of harm from the realization of a specified hazard while being exposed to that hazard. Although not mandatory, it has become common practice to take the level of harm as the risk of death and to express the risk on a per exposure-year basis.

Individual risk is location-specific, and can be graphically displayed as risk contours. Risk contours connect points of equivalent risk, as contour lines on topographical maps connect points of equivalent altitude. Since there is a wide range of potential release frequencies, risk contours often display order of magnitude individual risk. Figure 1 depicts hypothetical risk contours overlayed onto a "typical" offsite map for illustrative purposes. For example, Figure 1 shows that the risk to anyone in the vicinity of the school is about 10^{-5} fatalities/exposure-year (one fatality in 100,000 years).

While there may appear to be just one interpretation of risk contours, there are actually at least four. Risk contours can reflect an average level of risk over a long period of time, or a peak level of risk. In addition, risk contours can reflect the risk to the average individual, or the risk to a particular group of people. The type of risk contour that is produced is dependent on the study parameters.

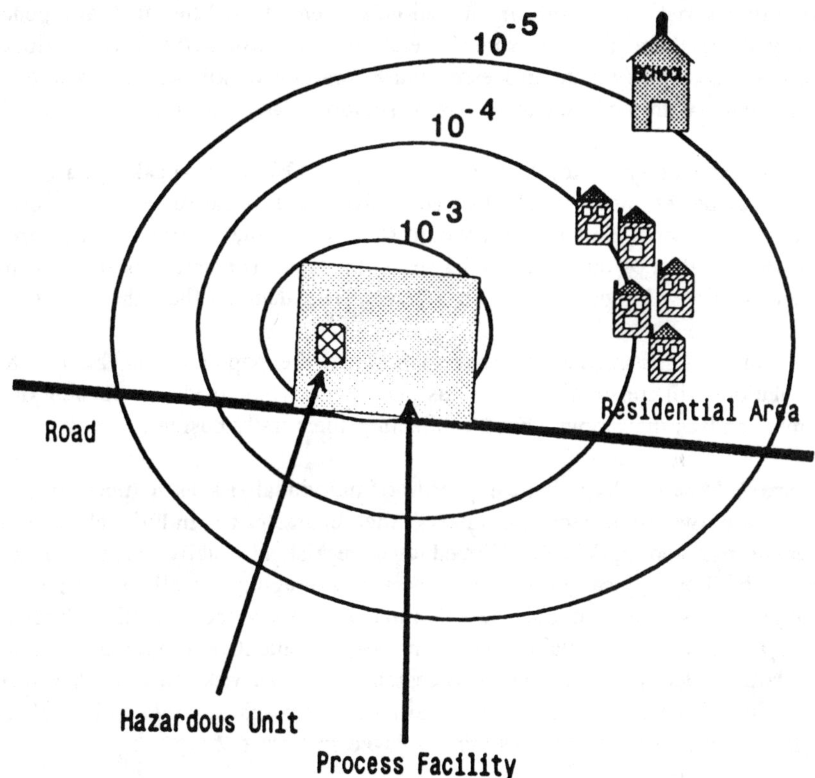

Figure 1. Hypothetical Risk Contours

Quantitative risk analysis, as a risk management tool, is still in its early stages. This paper points out one area, the application of individual risk, that requires further refinement. For those establishing risk policy, it is important to be aware of the possible interpretations of individual risk and to select those interpretations that are most appropriate. Regardless of which type of individual risk is selected, it is important that it is correctly interpreted. This paper provides discussion of the various types of individual risk and how they can be influenced by the study parameters.

Peak vs. Timed Average Risk

Individual risk is rarely a constant value and will fluctuate due to several factors. One factor is non-steady-state operations. Many processes take place batch-wise, exhibiting a risk only while in operation. Even in facilities that are generally steady-state, there may be some operations that cause the risk to fluctuate. Furthermore, steady-state processes must come down for periodic maintenance turnaround, with the associated risk of shutdown and startup.

Weather is another factor that influences risk. Generally speaking, a nighttime release disperses more slowly than an equivalent daytime release. Therefore, it would be expected that the risk would show daily fluctuations. Wind direction is another influence that weather has upon the risk. The areas located downwind of the facility will most likely face a greater risk than do the other areas.

One other significant factor is the mobility of the population. The risk to the population will constantly change as people go from inside to outside of their homes, and as they move about conducting their daily business.

Figure 2 depicts a hypothetical profile of individual risk as a function of time. There are two important parameters that characterize individual risk: timed average risk and peak risk. Timed average risk is usually taken over a long period of time, typically a one-year period, averaging out all of the peaks and troughs. This results in a single value that represents the overall individual risk from a facility. This value can then be conveniently compared with a risk standard to determine its overall acceptability. Peak risk, on the other hand, is the value of the momentary individual risk as it reaches a relatively high value. Timed average and peak risks are displayed in Figure 2.

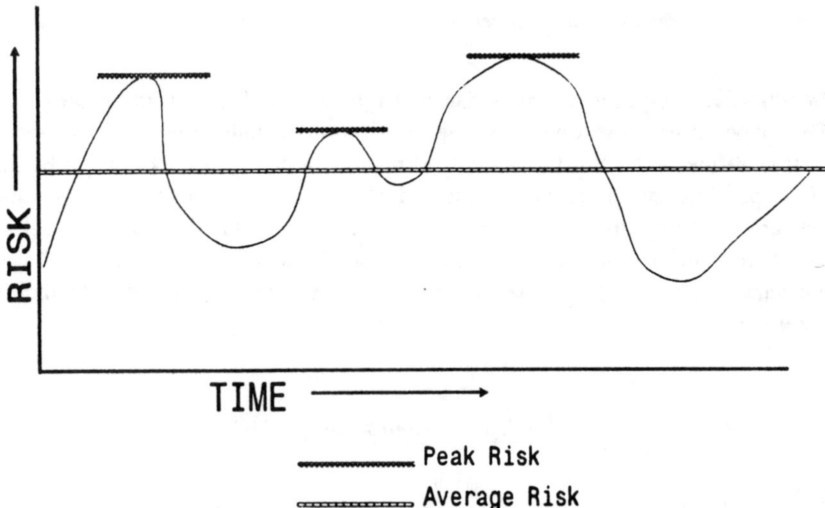

Figure 2. Fluctuations of Risk with Time

While much less frequently applied, peak risk can be very important in evaluating the acceptability of a facility, as some peak risks are deemed unacceptable, no matter how short the duration of exposure. For example, in comparison to not wearing seat belts at all, wearing seat belts *most* of the time substantially reduces the average level of risk, and wearing them *all* of the time only incrementally lowers the average further. Nevertheless, conscientious risk analysts would judge that the peak risks resulting from not wearing seat belts are unacceptable, and therefore always wear their seat belts.

Risk contours are frequently presented on an annualized basis. This, however, does not necessarily imply that the risk is evaluated on an annual timed average basis. In fact, the risk could be a peak risk, presented on an annual basis. For example, without knowing more about the study that produced Figure 1, it is impossible to discern whether the residential risk of 10^{-4} per year represents a timed average risk or a peak risk. The timed average risk would include time spent indoors and possibly away from the home, while the peak risk may be the risk to a resident as he ventures outside [1] to fetch his morning paper.

Timed average and peak risks are not interchangeable. For example, suppose two alternatives were being considered to mitigate the risk of a particularly dangerous operation. One alternative would be to increase safety measures to reduce the danger of the operation, while the other alternative would simply be to rotate in several operators to do the job. Both of these measures may adequately reduce the timed average risk. However, the latter option of "operator Russian roulette" does nothing to reduce the actual danger of the process. The reduced chance that any given operator will be injured has been balanced by the increase in the number of operators who will be exposed to this risk. The dilution of risk does not mitigate the risk.

Some residents rarely stray far from home and, for these people, their individual risk, given on an exposure per year basis, approximates their true risk on a calendar-year basis. This is contrasted with transportation risks where, owing to the very limited exposure time, the risk on a calendar-year basis is far less than the risk given on an exposure-year basis. Figure 1 shows that the risk level on the road approaches 10^{-3} per exposure-year. This peak risk is only imposed upon drivers for the limited time it takes to pass by the process facility and then their risk drops back down to much lower levels. The timed average risk could be estimated by reducing this peak risk by approximately two or three orders of

---

[1] It is tacitly assumed that the risks are greater outdoors than indoors. While this is largely true, it does not apply for explosions, where humans can tolerate high levels of overpressure, whereas structures cannot. Therefore, in situations where explosions create the greatest hazard, the reader is instructed to substitute "indoors" for "outdoors" for the remainder of the paper.

magnitude, thus accounting for the reduced amount of time spent on that stretch of road. Using the timed average risk, however, would negate the fact that, for a small period of time for each of the drivers, their risk is quite high. Neglecting this peak risk would be as illogical as lighting all but a short stretch of dark road for the reason that the time any one driver spends in that dark section would be small.

Average vs. Absolute Individual Risk

Every individual is unique and has a particular ability to tolerate and to escape the impact of a hazardous release. Heretofore, when discussing individual risk, the individual has yet to be defined. It is common practice to evaluate the risk to the average individual. In many applications, however, the risk to a particular group of people, or even to the most vulnerable individual, is of concern. This narrower application of individual risk is defined as absolute individual risk.

The issue of average individual vs. absolute individual risk is raised when evaluating the risk to residential areas. In Figure 1, the residential area abuts the plant fence line. One could argue that with the daily commute to work, shopping trips and other regular outings, the typical resident spends a considerable amount of time away from the home. As a result, their actual individual exposure level would be less than that indicated by the figure. While this would be true for the population at large, it may not be true for all individuals. The very young and the elderly tend to spend a great deal of their time at home and some may, in fact, be confined to the home.

In addition to the risk to individuals, absolute risk can apply to a group of people. For instance, due to their increased direct exposure, the individual risk to operators at a unit would be greater than the risks to the facility support staff.

With regard to industrial safety, most facilities take an absolute individual risk approach to safety. Certainly it would be unconscionable to have all employees use one-size-fits-all respirators, even if they do fit most of the work force. While the employees, on average, would be protected, some employees would not be adequately protected.

Establishing Study Parameters

Up to this point, the discussion has focused on the various possible interpretations of individual risk. This section provides a practical guide on how some of the study parameters can be adjusted to arrive at the desired form of individual risk. This is not necessarily a comprehensive list of all of the parameters that influence the risk contours. Nevertheless, the following four parameters will strongly influence the individual risk:

1. Personnel vulnerability
2. Personnel exposure
3. Mobility
4. Averaging time

Each of these parameters are discussed below, in turn.

Personnel vulnerability. Given a hazardous release, the QRA models determine the impact of the release onto the surrounding population. These models typically look at the impact from explosions, flash fires, pool fires, jet fires, and toxic gas releases. These models generally contain assumptions regarding the overall expected survival rate of people given various levels of exposure. These assumptions will then lead to the average individual risk. If, on the other hand, it was the objective to arrive at the absolute individual risk, then these vulnerability parameters would need to be adjusted to reflect the survivability rate of the weakest members of the population.

Risk analysis models often take into account whether a person is indoors or outdoors during the exposure period. With the exception of explosions, being indoors generally provides some protection against hazardous events. Depending on the climate and the general lifestyle of the local population, people typically spend about 90% of their time indoors. In order to evaluate the timed average individual risk to the population, it would be reasonable to take this high indoor fraction into account in the analysis. If, however the peak individual risk were evaluated, then it would be best to assume a 100% outdoor population. This does not imply that the population is always outdoors, which is highly unlikely for any individual. Instead, this can be interpreted as the risk to a person while outdoors, presented on an annualized basis.

Personnel exposure. The exposure parameters must reflect the population of concern. When evaluating the onsite risk due to pool fires, the company employees, on average, may spend 80% of the time inside, while the outside operators may spend 80% of the time outside. One set of parameters cannot reflect both of these groups of employees. If peak risk were a concern, then it should be assumed that the personnel are 100% outdoors, as that is where potential exposure to thermal radiation is greatest.

Mobility. As previously discussed, the actual mobility of the population will influence the risk contours. The movement of the population is not generally a study parameter. However, the effect of a mobile population can be accounted for by multiplying the risk by the fraction of time spent at each location. One example, which was already presented, was scaling down the individual risk to drivers on nearby roads to account for their short exposure durations. It has also been suggested that the risk to nearby residential areas should be attenuated to take into account that people spend much of their time at work, shopping, or otherwise away from the home. Both of these approaches would be correct if the desired outcome were a time averaged risk. For peak risks, however, the risks must be evaluated at maximum conditions.

Averaging time. The time averaging period selected for the study will also determine the type of individual risk that is derived. It is common practice to use yearly average operational and meteorological data to derive the risk. This, of course, will result in an annual average risk. By evaluating the risk over a year, it is presumed that any fluctuation to the risk, such as operational variances, weather changes and population movements will be accounted for. In that case, this analysis will calculate the average risk level to the surrounding population. This risk, however, may disguise the fact that some operations, such as the transferring of a highly toxic material, may be far more hazardous than others. If this operation is only rarely conducted, then it may not contribute significantly to the overall risk and, therefore, would not be flagged for consideration for risk mitigation measures. Nevertheless, this does not diminish its danger. By choosing the time period to coincide with this operation, and thus evaluate the peak risk, the relatively high risk of this operation over others becomes apparent.

As another example, if there were particular concern regarding the risk at a nearby school, then selecting the timeframe to coincide with school hours will provide the risk to the students, while they are at school, presented on an annualized basis. Table 1 summarizes how the above parameters can influence the interpretation of the risk.

Note that any other combination of the four parameters other than those given in the above table will lack interpretative meaning. For example, it is common practice to use yearly average data in evaluating the risk of a facility, while modeling the population as if they were always exposed. With possibly rare exceptions, it is a fair statement that all people spend some time indoors. Therefore, the modeling of 100% outdoors clearly does not represent even the most vulnerable in the population. It does, however, most likely represent a peak level of exposure, as the outdoor risk is generally greater than the indoor risk. However, by using annual average data, any other contributors to the peak risk from such factors resulting from non-steady-state operations or from adverse weather conditions would be averaged out in the analysis. As a result, the risk contours from this study would neither reflect the average risk nor the peak risk to the citizenry.

Table 1
Correspondence of Study Parameters to Individual Risk

	Parameters	Time Averaged Risk	Peak Risk
Average Individual Risk	Personnel vulnerability	Average population vulnerability.	Average population vulnerability.
	Personnel exposure	Actual population exposure.	Population assumed to be exposed to maximum potential risk.
	Mobility	Mobility of population accounted for.	Population is modeled as immobile.
	Averaging time	Long averaging time used to cover all potential variations (yearly is most common).	Limited averaging time selected to cover the particular interest of concern (e.g., batch process, risk to school children, etc.)
Absolute Individual Risk	Personnel vulnerability	Vulnerability representing weakest population.	Vulnerability representing weakest population.
	Personnel exposure	Actual population exposure.	Population assumed. to be exposed to maximum potential risk.
	Mobility	Mobility of population accounted for in transportation risk, population assumed to be stationary in residential environments.	Population is modeled as immobile.
	Averaging time	Long averaging time used to cover all potential variations (yearly is most common).	Limited averaging time selected to cover the particular interest of concern (e.g., batch process, risk to school children, etc.)

Conclusions

Individual risk can be interpreted several different ways. Individual risk can reflect the risk averaged over a period of time, or the risk at any given moment. Furthermore, individual risk can present the risk to the average person or to any unique individual or group of people. Each of these types of individual risk could vary considerably from the others. Therefore, in establishing individual risk standards, it is important to define what type of individual risks are to be evaluated. When evaluating risks, it is important to be aware how the study parameters affect the individual risk contours, so that they are correctly interpreted.

Although the selection of the most appropriate type of individual risk is dependent on the particular application, some general observations can be made. Timed average risk provides an overall representative measurement of the individual risk and is therefore useful for comparison to a standard. While it is desirable to keep the average risk below this standard, peak risk cannot be neglected, as high peak risks may also be unacceptable. Since timed average risk and peak risk are not mutually exclusive, it would be reasonable for risk rules or regulations to include a dual individual risk standard: one for timed average risk and one for peak risk.

The choice between average and absolute individual risk is not as straightforward. The intent of individual risk is often to assure that the safety of each individual is maintained, while managing societal risk. Following this through to the logical conclusion would indicate that the risk to the most vulnerable person must be protected. While this is fine in theory, it is fraught with practical problems. First, it must be determined who this vulnerable person is. Then, the study parameters must be developed specifically for this person, such as his (or her) vulnerability to toxic exposure and thermal radiation effects, and ability to take defensive measures in a given incident. The mobility of this person must also be defined. One can see that the task to evaluate the risk to the most vulnerable person would be monumental. In contrast, the same data for a "typical" person has been well developed and is readily accessible.

While evaluating absolute individual risk for the most vulnerable person is a daunting task, it may clearly be the appropriate course of action when it is applied to a specific group of people. One example discussed in this paper is the risk to process operators. This group of individuals will most likely have similar characteristics, and can be defined with one set of study parameters. Due to the training and work environment of these operators, their individual risks will be different from the general population and even from non-operations employees. Therefore, the individual risk for this particular group of people should be evaluated separately from the risk to other populations.

Managing risk is an important component of a process safety management program. This paper has provided discussion of several types of individual risk, along with suggestions on how they can be appropriately applied; first by determining the individual risks that are desired, and then carefully selecting the parameters that will provide that outcome. Taking this approach will result in greater understanding of the risks from industrial facilities, which will improve safety decision making.

REFERENCES

1. Dutch National Environmental Policy Plan (1989), *Premises for Risk Management*, Second Chamber of the States General, session 1988-89, 21137, no. 5.

2. HSE (1988). *The Tolerability of Risk from Nuclear Power Stations*, Health and Safety Executive, Her Majesty's Stationery Office.

3. HSE (1989). *Risk Criteria for Land-Use Planning in the Vicinity of Major Industrial Hazards*, Health & Safety Executive, Her Majesty's Stationery Office.

4. HSE (1989). *Quantified Risk Assessment: Its Input to Decision Making*, Health and Safety Executive, Her Majesty's Stationery Office.

5. Norwegian Petroleum Directorate, Oslo (1990). *Regulations Concerning Implementation and Use of Risk Analysis in the Petroleum Activities, with Guidelines*.

6. Federal Safety Standards, 10 CFR Part 50 (1986). *Safety Goals for the Operation of Nuclear Power Plants*, Federal Register.

Process Safety: Life-Cycle Quality Control during Design and Operations

R. Ellis Knowlton
Chemetics International Company Ltd.,
1818 Cornwall Avenue, Vancouver, B.C. V6J 1C7

EXECUTIVE SUMMARY

Process Safety is an on-going activity involving every facility within each company in the Process Industries. However, the methods and management of Process Safety can be illustrated by considering a single facility throughout its life-cycle. For every facility, Process Safety begins at the initial concept, the so-called "Preliminary Appraisal" stage. It continues through design, construction, commissioning, normal operation, maintenance, modifications, accidents and abnormal occurrences, to the final taking out of service.

At each stage there is a hierarchy of checks carried out both within a corporation and externally by regulators and insurers, which ensure the safety and environmental protection of the operators, the general public, the unit and the environment. This external hierarchy extends upwards to include national codes, code-writing bodies and process safety coordinating committees such as the Dutch Process Safety Committee.

This paper uses a matrix presentation of the methods and management of Process Safety with the sequential development and operation of the facility shown horizontally and the hierarchy of control and management systems, vertically.

The various cells of the matrix interact, with in general, information flowing up and control flowing down. For an example, consider a basic activity such as the production of a P&ID. There are primary safety checks on a skills basis. Above these in the hierarchy, are secondary checks such as a Hazop. These secondary checks can be subjected to a tertiary level of control such as a Corporate Audit on the quality of the Hazop. Further controls in the checking of the P&ID are exercised outside the corporation. For example, evidence of the secondary checks may be required by permit -giving authorities and insurance organizations.

In addition to these "feed down" controls, there are "feed forward" systems to influence future stages in the project and "feed back" systems to improve the quality of earlier stages in future projects.

The aim of the paper is to give in a succinct form, the control and management architecture. This matrix model has been used to illustrate the requirements of Hydrogen Safety in Canada and of Process Safety in Taiwan, India, China and Thailand. An analogy will be drawn between Process Safety and the Quality Assurance requirements for consumable products such as pharmaceuticals.

1.   INTRODUCTION

The Process Industries have been a vital component of the industrial base of Europe and North America since the beginning of the twentieth century. The progressive industrialization of many other parts of the world has been accompanied by a corresponding spread of the Process Industries. As the scale and complexity of operations has increased, so has the potential for disasters to people, property and the environment. This potential has been demonstrated with tragic results on a number of occasions.

Initially, the only method available to avoid accidents was experience, either individually or collectively, as expressed in codes and standards. Although experience is a valuable form of feedback, it is expensive in terms of human suffering and financial loss. Experience has therefore been supplemented by two further approaches. One approach has been to improve the engineering knowledge base so that designers can design a unit "right first time". The other, complementary approach has been to develop a variety of methods of identifying and assessing hazards so that they may be eliminated, controlled or contained. These methods are collectively known as "Process Safety" and their managerial and regulatory control is known as "Process Safety Management". Both the methods and the management systems have evolved to meet the needs of the Process Industries.

If a new industry were to be developed with similarities to the Process Industries, the equivalent to Process Safety could follow the same path and evolve by experience. Alternatively, the safety aspects of the new industry could develop more rapidly by making use of the accumulated experience of the Process Industries.

This situation has occurred in practice with the nascent industry of "Hydrogen Energy". Hydrogen Energy is based on the use of hydrogen as a secondary form of energy or "energy currency" in the same way that electrical energy is a secondary form of energy. Although hydrogen has a number of desirable characteristics, it does have its own distinctive safety problems which would need to be dealt with in a manner similar to the Process Industries. There is thus a concept of "Hydrogen Safety" which is analogous to Process Safety.

In the early 1980's the National Research Council of Canada initiated a project called "Hydrogen Safety". The aim of this project was to identify the problems associated with the extended use of hydrogen as a ground transportation fuel. Chemetics was awarded the contract for this project because of Chemetics experience in Process Safety, together with its background knowledge of hydrogen. Chemetics identified a number of problems which would need to be addressed, including the issue of how Hydrogen Safety was to be managed.

As part of the broad issue of the "Management of Hydrogen Safety", an overview was produced which showed the architecture of a management control system. The overview was based on the life-cycle of a single project to design, build, operate and finally shut down a facility, together with the hierarchy of checks both inside a company and outside by regulatory and insurance organizations to ensure the facility was safe and environmentally benign.

Although the overview was based on a single facility, the principles could be generalized to cover all the facilities within a company and for all companies within a country. This overview was a small part of a consultancy project. The final report is given in Ref. 1.

Although originally developed for Hydrogen Energy, the format for the overview has been found to be useful in discussing the "Executive Management of Process Safety" in seminars in Taipei, Beijing, Shanghai, Bangkok and Madras. Some elements of this overview have been mentioned in papers presented in London and Denver (Refs. 2 and 3) but this is the first time the overview has been the subject of a paper in its own right.

The aim is to give a simple model which can help answer:

- What is Process Safety?
- How is it managed?
- Are there opportunities for further improvements?
- Is there synergy between Process Safety and Quality Assurance?

The paper will start with chapters on the project sequence and the management hierarchy before describing the matrix model itself.

2.   THE PROJECT SEQUENCE

Typically, the need to create a facility is market driven. There is a need to build a plant to make a product, or a range of products. The methods of arriving at the decision to proceed are outside the scope of this paper. However, the project requires the following sequential steps:

- Initiation and design
- Engineering implementation
- Personnel and plant control
- Commissioning
- Operation
- Closure

Most of the above can be sub-divided to give the following 15 stages.

Initiation and Design (Stages 1, 2 and 3)

1.   <u>Preliminary Appraisal</u>

At this stage, only the nature of the product is know, i.e. chemical entity or formulation. The means of production have yet to be decided and no production site has been chosen. This stage will be relatively brief with commodity chemicals, but it could be lengthy in the fine chemicals field, particularly when a novel product is considered.

2.   <u>Project Definition</u>

At this stage, the main outline of the production process has been decided. A site has been chosen and a block layout has been prepared for the main items of equipment.

3.   <u>Engineering Design</u>

Detailed design leading to Piping and Instrument Drawings (P&IDs), tentative operating methods and control logic.

Engineering Implementation (Stages 4 and 5)

4.   <u>Procurement and Fabrication</u>

5.   <u>Construction and Assembly</u>

Plant Control and Personnel (Stages 6 and 7)

6.   <u>Detailed Operating Instructions and Software</u>

7.   <u>Personnel Recruitment and Training</u>

Commissioning (Stages 8, 9 and 10)

8.   <u>Pre-Commissioning Trials</u>

Water trials, machinery tests, etc.

9.   <u>Commissioning</u>

Starting up the unit.

10.   <u>Post Commissioning</u>

Steady running is established and the startup team withdrawn.

Operations (Stages 11, 12, 13 and 14)

11.   <u>Normal Production</u>

12.   <u>Routine Maintenance</u>

13.   <u>Incidents</u>

14.   <u>Modifications</u>

Closure (Stage 15)

15.   <u>Final Shutdown</u>

Clean up, dismantling and landscaping.

3   THE HIERARCHY OF CHECKS AND CONTROLS

The sequence of activities described in the previous chapter will need to be managed to make sure that the quality of the work will meet the increasingly high standards of safety and environmental protection demanded by society. The management of a single project can be generalized to the management of all projects within a company and for all companies within a country or a group of countries.

Safety and Environmental Protection start at the operational level with good design, sound engineering implementation and high standards of performance in production and maintenance. In order to make sure that high standards are maintained and improvements initiated, when necessary, there is a hierarchy of checks and control activities. These are shown as consisting of three levels of checking within a corporation and a further five levels which are shown as outside. The division is between inside and outside made partly on the grounds of simplicity, because some of the higher level activities shown as occurring outside the corporation also have their counterparts inside.

These eight levels move progressively from the skill and task-specific type of check to broader and more policy oriented controls. The operational level is designated row "0" and the checking levels are shown as rows which are identified by Roman Numerals, starting with the row closest to the operational level.

PRIMARY CHECKS LEVEL I

These are carried out on a skills basis and are closely linked to the operational level. For example, a mechanical engineer will check the strength of a proposed vessel against the anticipated maximum pressures.

SECONDARY CHECKS LEVEL II

These are usually carried out by multi-disciplinary teams, at meetings specifically designed for this purpose and which are outside the normal design and operating activities. Two examples are, a Guide Word Hazop Study at the "design freeze" stage, or a Safety Audit of a production unit.

CORPORATE CHECKS LEVEL III

These are checks which ensure that the Primary and Secondary checks have been carried out to the required standard. For example, a company could "audit" a Hazop Study. The checks at this level, ensure that the corporate policies are being followed and that corporate safety objectives will be achieved.

EXTERNAL CHECKS

These external checks are made to ensure that national standards are being met, that there are no undue financial risks and that improvements are initiated when necessary.

STATUTORY APPROVAL & INSURANCE APPROVAL LEVEL IV

This is the first level of contact between a corporation and the "outside world" with respect to a particular project. For example, an outline "planning permit" could be required for a new facility.

REGISTRATION LEVEL V

It would be very expensive and time consuming for every individual proposal to be checked for every aspect and for every organization which could be involved. Therefore, systems have been developed, which enable certain designs and certain organizations to be "registered". Such registration can be obtained by a suitable audit together with demonstrated satisfactory performance. For example, there are registers of "approved designs" and "approved fabricators".

CODES LEVEL VI

The combination of collective experience and theoretical knowledge, has led to the development of "codes of good practice" for designs and operations. Codes are more general than the registration of specific designs. Examples are "vessel codes" and "codes of good practice for construction".

CODE IMPROVEMENTS LEVEL VII

Experience, the advances in theoretical knowledge and the development of new areas of business will, from time to time, show the need improvements to existing codes and even the creation of new codes. This has led to the establishment of code-improvement bodies, e.g. code writers for piping codes.

INDUSTRY-WIDE PROCESS SAFETY COMMITTEES LEVEL VIII

The requirements of Process Safety and Environmental Protection, have shown the need for a forum in which broad problems can be addressed by representatives from a number of organizations. Typically, such committees are formed with members drawn from industry, academia and government regulators. These bodies are often advisory and operate on the collective reputations of the members. An example is the Dutch Process Safety (RIVEPRO) Committee. The Canadian Major Industrial Accident Co-Ordinating Committee (MIACC) is a similar body.

CORPORATE CODES AND CORPORATE COMMITTEES

Although the registration of designs, codes and safety committees have been described as activities outside an individual company, it is usual for larger companies to have their own standard designs, internal codes and policy making committees. These activities perform similar functions to their external counterparts. They have been omitted solely to simplify the description.

4   A MATRIX REPRESENTATION OF PROCESS SAFETY MANAGEMENT

One way of combining the time sequence of a project and the hierarchy of checks and controls, is to display these two variables in the form of a matrix, with "time sequence" flowing from left to right and the "hierarchy" of controls shown vertically. Within the cells of the matrix thus formed, it is possible to give an outline of the nature of the checks within the hierarchy which are appropriate to each stage.

The full matrix is large with nine rows and 15 columns. It will be displayed as a "poster" at the Conference. In order to reduce the size to that suitable for publication, it is shown in summary form in Figure 1. The details are then given in separate parts in Figures 2-7. These figures are shown with the rows lined up so that the complete picture could be formed by joining the Figures together.

The original matrix mentioned in Reference 1 was even larger, with 15 rows and 19 columns.

FIGURE 1
SUMMARY MATRIX

Row No.	Figure Numbers						Checking & Control Activities
	2	3	4	5	6	7	
VIII							Industry Process Safety Coordination
VII							Code Revision
VI							Codes & Collected Data
V							Registration & Statutory Files
IV							Statutory & Insurance Approval
III							Corporate Checks
II							Secondary Checks
I							Primary Checks
0							Ongoing Design & Operations
Col. No.	1 2 3	4 5	6 7	8 9 10	11 12 13 14	15	

FIGURE 2
INITIATION AND DESIGN

Rows	Preliminary Appraisal	Project Definition	Detailed Design
VIII	Overall Safety Coordination		
VII	Code - Writing Bodies		
VI	Material Codes	Spacing & Layout Codes	Functional Engineering Codes
V	Material Data	Planning Files. Insurance Records	Registers of Designs & Emergency Procedures
IV	Statutory & Insurance Checks on Materials	Zoning & Insurance Risk Management	Permitting Controls, Insurance Evaluations
III	Corporate Audit on Materials	Corporate Audit to Maintain Policy	Audit on Quality of Identification & Assessments
II	PHA on Materials e.g. Checklist Hazop	PHA on Siting, e.g. Checklist Hazop on Layout	Design Stage Hazard Identification & Assessment, e.g. Guideword Hazop & Quantified Analysis
I	Material Properties, e.g. Corrosion, Toxicity, etc.	Planning, Spacing, Effluent Checks & Environmental Checks	Material Codes Vessel & Equipment Checks
0	Analysis of Requirements & Options	Siting, Process Selection & Block Layouts	Materials, Equipment, P&IDs, Operating Methods & Instrumentation
Col.	1	2	3

FIGURE 3
ENGINEERING IMPLEMENTATION

Rows	Procurement & Fabrication	Construction & Assembly
VIII	Overall Process Safety Coordination	
VII	Code Writing Bodies	
VI	Codes of Good Practice for Fabricators	Construction Codes
V	Register of Approved Fabricators	Register of Approved Constructors
IV	External Inspection of Fabricators	External Inspection of Construction
III	Corporate Audit of Manufacturers	Corporate Audit of Construction Safety
II	Audit of Fabricators Quality Assurance by Clients' Q.A.	On Site Inspection by Client or Client's Appointees
I	Manufacturers & Fabricators Internal Quality Assurance	Checking by Constructors' Inspectorates
0	Purchase of Standard Components & Fabrication of Equipment	Site Preparation, Construction of Plant, Piping & Assembly
Col.	4	5

FIGURE 4
PLANT CONTROL AND PERSONNEL

Rows	Operating Instructions & Software	Personnel Recruitment & Training
VIII	Overall Process Safety Coordination	
VII	Code Writing Bodies	
VI	Codes for Production of Software & Instructions	Codes for Operator Selection & Training
V	Register of Software Producers & Instruction Writers	Register of Trainers
IV	Inspectorate for Software & Instructions	Training Inspectorate
III	Corporate Audit of Secondary Checks	Corporate Audit of Training
II	Validation of Operating Methods, e.g. Hazop & "C Hazop"	Internal, Operator Examinations
I	Review of Instructions "Program Validation" of Software	Operator Testing on Tasks
0	Preparation of Operating & Maintenance Instructions. Computer Programs	Manning Requirements Recruiting & Training
Col.	6	7

FIGURE 5
COMMISSIONING

Rows	Precommissioning	Commissioning	Post Commissioning
VIII	Overall Safety Coordination		
VII	Code Writing Bodies		
VI	Startup Codes	Commissioning Codes	"Sign Off" Codes
V	Statutory Insurance Files	Statutory & Insurance Reports	Registration of Production Unit
IV	Permit to Start Manufacture	On Site Inspection	Permit to Continue Manufacture
III	Corporate Audits		
II	Pre Startup Hazard Identification	Safety Audit	Hazard Identification of Operational Plant
I	Conformance of Plant to Design	Conformance to Commissioning Procedures	Conformance to Revised Operating Methods
0	Pre Startup Checks, e.g. Water Trials	Startup	Post Startup Checks. Revisions to Drawings & Instructions
Col.	8	9	10

FIGURE 6
ONGOING OPERATIONS

Rows	Normal Operation	Routine Maintenance	Accidents & Abnormal Occurrences	Modifications
VIII	Overall Safety Coordination			
VII	Code Writing Bodies			
VI	Codes for Production	Codes for Maintenance	Codes for Emergencies	Codes for Modifications
V	Statutory & Insurance Files	Register of Maintenance Methods & Contractors	Statutory & Insurance Files	Statutory & Insurance Files
IV	Routine Inspection	Routine Inspection	Statutory & Insurance Investigation	Permits to Modify
III	Corporate Audit	Corporate Audit	Corporate Investigation	Corporate Audit
II	Safety Audits Hazop or Similar	Safety Checks on Preparation, Execution & Recommissioning	Formal Accident Investigation	"Control of Change" Checks
I	Ongoing Supervision	Permit to Work, etc.	Containment & Restoration	Review of Change & Permit to Work
0	Production According to Instructions	Maintenance to Agreed Procedures	Accidents & Abnormal Occurrences	Execution of Change
Col.	11	12	13	14

FIGURE 7
CLOSURE

Rows	Final Shutdown	Checking & Control Activities
VIII	Overall Safety Coordination	Industry Wide Process Safety Coordination
VII	Code Writing Bodies	Code Revision
VI	Codes for Shutdown & Clean Up	Codes & Collected Data
V	Permit to Cease Operation	Registration & Statutory Files
IV	Final Inspection & Environmental Audit	Statutory & Insurance Approval
III	Corporate Audit	Corporate Checks
II	Hazard Identification of Dismantling	Secondary Checks
I	Permit to Work, Etc.	Primary Checks
0	Final Shutdown, Cleanout, Dismantling & Landscaping	Ongoing Design & Operations
Col.	15	

One of the aims of this matrix is to give an indication of the types of checks and controls which could be carried out at different stages and a different levels inside and outside a company. Thus a system can be checked for comprehensive coverage. However, the individual cells are only part of control architecture. In general, informarion and instructions have to flow from one cell to another. In some circumstances, the flow pattern can be extremely complex. Thus a complete system will require not only all the cells, but also all the necessary flow patterns.

To illustrate the types of flows that are possible, these are discussed as being "feed down" and "feed forward" controls as a project proceeds, together with "feed-back" controls, which can operate when corrective action is necessary. These are discussed in the following sections.

5.   FEED DOWN AND FEED FORWARD CONTROLS

As a simple example, consider a project which is at the detailed design stage and a designer is working on a single piece of equipment, say a vessel. This activity is within cell 0,3. The design of the vessel is partly controlled by the vessel code (cell VI,3). It could be a registered design (cell V,3). The design would be checked for adequate performance by the primary checking (cell I,3). The complete design, including the vessel would be subjected to the secondary checking by, say a Guide Word Hazop (cell II, 3). Any unusual deviations affecting the vessel could be identified leading to corrections being fed down to the basic design (cell 0,3). As part of the overall approval process by statutory and insurance bodies, the vessel could be checked again by outside bodies (cell IV,3).

In addition to the above feed-down controls, the vessel would be subjected to feed forward controls. The vessel would be fabricated (cell 0,4), and the fabricator's quality assurance would check the fabrication (cell I,4). The fabricator's quality assurance itself could be audited (cell II,4) and, in turn, this audit may be subjected to a corporate audit (cell III,4). The fabrication could be executed by a fabricator which could be registered as such by outside bodies (cell V,4).

The operation of the vessel will be controlled by operating instructions and may require special operator training, Column 6. These instructions and training could be specified by the vessel code (cell VI,3) and implemented at Column 6. Instruction and training may possibly be amplified by the secondary safety checks on the design (cell II,3). The vessel would be installed (cell 0,5) and the installation checked (cells I,5 and II,5).

The normal operation of the vessel (cell 0,11) could be checked by a safety audit (cell II,11). Maintenance and modifications (cells 0,12 and 0,14) could be carried out and checked by cells II,12 and II,14.

Even this simple example can be extended. For example, the process fluid could have been found to have unusual corrosion problems at the preliminary appraisal stage, say cell II, 1 which could modify the design and the location of the vessel changed to reduce hazards at the project definition stage (cell II,2).

6.   FEED BACK CONTROLS

In addition to the feed down and feed forward controls, there are also feed back controls. Continuing the example of the vessel, let us suppose that despite all the checks and controls, there is an accident or an abnormal occurrence involving the vessel (cell 0,13). This would be followed by an investigation (cell II,13), and steps would be taken to prevent an occurrence. The results of such an investigation could influence many of the earlier cells.

For example, incorrect operation or maintenance would feed back to the checking procedures for these activities, Columns 11 and 12. Incorrect fabrication would feed back to Column 4 and result in a strengthening of the audit of a fabricator's quality assurance (cell II,4).

The accident could have been caused because the initiating deviation was not identified during the checking at the design stage (cell II,3), or that the deviation was correctly identified, but the resultant design change was inadequate (cell 0,3).

There may be circumstances which were not adequately covered by the vessels code (cell VI,3) leading to suggestions for an improved code (cell VII,3). Finally, if such failures were starting to have a significant impact in Process Safety within a country, the policy making Process Safety Committee may become involved Row VIII.

7.   THE MATRIX AS AN AID TO PROCESS SAFETY MANAGEMENT

As mentioned earlier, although the original purpose in developing the matrix was to help define the role of the Canadian Hydrogen Safety Committee, it has also been found to be useful as a means of showing "the big picture" of Process Safety Management as a single illustration.

Management will already know the sequence of activities necessary to create and operate a new facility. It is also highly likely that all the "Primary" checks (Row I) are already in place, as these are an integral part of the individual skills of the engineers, scientists and other professionals involved in the creation and operation of the facility. The main use of the matrix is likely to be as a means by which management can assess its own programs from "secondary" checks upwards, and to show the relationships between the internal and external checking systems. It is a means of determining whether the programs are "comprehensive", i.e. all relevant cells are in operation and "integrated", i.e. all the relevant flows of information and instructions between the cells are taking place. It is not suggested that every cell is relevant for every organization, but that any omissions are by design and not by default.

In the more progressive companies, it is unlikely that any "gaps" will be found although the matrix format itself may be unfamiliar. However, a few companies may find some gaps. The following are a few examples where further action may be desirable.

In column 4, "Procurement and Fabrication", there is a secondary check, cell II,4, "Audit of Fabricators' Quality Assurance" and a corporate cell, cell III,4 to ensure the effectiveness of this secondary check. Recent experience in the Process Industries concerning a small but vital component of piping, seems to indicate that these secondary and corporate checks may need to be pursued more intensively.

The secondary checks such as a "Guide Word Hazop" during the design stage, cell II,3 are becoming very common and in an increasing number of countries, they are mandatory. However, a "Preliminary Hazard Analysis" using "Creative Checklist Hazop" or similar procedures at the "Preliminary Appraisal" and "Project Definition" stages, cells II,1 and II,2 are less common. Such checks could identify potential environmental and plant interaction problems at an early stage, resulting in smaller and less expensive design changes being required later in the design process.

One application which could benefit a crowded site is the use of a "Creative Checklist Hazop" at the Project Definition stage, cell II,2 to identify not only safety hazards during the construction phase, Column 5, but also identify any disruptions to "plant operability", which could be caused by the construction activity itself. Such a study has been found to be useful as an aid to site selection.

Accidents and abnormal occurrences are investigated promptly and thoroughly and the results are fed back into the system to prevent a recurrence. However, the secondary hazard identification checks carried out during design or operations (cells II,3 and II,11), often generate "pseudo accidents", i.e. accidents which have yet to happen. The feed back of information from such pseudo accidents can be almost as valuable as the feed back from accidents which have actually happened. This feature is likely to be of most value when novel technology is involved and experience of actual accidents is limited. This form of feed-back was suggested for Hydrogen Energy and could be useful in an emerging technology such as genetic engineering.

Finally, although it is possible to employ established secondary checks during "Final Shutdown", cell II,15, it is likely that little use is made of this option.

The aim of this chapter is to provide management with a few ideas to help decide:

i)   Whether the existing programs cover all the activities necessary within their own organization for effective Life Cycle Quality Control During Design and Operation for all projects.

ii)   Whether the information flows to and from these activities are adequate.

iii)   Whether there are any opportunities for improvements.

Process Safety Management programs are an "overlay" which ensure that the necessary activities and information flows take place. It is hoped this matrix will be a useful substrate for such an overlay.

8. PROCESS SAFETY AND QUALITY INSURANCE

The life cycle of a production unit, together with the requirements of process safety, has some similarity to the life cycle of a consumable product such as a pharmaceutical drug and the quality assurance during its life cycle. The time-scales of operation are very different; it takes many years of R&D to launch a drug. However, the various stages and checks have certain similarities.

For example the R&D stages of a drug are closely controlled by many requirements, including "good laboratory practice," which has some similarities to the primary and secondary safety checks up to and including engineering design.

The final stages of "commissioning" of a drug will be governed by good clinical trials practice and the requirements of a controlled release.

Ongoing use will be controlled by good manufacturing practice, a vigorous quality assurance system to ensure consistent quality of the product, and a vigorous reporting system for any adverse effects—the equivalent of supervision during production and the reporting of accidents and abnormal occurrences.

Not surprisingly, some of the methodologies of process safety can be applied to quality assurance. As examples, Hazop was used in the following two quality assurance applications:

1. To identify the hazards to the quality of x-ray film.
2. To identify any hazards to the quality of the toxicity data generated during the drug development phase.

Conversely, the quality assurance requirements during "fabrication," although part of process safety, are also part of the technology of quality assurance in general. There are, no doubt, many other examples of "crossovers" in both directions.

In conclusion, both process safety and quality assurance are components of what could be called "systems integrity," that is, the means by which any system can be tested and have its weaknesses identified and removed.

REFERENCES

1.　An Assessment of the Hazards Associated with the use of Hydrogen as a Ground Transportation Fuel. National Research Council of Canada 1983. Final Report, Project 07SX 31155-1-6622.

2.　Avoiding the Process Safety "Management Gap". R.E. Knowlton. I Chem E Symposium Series 110.

3.　Dealing with the Process Safety "Management Gap". Plant/Operations Progress (Vol. 9 No. 2) April 1990.

Using Feedback to Improve Your PSM

Mark Paradies
Ted Light
Linda Unger
System Improvements, Inc.
238 S. Peters Road, Suite 301, Knoxville, TN 37923

WHAT IS FEEDBACK?

The safety expert Calvin (of the comic strip Calvin & Hobbes) once said:

"It's true Hobbes, **Ignorance is Bliss!**
Once you know things, you start seeing problems everywhere …
… and once you start seeing problems, you feel like you ought to fix them …
… and fixing problems always seems to require personal change …
… and change means doing things that aren't fun! I say phooey to that!
But if you're willfully stupid, you don't know any better, so you can keep doing whatever you like!
The secret to happiness is short-term, stupid self-interest!"

Calvin doesn't want to learn from his experience. Feedback is learning from experience. In strictly engineering terms, feedback is a sampling of the output of a system and feeding a sample of that output back to the system and using that signal to control the process. This "feedback" process for Process Safety Management (PSM) is represented in Figure 1.

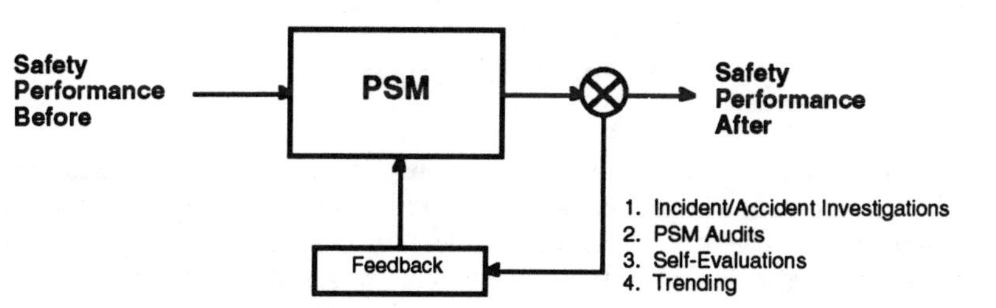

Figure 1. PSM Feedback

USING FEEDBACK TO IMPROVE PSM AND BECOME PROACTIVE

OSHA doesn't allow companies who handle significant quantities of hazardous chemicals to behave like Calvin. The OSHA PSM Regulation requires companies to find and correct the root causes of accidents and incidents. Some companies want to do more than just comply with OSHA regulations. They want to become *PROACTIVE* and find the root causes of problems before accidents occur. How can a company move beyond the reactive mode of responding to accidents and incidents to a proactive mode of fixing problems before an accident happens? They need to use all their sources of PSM feedback. (See Figures 2 and 3.)

When a company looks at accident statistics, they are looking at the tip of an iceberg. Most of the problem is below the waterline - undiscovered. To have an effective PSM program, a company must look beyond the obvious accidents and start looking at the statistics from incidents, near-misses, and unsafe conditions (that can, under the right conditions, cause accidents). This requires programs that look for the bigger portion of the iceberg and get at the causes of the more prevalent incidents, near-misses, and unsafe conditions.

Therefore, we will discuss four methods of uncovering problems that are "below the waterline":

1) **Incident Investigation** of events for which consequences weren't as serious as an accident;
2) **Near-Miss Programs** for self reporting of problem situations that didn't cause an accident but were recognized as potentially hazardous by the personnel involved;
3) **Audits and Self-Evaluations** to look for unsafe conditions before they can cause accidents; and
4) **Trending** of safety statistics to identify programmatic problems that, when corrected, will fix the causes of whole classes of problems. Trending provides management with the information needed to control the PSM process.

Incident Investigations

An incident is an accident with less serious consequences. One usually calls an incident an accident if a death or serious injury occurs (Bhopal), if extensive environmental damage occurs (Exxon Valdez), or if expensive equipment damage results (Three Mile Island). If the event had causes identical to the accidents mentioned before but, by some good fortune, the damage or injuries were less, it would be called an incident. Therefore, incidents are at, or just below, the waterline of the iceberg. With slightly different circumstances, they could have been accidents.

Therefore, one way to prevent accidents from occurring is to find and correct the root causes of incidents. This can be done with systems similar to those used for accident investigation. The key differences are that incidents are much harder to

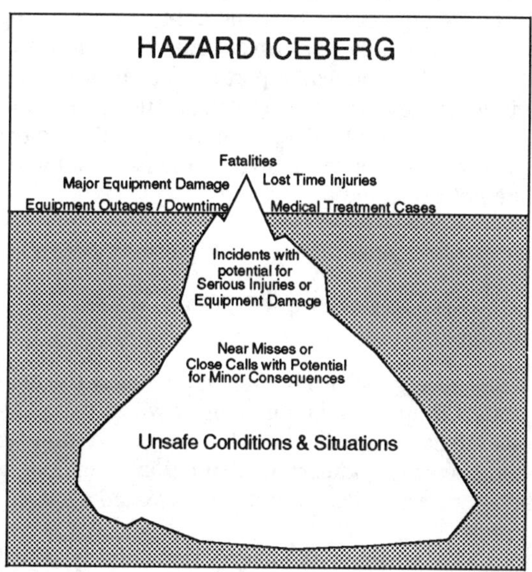

Figure 2. Hazard Iceberg

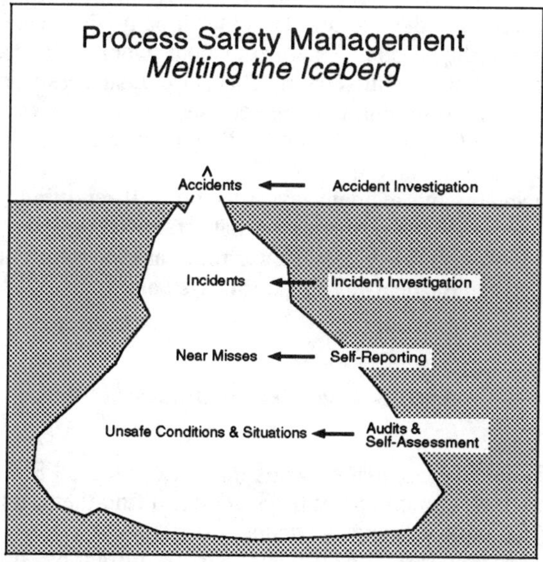

Figure 3. PSM Feedback - Melting the Iceberg

detect (an accident that causes a fatality will obviously be reported - a similar incident that causes a bump on the head probably won't be reported) and that incident investigations are usually performed using fewer resources. Therefore, to avoid inconsistent, incomplete incident reporting, a company should pay particular attention to defining what they are going to call an "incident" so that everyone can understand when one has occurred. Also, a company needs to pay special attention to the system they choose for root cause investigation so that the resources required to use the system are not excessive.

Self-Reporting of Near Misses

Further down the iceberg (occurring even more frequently) are near-misses. The term "near-miss" comes from the aviation industry where two planes that come too close together are said to have experienced a "near-miss". A more accurate term may be "near-hit". In the chemical industry this situation is analogous to an operator almost opening the wrong valve that would have released a hazardous substance to the environment.

Some industries have government/industry supported systems for identification of near misses. Examples include the Aviation Safety Reporting System (ASRS) run for the FAA by NASA and the Human Performance Enhancement System (HPES) run by the Institute of Nuclear Power Operations.

Because a near-miss is seldom obvious to anyone except those involved, a successful system requires the operator (whether it be a pilot or a refinery outside operator) to self-report the problem. Operators would seldom self-report these events if reporting them resulted in punishment. Therefore, a successful system requires confidentiality of the reporter and a guarantee that the reporter won't be punished. Both ASRS and the HPES are similar in that they have such a confidentiality/non-punishment agreement.

Similar to accident and incident investigation, self-reporting systems need to identify the root causes of the problems so that effective corrective actions can be implemented. Therefore, a fairly easy to use root cause system could be used for accidents, incidents, and the near-miss reporting systems.

Audits and Self-Evaluations

The most frequently occurring events that make up the PSM Iceberg are the unsafe conditions. Unless a company takes action to find them, unsafe conditions usually go unnoticed until an accident occurs.

Conducting a one time hazard analysis (required by the OSHA PSM regulation) is one way to look for these types of problems. Unfortunately, even when these types of analyses are coupled with change control, they don't find developing problems. For example, they don't find less than adequate preventive maintenance that is letting the plant's material condition degrade or procedures being disregarded for more expedient but less safe practices. They can't observe operations with an eye for making improvements. However, two other techniques can - Audits and Self-Evaluations.

Audits and Self-Evaluations are the process of having trained observers watch an activity and suggest ways in which it might be improved. The difference between Audits and Self-Evaluations is that Audits are performed by someone outside the organization being observed and Self-Evaluations are performed by people from the organization being observed. Each type of observation has its own advantages. A combination of Audits and Self-Evaluations can be more effective than either technique alone.

Even with Audits and Self-Evaluations, the evaluator needs to determine the root causes of problems that are observed. If the system for root cause analysis is chosen wisely, the same system that is used for accidents, incidents, and near-misses could be applied to Audits and Self-Evaluations.

Trending

A company can become even more effective in preventing accidents if instead of looking at each accident as a unique event that has specific causes, they instead view the accident as the result of several specific causes that are the result of more generic, programmatic shortcomings. The generic/programmatic shortcomings allow the specific causes to occur. The programmatic shortcomings can be more clearly seen and demonstrated to others using statistics obtained through trending incident causes.

For example, one might view the failure of a single operator to use an important procedure as a single operator's attitude problem. However, reviewing the accident, incident, near-miss, and audit/self-evaluation data can show that others are also disregarding procedures and that this disregard is increasing. This programmatic problem requires broader corrective action.

Management can use the trending information to decide where programmatic improvements are needed and how much improvement can be expected. Management can also use the information to review the effectiveness of the PSM program. For facilities that have very few explosions and fatalities, data from incident investigations, near-miss investigations, and audits and self-evaluations are the only source of information providing feedback on the PSM system and what needs to be improved to improve safety.

To effectively identify and use trends, a company needs a consistent accounting of problem root causes. This includes problems that are identified during accident and incident investigations, reported from the self-reporting system, and identified during audits and self-evaluations. Therefore, one needs to carefully choose a root cause analysis system that will work on all the parts of the PSM Iceberg and is designed to trend the data.

CONCLUSIONS

The causes of most accidents are the same as the causes of most incidents and near-misses - unsafe conditions or situations in the system. To measure the effectiveness of a company's PSM process, management needs trending information from incident investigations, near-miss investigations, audits and self-evaluations. By using this feedback, a company can improve their PSM and go beyond responding to accidents to proactively preventing accidents. However, to be efficient and effective, the approach must be coordinated.

If a root cause analysis system is designed using sound equipment reliability principles and human performance theories, it will be applicable to all four methods of operating experience feedback. In addition, if a root cause analysis system is designed to be trendable (having no overlapping categories), statistics from all four feedback methods can be used together to look for generic, programmatic areas that need enhancement within a company. These statistics can provide management with a powerful tool in controlling safety performance. Therefore, especially important is the selection of the right root cause analysis system that is designed to work not only for accidents but also for everyday incidents, near-misses, and unsafe conditions.

Process Safety Management and Product Stewardship: Re-engineering Engineers

Mary L. Woodell
Controversy Management and Risk Communications, Washington, DC

Barbara Ex
Arthur D. Little, Inc., Cambridge, UK

INTRODUCTION

Process safety management (PSM) and product stewardship standards govern different dimensions of the manufacturing process. However, they represent a significant common shift in emphasis from predecessor standards and practices.

In essence, these bellwether standards for industry practice have changed the definition of the problem for managing product and process safety. In effect, they actively promote compliance with the <u>intent</u> of the standards -- not merely with the letter.

The authors believe that redefining the nature of the problem requires industry to redefine its approach to problem-solving, and that this urgent management challenge has been largely underrecognized. For engineers, safety professionals, and other technical experts, this challenge involves rethinking their roles, and redirecting their energies. For example, where it was once sufficient to accomplish a task, engineering and safety professionals must now contribute to a process, to assure the integration of product and process safety issues into core business priorities.

This paper explores the re-engineering of problem-solving to reflect PSM and product stewardship requirements, and to maximize the benefits of the process-oriented approach they represent, focusing on the evolving roles and responsibilities of safety, engineering, and other technical specialists.

COMMON THRUSTS AND TRAITS

The Center for Chemical Process Safety has defined process safety management (PSM) as the application of management systems to the identification, understanding, and control of process hazards to prevent process-related injuries and incidents. Implicit in this definition is the concept that PSM, to be effective, must be both systematic and systemic. PSM expands the purview of traditional process safety activities to emphasize foresight and prevention, in addition to reactive intervention.

Product stewardship similarly broadens the definition of traditional product safety efforts, by applying these disciplines to the entire product lifecycle. The Chemical Manufacturers Association has adopted product stewardship as one of the codes of practice under its Responsible Care program. Its goal is to continuously improve product safety performance as it may affect human health and safety and the integrity of the environment.

Both product stewardship and PSM require that engineers redefine the way they look at problems. Often the shift is from solving problems to finding them. Under these concepts, the role of engineering is expanded beyond traditional boundaries to include health, safety, and environmental concerns associated with the product under customer use and disposal.

Product and process safety traditionally have been separate functions within the process industries, residing in different organizations and relying on different personnel. Both conceptually and practically, these functions tend to operate largely in isolation, from each other and from the outside world. At the same time, they share a number of common traits and goals. Understanding them can help create the opportunity for efficiency and for overall safety improvements.

For example, a company that has recently undertaken a global product stewardship initiative can apply many of the lessons learned to the renovation of its PSM systems. Similarly, many of the tools, techniques, and internal structures required for implementing PSM can be readily adapted to product stewardship programs.

Taking advantage of these potential efficiencies begins with the question: What do process safety management and product stewardship have in common? Five interrelated characteristics suggest themselves.

Process-oriented, not task-specific, approaches

Unlike precursor regulations, new PSM standards emphasize process and results, rather than mandating specific, discrete tasks. Similarly, product stewardship addresses the entire product lifecycle, from concept through manufacturing, distribution, disposal, and recycling. Each of the steps along the way may involve a number of specific tasks that can and should be adopted as standard operating procedure. However, the <u>overall</u> thrust of PSM and product stewardship encourages an understanding of how these steps interrelate and reinforce each other.

Taken together, the steps required for achieving effective product stewardship or compliance with PSM regulations create a continuum, or process, whose broad goal is the <u>systemic</u> improvement of product and process safety. This process-oriented approach helps minimize margins for error by eliminating many of the "hand-offs" from one task to the next, and by providing a common context for a range of very different activities.

Principles of continuous improvement promoted

PSM and product stewardship principles and practices share much in common with Total Quality Management and other quality initiatives, in that all are concerned with continuous improvement. As applied to product and process safety, this concept recognizes that risk reduction is a process, not an event, and that safety performance can always be improved, regardless of the starting point.

By these standards, process and product safety are never "finished;" as a result, continuous improvement is literally a moving target. With goals that are evolutionary, not finite, assessing progress presents a key challenge. Whether internal (as in self-evaluation) or external (where regulatory or industry oversight may apply), establishing meaningful measurement criteria presents a key challenge in any continuous improvement initiative.

Interdisciplinary teamwork required

Traditional process and product safety programs emphasize individual, sequential responsibilities for specific functions. Under this approach, research and development (R&D) specialists and plant managers, for example, may rarely even meet, much less collaborate in any meaningful sense. Although they share the common goal of producing a marketable product that performs to specification, their traditional roles are defined at opposite ends of the product lifecycle, with the result that their functions rarely interact. Similarly, the maintenance engineering and other safety functions in traditional companies often do not routinely interact with design engineering or process R&D.

PSM and product stewardship principles seek to change these traditions, and the work habits associated with them by encouraging interaction and collaboration among the relevant professional specialists. Rather than focusing solely on a given task series, professionals in related fields contribute more broadly to the entire manufacturing process.

Management systems emphasized

As defined by CCPS in its 1989 volume, "Guidelines for the Technical Management of Chemical Process Safety," management systems "consist of explicit sets of arrangements for planning, organizing, implementing, and controlling work within complex organizations." As applied to process safety and manufacturing operations, management systems are increasingly recognized as a critical dimension. For example, accident investigations, which traditionally have focused largely on mechanical failure, today often involve parallel studies of the management systems that underlie such failures and may have contributed to them.

In a similar vein, many facility managers may view maintenance requirements as a reaction to specific, high-profile failures rather than as a systematic program that improves overall reliability. In contrast, a contrast, a company with a well-designed preventive maintenance program will help assure consistency, minimizing the chances that critical improvements may fall through the cracks and improving accountability for overall performance.

Safety issues integrated into ongoing business operations

One important goal for both PSM and product stewardship--and a desired result for initiatives in both areas--is better integration of safety functions into broader business decisions. Systems such as these derive from the recognition that safety cannot be effectively "delegated," nor can it reside in a single box in a company's organization chart. Rather, safety must be seen as an integral concern in business operations and a priority for business success.

This perspective applies equally to product safety and process safety. Companies increasingly recognize that failure to assure employee, customer, and community safety can quickly put them out of business, either through exposure to legal liability or because of damage to corporate image and reputation. On the upside, early and consistent attention to product and process safety issues often creates cost-efficiencies; building product and process integrity into manufacturing tends to cost less than retrofitting to correct a problem after the fact.

IMPLICATIONS FOR SAFETY AND OTHER TECHNICAL PROFESSIONALS

The safety specialist's role has never been more critical, nor has it ever been more complex. Both product stewardship and PSM stndards demand the evolution of numerous industry responsibilities, both technical and managerial. For example, where once it may have been sufficient for safety professionals to consider process risk only within traditional plant boundaries, product stewardship encourages the participation of both customers and suppliers. Moreover, a given process, such as waste minimization, may comprise a number of non-sequential activities, and involve personnel from outside the traditional experience of environmental engineering.

Similarly, PSM takes a management systems approach to a range of safety-related activities, many of which have historically been handled independently. As a result, the implementation of PSM at the facility, divisional, or corporate level may require technical and managerial professionals to consider--and perform--these activities in a different light. Where safety professionals were once perhaps solely responsible for supervising and documenting risk-related activities, PSM involves other groups, including maintenance, engineering, and operations, all of whom share accountability for the process.

The changing roles of safety professionals and other specialists in product and process design are illustrated in Figures 1 and 2, respectively. These figures contrast traditional approaches to stages of these processes with those contemplated by product stewardship and PSM, and highlight some of the issues these management systems are designed to address.

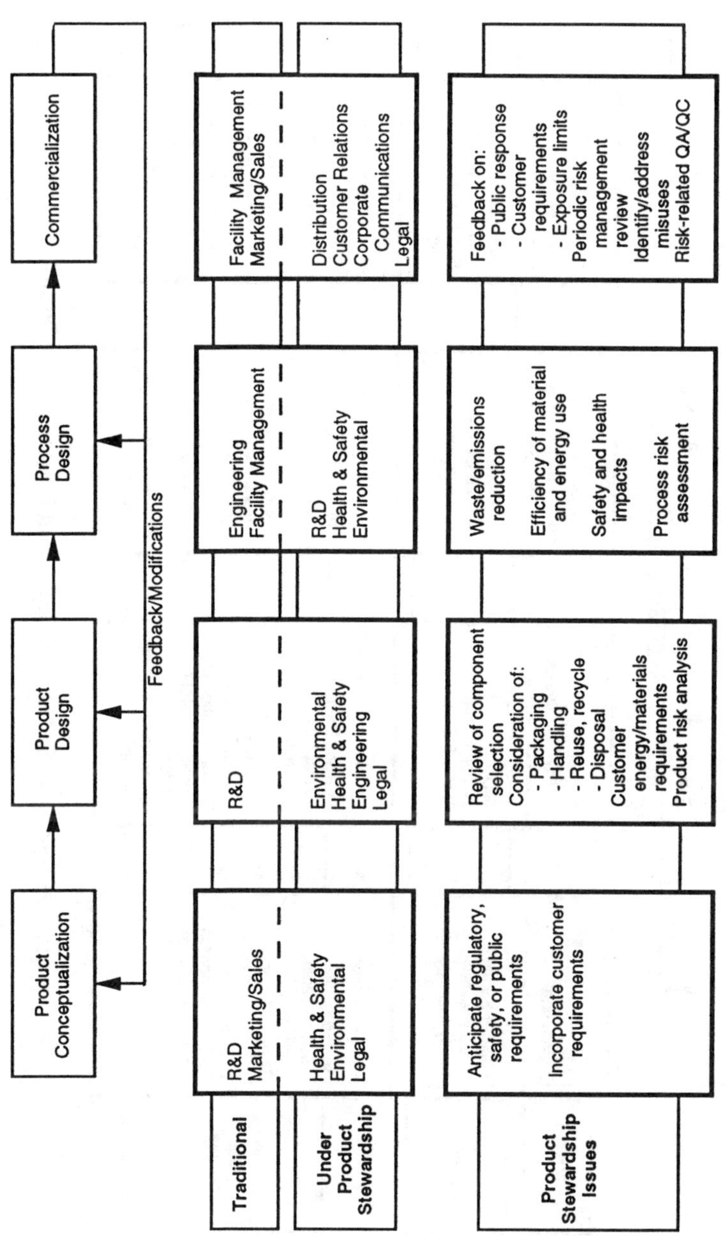

Figure 1: Product Stewardship Issues in the Design Process

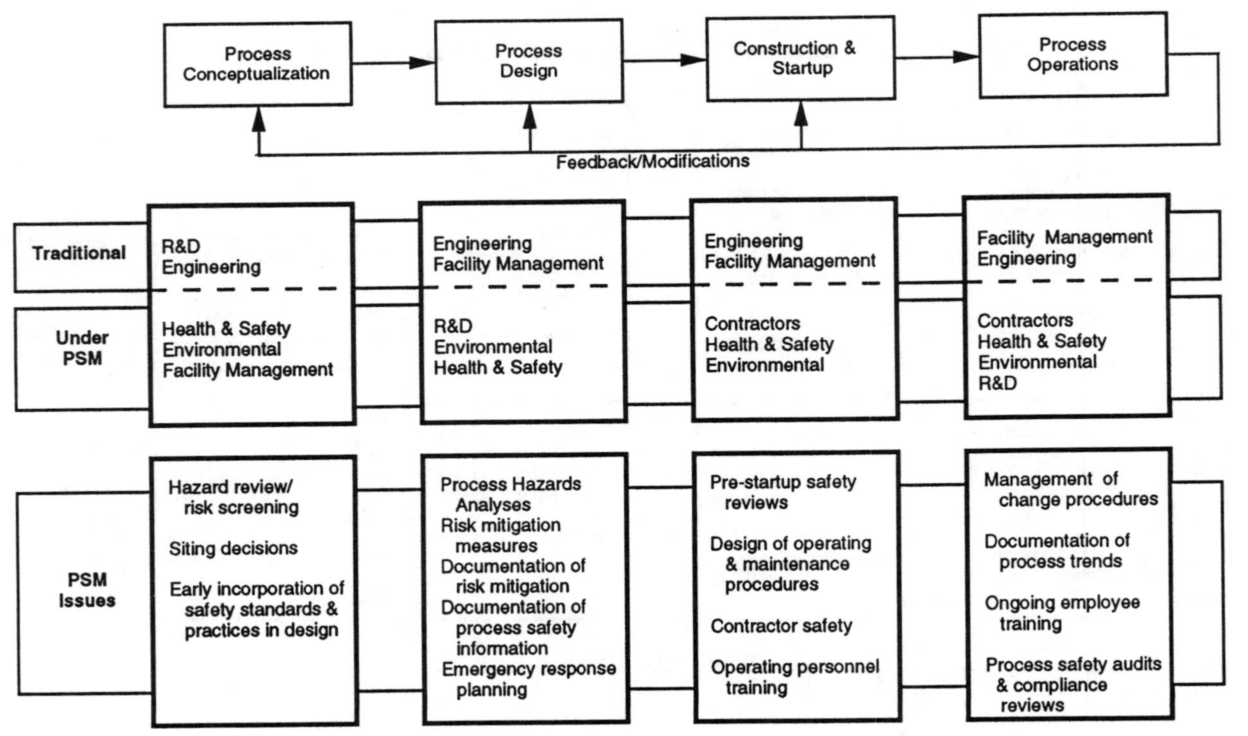

Figure 2: Process Safety Management Issues in the Design Process

Lessons from the quality movement

For companies in which teamwork principles have become the norm in a range of business operations and support functions, adapting them to product and process safety questions may be relatively straightforward.

In others, these concepts may be less mature, suggesting both the need and the opportunity to focus on the management issues arising from product stewardship and PSM. Regardless of the starting point, the following lessons from Total Quality Management and other quality initiatives may apply:

1.　　Involve people early. Both product stewardship and PSM take a lifecycle approach to safety, consistent with principles of continuous improvement. In keeping with this approach, definitions of specific tasks expand to include their integration with other activities. For example, product stewardship contemplates addressing customer usage, disposal, and recycling issues as early as the preliminary research and development stage. Similarly, PSM considers process hazards analysis as part of preliminary process design, rather than as an "add-on" once design is complete.

The kind of foresight these examples imply requires the early input of a range of specialists--many of whom have traditionally become involved much later in the product or process cycle. Getting them on board early often requires the active effort of a PSM or product stewardship "champion" to encourage their paticipation.

2.　　Gain top level support. Commitments to PSM and product stewardship, like other broad industry initiatives, require the endorsement and active support of senior-level management. Initiatives such as these require resources, and they are inherently long-term, which means that consistency and follow-through are essential to their success. Without the continuing, explicit support of senior management, they can all too easily fail.

Gaining this support requires identifying the benefits of product stewardship and PSM in terms of the broad corporate goals and business interests these initiatives promote. For example, a strong product stewardship program may have the effect of minimizing liability arising from customer misuse or disposal of hazardous products, while at the same time strengthening customer or supplier relations. Similarly, PSM, when properly and consistently implemented, can improve productivity by reducing lost-time accidents and other downtime, which in turn enables the company to be more responsive to customer requirements.

3.　　Establish clearcut mission and goals. Complex, multidisciplinary initiatives like PSM and product stewardship require focus and discipline to succeed. Unlike more task-oriented programs, these systems allow considerable latitude for adaptation and interpretation by the individual companies undertaking them. Moreover, continuous improvement by definition has no end-point; it is never "complete." For these reasons, it is critical that missions and goals be determined and articulated on the front end of the effort, to assure that everyone involved understands both the broad objectives and his or her individual responsibilities for meeting them.

As a practical matter, this means taking the time to analyze the requirements of product stewardship and PSM as they apply to specific company needs, practices, resources, and goals. This provides a sound basis for key decisions, such as staffing, funding, and performance measurement, that will determine how these initiatives will be implemented.

4.   Seek out new information sources. Hazards that are not identified cannot be mitigated; identifying them requires knowledge of the product or process and its design basis. Both PSM and product stewardship call for expanding the base of information and understanding of inherent hazards, as a means of improving risk mitigation. This also means assuring that relevant information is shared among members of teams focusing on these initiatives. As a matter of practice, many professionals keep current on developments in their own immediate field of interest, but often the information is not organized and shared.

The more information that can be made available to the entire team, the stronger the framework for building process and product safety improvements. Sources can be both internal and external, formal or informal. For example, operator experience, customer feedback on product performance, published equipment reliability data, public research on product hazards, and process design documents all contribute to the knowledge base.

Assembling all this information does not necessarily require a large, centralized information center. Decentralized organizations may choose to have each function compile and maintain information relevant to its own specialty. What is important is that there is a breadth of available information, that it is kept current, and that it is easily accessible across organizational boundaries.

5.   Develop meaningful benchmarks and milestones. Effective PSM or product stewardship requires a series of reviews at different stages of the product or process lifecycle. For example, some companies might choose to carry out risk assessments at three points in the design of a new product: 1) at the experimental/research stage; 2) at the development/test market stage; and 3) at the commercialization/full production stage. Expected results from each stage can include identification of additional data required, design modifications, limitations on the use of or markets for the product, or the decision to discontinue development. In each case, the criteria for risk management decisions should be in line with company policy.

In terms of PSM, the challenge is twofold: to organize diverse (and often scattered) safety-related activities into a coherent system, and to devise measurement mechanisms that apply to both the individual activities and the system comprising them. For example, the broad goal may be to reduce lost-time accidents by a certain percentage within a certain timeframe. Achieving this goal may require such measurable efforts as streamlining documentation of accident investigations by a specific factor, consolidating safety training programs to expedite delivery by a certain date, and assuring that process and instrumentation diagrams are updated at each facility.

6.   Encourage inter-company cooperation. No management structure would knowingly introduce an unreasonably dangerous process, or a product that exposed customers to needless risk. However, process design and operation are often directed by individuals and units spread throughout at company, just as product-related R&D, safety, manufacturing, and customer support often do not interact on a consistent basis, separated both by functional specialty and by organizational structure. The increasing complexity of modern processes requires that these diverse individuals and units work together effectively to identify safety issues related to the process as a whole and the product throughout its lifecycle.

For example, a risk assessment task force might include researchers involved in research and development, the product manager, an operations manager, specialists in various environmental, health, and safety disciplines, and others with relevant expertise. Special circumstances may also call for non-traditional members to be included, such as legal counsel where there may be potential for product misuse, or marketing specialists who can shed light on customer usage patterns. If product packaging is involved, personnel from distribution or marketing may also contribute.

One effective way to encourage inter-company cooperation is through training. While engineers and other professionals commonly learn new skills through formal classromm training, operations and maintenance personnel tend to receive on-the-job training. In fact, both types of training have value and can be productively combined and applied at all levels. Engineering and safety professionals in particular should work at "switching hats" with other functions. This helps ensure not only that these specialists develop additional skills, but will also provide them with a broader perspective on the kinds of issues faced by their colleagues in other business areas.

7.   Inspire a sense of accountability. The single most important concept underlying both PSM and product stewardship is accountability. Every employee at every level must recognize that, regardless of job title, he or she has a responsibility for health and safety: of co-workers, of neighbors, of customers. PSM and product stewardship programs offer effective means of establishing and directing that responsibility, and monitoring its results.

Technical professionals must learn to broaden their sense of responsibility for safety beyond traditional parameters such as the safety margin in design specifications. If engineers take the attitude that safety problems are being handled by "someone else," or simply a matter of competent engineers complying with the correct standards, then complex or subtle hazards will not be identified, and processes will not be redesigned and retested to assure that they have been properly dealt with.

For example, one of the greatest pressures during design and construction is to get the process ready, on time and on budget. Actions that either add to direct costs or require additional time will tend to be ignored or given limited priority. Accountability for safety performance is necessary to ensure proper regard for process safety during rushed startups.

8.   Recognize that the process never ends.   Product and process lifecycles are by definition dynamic, not static. PSM and product stewardship initiatives must reflect this kind of evolution to accommodate constant change--regardless of whether it is prompted internally or by market demand. For example, most major companies have adequate safety review procedures for major capital projects. However, facility-level projects involving changes are often performed as "maintenance," implying a constant status quo.

The problems created by process complexity are magnified by changes during the process lifecycle. Even where the process is modified in a structured way, the exponential impact of the change on potential process hazards can overwhelm the engineer's capacity. Product stewardship and PSM, which are based on explicit management systems, are the only way to deal effectively with the continuing evolution of a process.

9.   Use your imagination. TQM and other, comparable programs, have prompted radical changes in the definition of quality and the ways it can be achieved. The most effective of these programs derive from creative application of existing expertise. Similarly, PSM and product stewardship initiatives invite a rethinking of "the way we've always done it," and offer technical professionals the opportunity to challenge traditional product and process safety programs. Accepting this challenge requires imagination: could we do this better?

The result need not be drastic; it may, in some instances, be a relatively minor modification to an inherently sound existing system. What is important is that the process that produced that modification is sufficiently broad and imaginative to assure that this is, in fact, the best solution.

CONCLUSION

Market demand--societal, political, and economic--for credible, continuous improvement in safety performance has never been more clear or more urgent. Companies that do not respond effectively put their future value at risk. Engineers and safety specialists, who understand process hazards and product risks and the tools required to manage them, are critical to their companies' ability to address these demands credibly and responsibly, through initiatives like PSM and product stewardship.

Implementing them properly requires time, effort, and imagination, and the challenges are real. The results, too, are real--measurable improvements in product and process safety, that in turn translate into improved productivity, profitability, and corporate reputation. By applying their expertise in new ways, and by re-engineering their approach to problem-solving, technical professionals play a key role in achieving these results.

Designing, Developing, and Operating Crisis Management Training Systems

P. Schulein

D. Kloet

D. J. Stolk

Netherlands Defense Research Organization, Physics and Electronics Laboratory TNO, P.O. Box 96864, 2509 JG, The Hague, The Netherlands

SUMMARY

The TNO Physics and Electronics Laboratory (TNO-FEL) defines Crisis Management as controlling an organisation during times of crisis, when procedural measures alone do no longer guarantee a satisfactory solution to problems. Crisis management addresses handling complex decision problems without an obviously best solution, within a short period of time, often based on incomplete and/or unreliable data.

The TNO-FEL concepts that were used to build crisis management support and training systems both for military and civil applications will be addressed in this paper. Training goals, management modelling and crisis game principles will be introduced as a way to structure a crisis management training problem. Some projects in the area of Disaster Control will be addressed too. Experiences with projects on behalf of the Netherlands Ministries of Defence, the Environment and Internal Affairs will be shared. Finally, current activities in the area of Disaster Control Crisis Management Gaming will be presented.

INTRODUCTION

The Netherlands Organisation for Applied Scientific Research (TNO) currently operates 30 laboratories with a staff of about 5500 employees. With nearly 600 employees the Physics and Electronics Laboratory (TNO-FEL) is the largest of these laboratories. The TNO-FEL field of work is divided into four research divisions and a division for technological development.

The five divisions are:
1.   Operational Research,
2.   System Development and Information Technology,
3.   Radar and Communication,
4.   Physics and Acoustics,
5.   Technological Development.

TNO-FEL, located in The Hague, is one of three institutes belonging to the Defence Research Group of TNO. Although most research is done on behalf of the Ministry of Defence, projects are being carried out for other authorities like trade and industry.

The Operations Research division of TNO-FEL is, amongst other things, concerned with the aspects of emergency planning. Within the framework of emergency planning decision support systems have been developed during the last five years of the 1980s. These relate to incidents involving hazardous materials, public information systems and evacuation planning.

Currently TNO-FEL concentrates on future developments in crisis gaming which concerns the simulation and training of controlling of an organisation in times of a crisis. Based on experience with military applications, a conceptual crisis management model has been developed. In this model the characteristics of a crisis team and its decisions can be addressed. Using this model crisis management training systems can be built both for military and civil applications.

In case of an emergency a crisis team is formed at a very short notice. Members of such a team often have other primary jobs. Therefore, the availability of good training facilities and the regular repetition of training is necessary. The development of an adequate organisationstructure for this team and the training of its participants is the purpose of crisis gaming. The game presents situations that require handling complex decisions within a short period of time, based probably on incomplete and unreliable data. Players of such a game do not play against an opponent, but try to survive in a supervisor-controlled environment. In this paper the general structure of such a game will be presented and some indications about future use will be given.

CRISIS MANAGEMENT

Crisis Management (i.e. controlling resources to reach the most effective performance in any feasible situation) is a demanding job. To use   current technological and organisational resources effectively, detailed, complete, accurate and real-time information concerning the (probable) emergency area and all included participants is needed.

Technological advances in sensor systems, reporting systems and communication networks have enlarged the available amounts of data. Data quality and reliability however vary significantly. Data fusion becomes mandatory to eliminate uncertainties and ambiguities in the data available.

To produce intelligence from available data, and use that intelligence to allocate emergency control assets, is a time and resource consuming affair. Two important solutions are promoted to alleviate this emergency management problem:

1   Numerous Command, Control and Information Systems (CCIS) and Decision Support Systems (DSS) have been designed, developed and fielded.
2   Crisis Management Training methods are developed with extensive use of simulation systems.

A large part of the CCIS/DSS system designs mentioned are technology driven. Obviously all design methodologies that are geared to produce an accurate, logical and consistent system are mandatory in designing complex systems like CCIS's. Furthermore, a close match between the organisational behaviour and any CCIS/DSS is required to gain acceptance and participation from the (future) users and insure effective use of the system. To structure and solve this problem a management model has been introduced.

BASIC MANAGEMENT MODEL

During contacts with the Netherlands armed forces and civilian organisations it turned out to be very difficult to define management tasks. Members of those organisations had difficulty stating clearly what activities were actually performed while executing their management tasks and why they did them anyway. Goals, tasks and procedures were often used without discrimination to describe the management activities. To offer some help to this problem a management model has been introduced.

This model distinguishes 4 management levels:

1   Execution Management: management of daily work and routine tasks;
2   Procedural Management: management of procedural changes and evolution;
3   Emergency Management: management of events that are not routine but can be dealt with within the (sub)organisation involved;
4   Crisis Management: management of events that are unusual and require the involvement of bordering parts of the organisation or higher level decisions and/or resources.

Depending on the organisation each of these levels can be implemented by one or more hierarchical levels.

These levels interact. Basically, at each level 3 alternatives can happen:

1   involvement of a higher level management;
2   regular management at that same level;
3   delegation to a lower level management.

Whatever alternative is invoked, the basic steps within each management level are the same, only the scope is different:

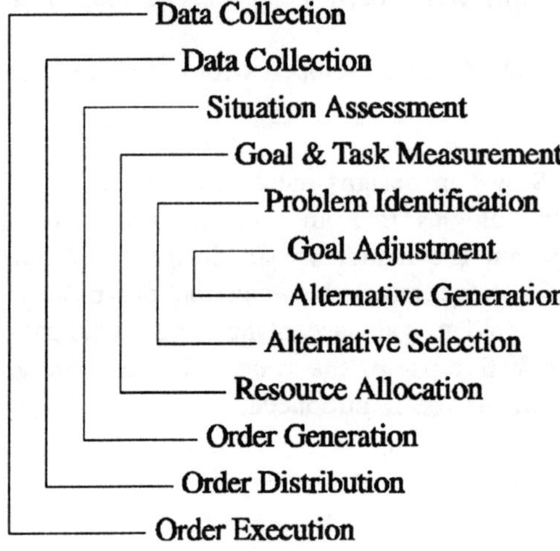

Figure 1. Management Steps

This model can be used to structure an organisation and its management problems in order to design either a CCIS or a Crisis Management Training System.

TRAINING GOALS

Most important for any training system are its goals. Without a precise definition of the goals of a training system it is extremely difficult to design, develop and validate such a system adequately. At the same time it becomes almost impossible to assess learning curves for people using the system and to grade their performances. Goals can be set at a number of levels. Levels that are currently used are:

- Level 0 : Personal Skills
- Level 1 : System Components (organisation, infrastructure)
- Level 2 : System Functions (procedures, regular operations)
- Level 3 : Emergency Assessment and Containment
  (local problems)
- Level 4 : Crises Assessment and Containment
  (large problems, uncertain environment, restructuring, retasking)

Training systems can be developed for all levels. The management level increases with each training level and with each additional level it becomes increasingly difficult to model both the system and its management.

TRAINING METHODS

Training can be implemented by a number of methods. Passive training by reading and listening to presentations and more active training by writing essays and doing case studies.

Training systems can be divided into two categories. One category offers training opportunities that can be achieved by other means such as exercises, but that can be performed less expensive or less dangerous on simulation systems. The other category contains systems that provide training opportunities that cannot be achieved by any other training methods short of actual combat operations.

This category contains Crisis Management Training Systems. The use of automated systems and computer support turns out to be mandatory for training systems that model an organisation and its activities timewise. To model and present an organisation and to extrapolate a current situation into the future by relating the situation to all actions that have been initiated by the trainees requires manipulation of a huge amount of data. Real-time manipulation for training purposes is beyond human capability without computer support. So Crisis Management Training Systems are in essence highly interactive and flexible computer simulations. Due to developments in computer sciences in both hardware and software areas it became feasible to build these systems.

TO CON OR NOT TO CON

Building a simulation model of a system is not always the most effective way to reach training goals. If an application is very close to handling resources in great detail then a simulation model is indispensable. This is also the case when not only a result of an action is important but also the way in which the result was achieved: "What did where why and how well?" Examples of these types of models are decision support systems and simulations with which experts can develop new strategies to counter a problem.

The drawback of a simulation model is the amount of work that is needed to built one, and the time people need to invest to be able to use one profitably. On the other hand, if the goal is to train managementskills, especially high level management, it is acceptable to sacrifice some detail in the model to place more emphasis on the decision-making process. In this case the model just delivers a situation description to the player and the responses of the player are limited to a predefined set of possibilities, one of which is of course doing nothing. The system therefore is just an information management tool, using a decision tree based on the actual decision space of the managers. Players are conned into believing that they are free playing the system, when in fact the system is playing snapshots of a given situation and chooses the next snapshot based on time constraints and player actions.

The decision tree system allows the building of standardized gaming systems and to fill them with information on the current situation in training. It is cheaper to build, easier to use, but gives less detail. Both systems will be explained below.

CRISIS MANAGEMENT TRAINING
SIMULATION SYSTEM COMPONENTS

The basic layout of a simulation system includes the following components:

The situation model describes the actual system that is simulated.

The gaming system has the basic gaming structures that allow the situation model to be run as an interactive game.

The presentation manager generates the user interface and controls the information and command flows between the players and the gaming system.

The reorganisation tool allows the staff to change the Presentation Manager and Gaming System so that other management or interface structures can be tried.

The control tool allows the staff to run the simulation.

In these systems only one playing team is indicated and the environment is played by the scenario and/or staff. This is important to be able to precisely address management problems or decision moments, since in this situation the flow of the exercise can be more tightly controlled than in a two-sided, free play game.

<table>
<tr><td rowspan="4">R
e
o
r
g
a
n
i
s
i
n
g

T
o
o
l</td><td>Situation
Model</td><td rowspan="4">C
o
n
t
r
o
l

T
o
o
l</td><td rowspan="2">S
c
e
n
a
r
i
o</td></tr>
<tr><td>Gaming System</td></tr>
<tr><td>Presentation
Manager</td><td rowspan="2">Staff</td></tr>
<tr><td>Players</td></tr>
</table>

Figure 2.   Crisis Management Simulation System

The basic principles guiding a model like the one presented above are the following:

A crisis management training system contains a nucleus actually simulating the target organisation, a decision space and action control model and a Command and Control model.

NUCLEUS

A simulation system for a unit at operational level should contain a nucleus that is able to simulate the regular activities of that unit. This nucleus contains triggers that will respond to outside influences or control measures from its management.

Outside influence is governed by a scenario or an opponent, depending on the type of unit and the type of training demanded.

The triggers control the units reflex actions and imply the control level of the managements. A large number of triggers and reflex actions requires high level managementskills of the players, far beyond the actual execution of activities in the unit. This restricts the number of automated options that are available at execution level, but lessens the workload of the players .

A small number of triggers and reflex actions require low level managementskills by the players with direct control of basic units actions. This facilitates a flexible execution of basic actions but imposes a higher workload on management.

DECISION SPACE AND ACTION CONTROL

To implement a training system one should define a decision space and include action control functions.

The decision space of the management in a system can be controlled by controlling both the information flow from the system to the management and the command flow from management to the system.

Information flow control is accomplished by designing different levels of information about the items found in the system. This difference can apply to the amount of information available, the detail level of that information, the timeliness of that information and the reliability of that information.

Command flow control is accomplished by defining a language that can be restricted in type of commands, application of commands to certain elements of the system and accurate delivery of those commands to elements of the system in both time and reliability.

Action control should be implemented at two levels. On the management level all commands given should be checked against the restrictions imposed on the management at that moment. On the execution level no element of the unit should be allowed to perform an impossible action. In this area an accurate set of triggers as mentioned before is extremely helpful.

COMMAND & CONTROL AND MANAGEMENT

As stated previously, Command and Control (C2) is an important part of battle management. Modelling C2 in a simulation system requires a three level approach:

To simulate different management functions the system should be able to define the information items that go to each function (knowledge) and the commands that are available to each function (authority).

To simulate uncertainty the information items from the system to the management and the orders from the management to the system should be subject to changes and delays. And the interpretation of the orders by the system should be imperfect as well.

To be able to examine different management structures, the information and order flow and the quality of the information and the orders should be reconfigurable for training sessions. This is performed by a special games tool (the Presentation Manager), which is also able to change the layout of the presentation at will.

Both from a conceptual and a technical point of view designing Crisis Management Training facilities and CCIS/DSS are largely equivalent (i.e.

application, design methodologies, technology, user interface, functionality) [Schulein (2)].

So two applications of a Crisis Management Training System emerge: on the one hand it can be used to perform its primary goal of providing Crisis Management Training, on the other hand it can be used as a testbed to design and develop CCIS/DSS's.

CRISIS MANAGEMENT TRAINING DECISION TREE SYSTEM COMPONENTS

The basic layout of a decision tree system includes the following components:

The database includes the starting point of the exercise and all the situation descriptions for each timestep and combination of user actions.

The Decision Tree Engine has the basic decision tree structures allowing the model to choose the next step in the decision tree, based on the decision taken.

The Action Evaluator combines all player actions and rates them in such a way that the next decision can be made.

The presentation manager generates the user interface and controls the information and command flows between the players and the gaming system.

Reorganising Tool	
	Database & Scenario
	Decision Tree Engine
	Action Evaluator
	Presentation Manager
	Players

Figure 3.　Decision Tree System

The reorganisation tool allows the staff to change the Presentation Manager and Gaming System in such ways that other management or interface structures can be tried.

In this system the flow of the exercise can be tightly controlled by creating decision spaces beforehand. The scenario that is included in the database and has been prefabbed has a profound impact on the decision space of the players.

BUILDING TRAINING SYSTEMS

To be able to build systems like those presented above, a number of strategies need to be followed.

To be able to produce systems in a relatively short time hardware, software tools, concepts, design methodology, document layout and user interface are standardized.

To improve quality of software a large part of the systems are built in toolboxes. To facilitate maintainance and expansion the systems have a layered construction (see figure 4).

To improve project planning and user involvement system development is divided in parts of 6 months at the most, and at the end of each period an upgraded system is delivered to a customer. A number of systems (both simulation models and games) have already been built using this approach.

User Interface
I/O Toolbox
Reporting Functions
Process Model
Process Modelling Functions
Process Modelling Structures
Data Manipulation Functions
Datastructure Defenitions
Pascal Runtime Library
Operating System
Hardware

Figure 4. Layers in software

APPLICATIONS

The concepts presented herein have been applied already in military and civil systems. A major application will cover the area of emergency planning.

By law, every community in the Netherlands has to have a disaster control plan which contains the organisational and informational structures of disaster control resources. These contingency plans need to be tested and members of the organisations need to be trained regularly.

Tree main types of emergency planning training are:

1   Operational training for fire-brigade, police etc.
2   Command post exercises for the service commanders;
3   Training of mayors and their emergency planning staffs.

Current developments of crisis gaming mainly concern type 3 training, modelling industrial accidents. For such situations the mayor acts as the supreme commander of all operational units. He is assisted by an emergency planning staff in which all operational services are represented by officers. In case of an industrial accident the main goals of the mayor and his staff are to minimize the number of casualties and the financial and environmental damages, given a limited set of resources.

This task contains all elements of crisis management. Furthermore, since all participants have "normal" jobs to attend to, a number of strict requirements have been placed on training systems:

-    Training systems should use the same means and environment as in a real disaster situation (maps, books, symbols);
-    The same organisation should be used (same people same responsibilities);
-    The system should be low-cost, easy to use, flexible.
-    The system will not require any preparation or familiarization;
-    The system should be transportable to all cities.
-    Running the system should be instantaneous. No waiting.

Looking at the training levels the opportunity arises to build systems for level 1 and 2 using simulation models, because detailed knowledge of resource allocation and operational procedures is needed.

Fortunately at level 3 decision making is of paramount importance and the details of the accidents themselves can be much more abstract. This allows models to be built on PC systems, which are indeed cheap, fast, reliable and flexible.

CONCLUSIONS

Training with computer support is extremely dependent on the training goals defined. Decision tree models offer high flexibility and close control of the exercise, but sacrifice detail.

Simulation systems can be used for problem solving and training but can be very expensive to build and to maintain.

In both cases standardisation on concepts, design, development, software components and documentation is needed to lower costs and increase quality.

A crisis management simulation system for the Royal Netherlands Airforce, as described in appendix B, is already in use with the Royal Netherlands Defence College.

In the near future the management model and its associated crisis management training system structures will be applied to an Emergency Planning situation. Training emergency planning crisis teams, but also testing and evaluating contingency plans will be addressed by the system. An Emergency planning pilot system is scheduled for the summer of 1993.

Future systems will be based on the same concepts, software modules, documentation and training techniques, to be able to produce low-cost, high quality, flexible, standardized systems.

REFERENCES

1.	P. Schulein, "Crisis gaming for research and training", In: Simulation-gaming: On the improvement of competence, 19th ISAGA Conference, aug 1988, Utrecht, Netherlands, Paper

2.	P. Schulein, "Design, Development, and Application of a Crises Game", TNO Workshop "Human Failure in Process Industry", 1990, Paper

3.	E.A.M. Boots, "User Manual Airbase Operations Wargame", FEL-TNO report FEL-91-A083.

4.	W.C. Borawitz, E.A.M. Boots, P. Schulein, "Battle Management Training", The Sixth European Forum for System Simulation and Management Gaming, Paper, Nov 1990

5.	P. Schulein, E.A.M. Boots, "Training Airbases at Operational Level: Airbase Operations Wargame", The Seventh International Symposium on Military Operations Research, Shrivenham, UK, September 1990, Paper

APPENDIX A:
DECISION SUPPORT SYSTEMS
REM & IRIS

On behalf of the Netherlands Ministry of the Environment a Decision Support System (DSS) has been developed which will be used should an accident with a nuclear power plant occur in the Netherlands, including the border areas. The DSS is named REM (Radiological Emergency Manangement system).

REM supports the decision making of the authorities during the first hours of an accident. It will do so by:
- giving a quick assessment of the extent of the accident, assessing the size and the position of the area affected and calculating the number of people involved or affected;
- analysing the effects of possible countermeasures (shelter, evacuation, etc.) aimed at casualty reduction.

Another DSS is under development : IRIS, the information and computation system for incidents with hazardous materials. This system is being made for the Ministry of Internal Affairs .

IRIS supports the decision making process in case of an accident involving toxic or radioactive materials. Aspects of chemical and nuclear warfare will also be incorporated. Although IRIS handles more types of accidents than REM, thus resulting in more submodels, and more and larger databases; the basic principles are the same as those of REM.

Computer-aided decision making has great advantages:
- more standardization in obtaining and evaluating data,
- better communication facilities (output can be shown to everybody concerned),
- shorter calculation times,
- less calculation errors,
- more trace-back facilities which can be of use for evaluation.

It is necessary for the users of the DSSs or their results to be familiar with the shortcomings of their specific DSS (e.g. incomplete data, validity of models). Therefore training is as necessary for the decision makers and advising authorities, as it is for all other levels involved with crises management. Both in training sessions on learning how to operate IRIS and REM and in training sessions concerning emergency management, a strong demand has risen to expand those DSS systems into full Crisis Management Training Systems.

APPENDIX B:
AIRBASE OPERATIONS WARGAME

The main operating airbases are vital to NATO's fixed wing air operations both in the central region and in staging roles. No viable alternatives are expected to emerge in the next two decades. The military requirements towards air operations demand a flexible, interchangeable and cost-effective way of operating airpower. All functions are performed at the same geographical location, in a high risk environment, with few back up facilities. Thus an airbase is a prime candidate for Crisis Management Training and Crisis Management Support Systems.

The Royal Netherlands Airforce (RNLAF) is conducting a course for the Royal Netherlands Defence College to facilitate Battle Management training for RNLAF officers. TNO-FEL was tasked to design and develop a training aid with as primary goal to address Battle Management at airbase level.

The Battle Management Training System that was designed, the Airbase Operations Wargame (AOW), is in service with the Royal Netherlands Defence College. It includes all aspects of operating from a main operating base: Air Operations, Sortie Generation, Logistics and Support, Infrastructure, Passive Defence and Active Air/Ground Defence [Boots (3)].

The system models the airbase organisation below the level of the management team. The users of the system play one or more members of that team. Players communicate with the system in the same way a computerized Crisis Management System would interact with real life management teams [Schulein (5)].

The AOW System 1 is operational at the Royal Netherlands Defence College. The AOW System 2 is under development on a PC network and is scheduled to be completed in spring 1993.

When using the Airbase Operations Wargame the officers playing part in the simulation management teams addressed a number of points concerning availability of information, presentation, decision aids and came up with management alternatives.

Because changes in the training system are fairly easy to make and the system is built to simulate different management structures, in-depth discussions about the most feasible combination of management organisation and Battle Management Support Tools were stimulated. The Airbase Operations Wargame became a building box to express new ideas about Airbase Management and its Battle Management Support Systems.

Safety Analysis for Existing Facilities

J. A. Hoffmeister *
Martin Marietta Energy Systems
P.O. Box 2003
Oak Ridge, TN 37831-7212

Martin Marietta Energy Systems, Inc., manages five facilities for the U. S. Department of Energy (DOE). These facilities are the Oak Ridge K-25 Site, the Oak Ridge Y-12 Plant, and the Oak Ridge National Laboratory in Oak Ridge, Tennessee; the Paducah Gaseous Diffusion Plant in Paducah, Kentucky; and the Portsmouth Gaseous Diffusion Plant near Piketon, Ohio. Because of the recognized hazards involved in the processing and manufacture of nuclear materials, safety analysis has been performed throughout the life of these facilities. However, beginning in the mid-1980s, growing DOE emphasis on very rigorous safety analysis has resulted in the need to develop more comprehensive and rigorous safety documentation for each of these facilities. This factor combines with an increasing national focus on chemical process safety to necessitate a program to make significant improvements to the formal documentation evaluating and implementing the process safety programs at these facilities.

Energy Systems has implemented a program to provide the revised documentation using a systematic approach to evaluate these facilities which were built 35 to 50 years ago and which have undergone modifications in the intervening years. The facilities were built and modified according to prevailing engineering practices in compliance with the building codes at the time, but assembling a complete design record for these facilities is extremely difficult. Therefore, the program to develop an acceptable safety analysis documentation package for these facilities is based on a comprehensive evaluation of the existing facilities including structural evaluations and process safety evaluation. This approach should be valid for existing facilities in the chemical industry for which a set of safety measures is available.

---

*Managing contractor for the U.S. Department of Energy.

PHASE 0

The program was designed to have three phases, but on review, a preliminary phase was added to provide immediate qualitative review to select facilities. For the preliminary phase, plant management and operators are requested to identify the most significant hazards for the site and teams are assembled to review those hazards. The review is structured to produce a risk matrix which identifies the hazard and develops a qualitative estimate of the severity of potential consequences identified for the hazard. A qualitative probability estimate is then developed and a risk class is determined. The process is shown schematically in Figure 1.

The likelihood of occurrence of the accident/initiating event is ranked as "1" for low likelihood and "2" for high likelihood. A score of "2" implies significant potential for the event to occur during the next ten years. Although not all inclusive, attributes denoting high likelihood include: (1) the event has occurred at the plant before, and no changes have been made to prevent the initiator; (2) the event has occurred at similar plants or in similar processes, and there are no preventive measures in place at this plant; (3) there is a high probability of equipment failure and operators do not have training or procedures to respond to such a failure; (4) passive failure modes exist where there are no reliable procedures for inspection or testing to determine (or confirm) that the component is safely operable; or (5) a phenomenon or event that has caused an event in the past is not resolved or understood, yet conditions might still exist to cause a similar event at this facility.

Attributes denoting low likelihood include: (1) passive failure modes, such as pressure vessel rupture, piping leaks, and corrosion of DOT-approved cylinders where plant records and generic data indicate a low probability of occurrence; (2) operator-related events for which operators have good training, procedures, and operational experience; (3) cases where there were previous precursors or initiators, but modifications of equipment or procedures have been made to prevent recurrence; (4) events that are prevented by redundant control systems or that require multiple failures before the hazard will occur; or (5) events that have not occurred at this plant or similar facilities (assuming that these facilities have a significant time of operation.)

Potential consequences for accidents (or events) are ranked as "1" for low potential of the event or accident to have serious impact on the health and safety of the public or the site workers. Significant or high potential to have serious impact on the health and safety of the public or workers is ranked as "2." These consequence rankings are based on the unmitigated effects of the hazards. Attributes denoting high-consequence events involve releases of significant

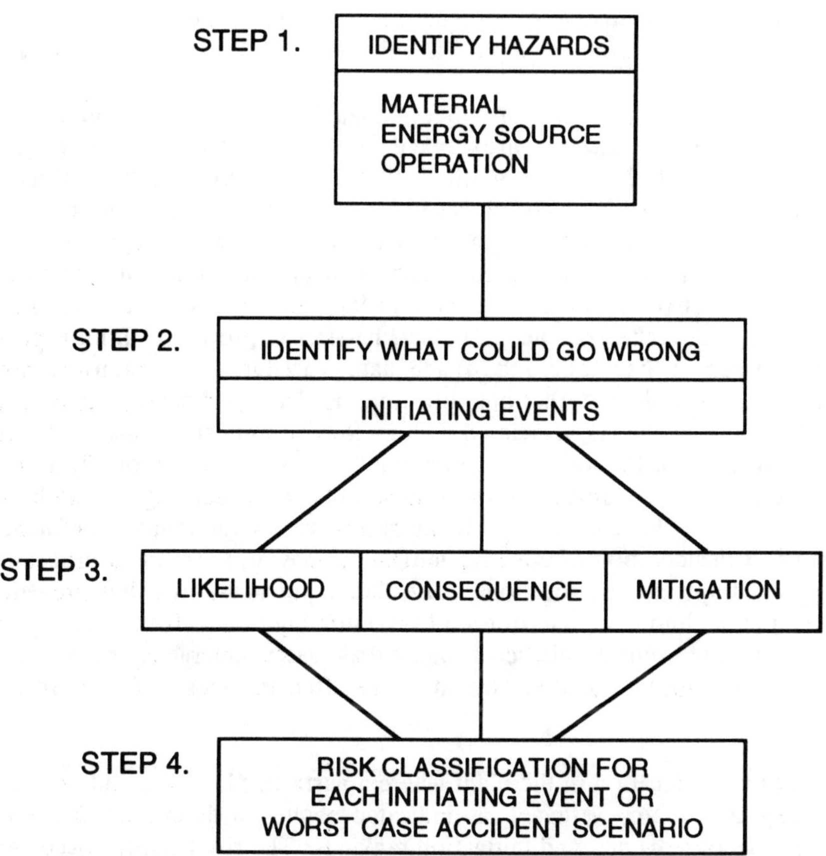

Figure 1. Risk Evaluation Process.

amounts of radiological materials, unique toxic materials, or high-level energy in the proximity of workers or the public. Attributes denoting low-consequence events include (1) release of insignificant amounts of hazardous materials such that radiological/toxic exposure limits for on-site personnel and persons at site boundaries would not be exceeded, (2) release conditions where serious exposure is unlikely due to limited access or length of exposure to the hazard, or (3) low-level energy release from the hazard is only likely to affect nearby workers.

Effective mitigation measures that can significantly reduce the consequences associated with the hazards are ranked as "1." Ineffective or nonexistent mitigation is ranked as "2." Attributes associated with effective mitigation include (1) the availability of active systems that are designed to isolate or contain releases or limit propagation of the hazard and are regularly tested, inspected, and maintained; (2) the ability of personnel or instrumentation to promptly detect the event through monitoring and alarm systems and to limit event severity; (3) the existence of natural barriers to prevent or limit exposure to site workers or the public; and (4) adequate procedures for operator response to the event with evidence that operator training, drills, and emergency exercises are in effect. Attributes associated with ineffective mitigation include (1) lack of one or more of the above measures for effective mitigation or (2) accident conditions that may result in exposures, fires, or other highly exothermic reactions that could defeat or disable the mitigative or containment/confinement systems. Consideration of effective mitigation may also extend to procedures, training, equipment modifications, surveillance practices, etc., that prevent or reduce the likelihood of occurrence of the initiating event. If the consequence of an initiating event is release of highly toxic gases, emergency response and training and procedures attendant to it are often the most effective form of mitigation.

For every location in the plant where a particular hazard could occur, the initiating events are evaluated, as indicated above, to determine the hazard likelihood, consequence, and mitigation rankings. The risk from the occurrence of a specific hazard can be classified by using the risk matrix shown in Table 1 based on the rankings for likelihood, consequence, and mitigation relating to that particular initiating event. It may be possible to define a worst-case accident scenario for a hazard in a particular location where the risk from such an accident scenario bounds consideration of a similar hazard in other locations. Three classes of risk are considered: (1) Class A - serious risk to public or workers identified which requires immediate correction or suspension of operation; (2) Class B - no serious risk identified, but opportunities for risk reduction or safety enhancement exist and operations are continued; and (3) Class C - no serious risk identified, but the information is preserved for possible improvement actions.

TABLE 1

Risk Matrix

Likelihood	Potential Consequence	Mitigation	Risk Class
2	2	2	A
2	1	1	B
2	2	1	B
2	1	1	C
1	2	2	B
1	1	2	C
1	2	1	C
1	1	1	C

PHASE I

Phase I is a comprehensive evaluation of every facility on the site. This evaluation is performed by facility personnel assisted by safety professionals as required to assist in interpretation of the guidance. The initial step is to fill out the Preliminary Hazard Screening (PHS) Worksheet displayed in Figure 2. This form is constructed to meet the concerns related to DOE activities, but a similar form could be constructed to fit almost any type of chemical operation. For each hazard, a measure is identified along with the action prescribed if that measure is exceeded. An action of "keep" for one or more of the hazard categories means that the facility must be subjected to a more formal evaluation to determine the safety requirements for the facility. If no "keeps" are found, then the facility is determined to have a generally accepted classification and no safety requirements are noted beyond those prescribed by OSHA or by applicable standards and codes. The completed worksheet for a generally accepted facility is combined with a brief facility description to provide the safety basis for operation of that facility. Any change to that facility or to the operation of that facility must be evaluated against that basis prior to the modification and the safety basis, i.e., PHS Worksheet and description, must be updated or the next phase of analysis must be performed if a "keep" is identified as a result of the modification. Records for these facilities will be maintained for inclusion in the complete site document packages generated in Phase III.

HAZARD	MEASURE	ACTION[*]	ACTION DECISION	ACTION DECISION BASIS
Rad Waste	$>0.002\ \mu Ci/g$	Keep		
Radioactive Materials	>levels in 5480.11	Keep		
X-ray	Doesn't meet ANSI x-ray standards	Keep		
Toxic Material	Mass/4000 ft^3 > IDLH/2	Keep		
Reactive Material	Hazard Level $\geq$ 2 per EPM-12.0	Keep		
Flammable Material	>110 gal (TWO 55-gal drums)	Keep		
Explosives	Any High/Class A or B	Keep		
	>10 oz. of low in one area/Class C	Keep		
Chemical Incompatability	$\geq$2 incompatable chemicals in same area	Keep		
Lasers	Class I or II	Screen Out		
	Class III w/ beam enclosed	Screen Out		
	Class III non-enclosed beam & Class IV	Keep		
Electrical Voltage/Current	600 V and >25 mA output 50 J stored energy @ 600 V	Keep		
Kinetic Energy	"Unique" (i.e., high energy fly wheel)	Keep		
Pressure	>3000 psig	Keep		
	>0.1 lb TNT equivalent energy	Keep		
Temperature	Results in unacceptable situation or byproduct	Keep		
Biohazard	Special Controls Required	Keep		
Asphyxiants	Affect large number of people	Keep		
Other or Unknown				

[*]If screened out because it is governed by a consensus standard - indicate the standard in the action decision basis section.

Figure 2. Preliminary Hazard Screening Worksheet

In the case of facilities identifying one or more "keeps" in the PHS Worksheet, Phase I continues with the hazard identification process which has the objective of documenting the process and logic by which: all hazards associated with each facility, process, or modification are systematically identified; significant hazards are identified for further evaluation; and a determination is made whether a credible initiating event exists for each significant hazard. To accomplish this purpose, the hazard identification process involves the following steps: division of the facility into systems and subsystems, determination of applicable hazards, initial postulation of consequences and initiating events, and identification of insignificant events and standard industrial hazards. Hazard consequence calculations are expressed in terms of possible doses, temperatures, pressures, etc. at various distances from the source.

For those facilities requiring further analysis, the next task is the production of logic models (consisting of top-down logic diagrams and event trees) developed around those accidents which pose the potential for significant consequences. Analytical tools such as FMEAs, HAZOPS, fault trees, and event trees are employed to evaluate the potential events which might result in the release of the hazards identified. Logic models are assembled by combining the results of the accident sequences identified through the use of the analytical tools with qualitative estimates of consequences resulting from the sequences.

Note that the exemption of standard industrial hazards from the detailed analysis is not intended to imply that these hazards are trivial, but rather to note that additional analysis is not likely to improve upon the ability of existing codes and standards to provide appropriate protection. For those facilities identified as having only standard industrial hazards at this point, the treatment is similar to those so-identified in the PHS Worksheets. This more extensive analysis becomes the safety basis for operation of the facility.

The nature of the facilities operated for DOE and the potential consequences identified for certain types of accidents has resulted in the specific application of natural phenomena criteria for facility designs. The events of concern include seismic, flooding (including rainfall events, dam failures, tsunami, etc.), and wind (including tornadoes and hurricanes.) A team of natural phenomena analysis specialists is employed to provide the specialized knowledge and experience to identify natural phenomena initiating events to be considered in hazard screening and logic model development. The approach used by this team is to perform a "walkdown" of the facility using the identified list of hazards of concern. This activity is an organized review of the facility using a checklist oriented approach to address natural phenomena effects. Examples of such activities for earthquakes are listed as follows: some aspects of "individual Plant Examination of External Events (IPEEE) for Severe Accident Vulnerability,"

10CFR50.54f; NRC Generic Letter 88-20; NRC Unresolved Safety Issue A-46, "Seismic Qualification of Equipment in Operating Plants," NRC Generic Letter 87-02; and recent experience in the activities conducted by Energy Systems and EQE, Inc., at the Oak Ridge National Laboratory (ORNL) High Flux Isotope Reactor (HFIR) prior to restart.

Several factors were noted in a review of this material. First, there is no one single approach to performing a "walkdown". Secondly, that the above examples of "walkdowns" address seismic issues and exclude almost all other natural phenomena. These findings led to the adoption of a methodology that was developed in the mid 1970s as the principal approach to be taken in Phase I. The adopted approach is reported in "Natural Phenomena Evaluation of Existing Buildings," Building Science Series 61, 1975, U. S. Department of Commerce National Bureau of Standards, BSS61. This methodology was chosen because of its inclusion of both seismic and wind phenomena. Also it is essentially the same in its implementation as many recently published "walkdown" methodologies.

Another factor to be considered is the interactions with adjacent facilities. Event scenarios from Phase I which might impact other facilities are provided to the evaluation teams for the potentially affected facilities for use in the detailed Phase II analysis. While the effects (if any) in terms of human health will be treated in the analysis of the facility releasing the materials, it is important to note that it is possible that the event could serve as an initiator for events in adjacent facilities.

To assure uniform implementation of the phases of the update program , a team has been created to oversee implementation of the approach at the Energy Systems level. This group, the Central Safety Evaluation Team (CSET), assembles the interpretation of the DOE Orders applicable to the facilities and provides guidelines for the documentation generated to comply with the program. This team is composed of managers from the sites as well managers from the technical organizations performing or supervising the analysis. In turn, each facility has a team, the Plant Safety Evaluation Team (PSET) which implements the guidance and monitors the site activities for consistency and accuracy.

The results of the process to this point are used to set a hazard level for each facility which is used to identify the level of safety analysis for that facility. The levels include designations such as: generally accepted, which has been previously discussed; low, which indicates the potential for minor on-site and negligible off-site impacts to people or the environment; moderate, which indicates the potential for considerable impacts on-site for people or the environment and no more than minor impacts off-site; and high, which indicates the potential for considerable impacts on-site or off-site to large numbers of people or for major impacts to the environment. This hazard level is the basis for determining the depth of treatment for the safety documentation of the facility. For those facilities identified as generally accepted, the previously described documentation is all that is required. The remaining facilities, identified as having one of the three higher hazard classifications, require a more structured form of analysis identified by DOE as a Final Safety Analysis Report (FSAR) which becomes increasingly quantitative and which is judged to increasingly stringent probabilistic standards as the hazard classification increases to high. The FSAR resembles the content of the documentation required by the OSHA Process Safety Management rule (29CFR1910.119) in every respect except for the comprehensive level of employee involvement. This particular program, because of the emphasis on involving facility personnel provides a greater degree of such involvement than would typically be incurred in the FSAR.

The results of this Phase I analysis may used to generate or update the set of Technical Safety Requirements (TSR) which set forth specific requirements on the operation of facilities required to control or mitigate specific accidents identified in the analysis. These requirements typically impose administrative controls on the maintenance and operation of the facility, including maintenance intervals, operating ranges, and safety-mandated operator actions. Phase I will also provide a listing of candidate safety class items (SCI), configuration management requirements and any needed risk reduction action plans (RRAP). SCI can be defined simply as hardware with the specific purpose of control or mitigation of accidents. Risk reduction action plans recognize immediate hardware or procedure changes which will result in a lowered perceived level of risk. The risk reduction actions can include removal of the hazard if the material is on longer needed or reduction of inventories to lessen the potential risk.

PHASE II

Phase II of the Program is a direct continuation of Phase I. The event-tree branches identified in Phase I as involving significant consequences will be analyzed quantitatively. The analyses will examine both probability and consequences of those accident sequences. In this manner the facilities determined to present the greatest potential risk may be analyzed on a priority basis with additional flexibility to set priorities within Phase II to account for limited resources. Facility and process descriptions will be prepared. SCIs and their safety functions will be identified and operational safety requirements will he created or modified as required. The organizational structure used in Phase I will continue with the same roles and responsibilities. Since the emphasis of the work is on specialized quantitative analysis, the program necessarily involves a greater participation of the safety analysts.

The products of Phase II are facility safety evaluation (FSE) documents which will include TSRs and lists of SCIs and administrative controls for safety (ACSs) that require configuration management, roll-up plans showing which FSEs are to be combined in the Phase III SAR documents, and, as necessary, risk reduction action plans. The roll-up plan for each site will define the facilities to be combined in FSEs and the FSEs to be combined in the SARs to be produced in Phase III. To avoid duplication of efforts and ensure consistency during Phase II for issues such as adjacent facility or natural phenomena impacts, the roll-up plans will be developed by the site Facility Safety Managers early in Phase II.

FSEs will be abbreviated safety analysis documents that provide an early focus on significant hazards and the analytic bases for TSR. The consequences considered in the accident analysis will be limited to human health effects with any SAR level evaluation of environmental effects deferred until Phase III. Facility and process descriptions including adequate depth of detail to support independent review of the accident analysis will be produced. SCIs and ACSs will be listed in the FSE documents. Configuration management will be required for those SCIs and ACSs.

Action plans prepared by the FSETs will focus on facility-specific actions or modifications necessary to reduce operating risk. Descriptions of any modifications proposed in the action plans will include approximate cost, schedule, and priority of the modification. The Phase II executive summary for each site will identify safety class items and administrative controls, discuss interactions between facilities; identify documents produced in Phase II and the hazard levels determined for all facilities; and define more detailed plans for Phase III.

The FSE will reflect the DOE guidance in both format and content. The chapter and subchapter numbers and titles for FSEs will exactly match those described in DOE guidance for SAR documents for these reasons: (1) to minimize the potential for any concerns that a reviewer might have that something may have been inadvertently left out, (2) to avoid the need to interpret where a section of FSE text should be incorporated in the eventual SAR document, and (3) to minimize rework during Phase III SAR preparation.

Although the format of an FSE will exactly match DOE guidance, the text contained within this format will be a previously agreed-upon subset of the content specified. As indicated in Figure 3, not all of the information expected in a 14-chapter SAR is required in an FSE. All chapter titles will be included in the FSE document but, instead of the text specified by DOE, the chapters and subchapters not required will merely contain the statement <u>not applicable to an FSE</u> or "not generally applicable to an FSE - limited information is incorporated in Sect. x.x," as appropriate. Figure 3 provides additional guidance on the FSE subset of DOE content requirements.

The descriptive text will characterize the facility, the equipment, the materials, and the operational processes. This information will be introduced and summarized in Chapter 1, and presented in detail in Chapters 5 and 6.

Facility and process descriptions must provide ample details to support the accident analysis and TSR sections. The functions and capabilities must be described as well as the physical attributes of the equipment and structures (i.e., need to know that pump X must supply adequate cooling water to heat exchanger Y, but may not need to know it is Ingersoll-Rand Model 303, Serial number x67809). Equipment capacities, nominal flows, and operating modes must be described.

The facility description in FSE Chapter 5 should include major buildings, structures and utility systems with emphasis on those features which could impact safety. The location of the facility should be indicated on a map which also depicts the site boundary. The boundaries of the facility (i.e., building, wing, room, individual process area with separate ventilation system, etc.) encompassed by this FSE should be clearly identified. Building plan, elevation and section drawings should be used to show major equipment locations and functional features for the process areas covered in the FSE. FSETs are encouraged to have all necessary drawings prepared by Engineering on the computer aided design (CAD) system to facilitate future drawing revisions and for incorporation into the final SAR documents. Special safety feature such as confinement, shielding, criticality alarms, or ventilation system design should be described in this chapter. Utility systems should be described to the extent relevant to the accident analysis.

Process/operation systems text in Chapter 6 should provide detailed descriptions of all primary and auxiliary processes and the equipment and controls involved in these processes. Aspects of the process or equipment required to maintain safety and to ensure nuclear subcritical conditions should be included. The process description chapter must also include descriptions of

CHAPTER 1.   INTRODUCTION AND GENERAL DESCRIPTION OF THE FACILITY

   1.1 INTRODUCTION
        Content will be as prescribed in DOE guidance.
   1.2 GENERAL SITE AND BUILDING DESCRIPTION
        Provide summary description of site and building. Meteorological and demographic
        information specified in DOE/OR-901 is applicable to an FSE only to the extent
        needed to support accident consequence analyses.
   1.3 GENERAL PROCESS DESCRIPTION
        Provide summary description of the process. Principal design criteria, operating
        systems, waste management system information, etc., are not applicable to an FSE.
   1.4 REQUIREMENTS FOR FURTHER TECHNICAL INFORMATION
        Not applicable to an FSE.
   1.5 COMPARISON OF FINAL (FSAR) AND PRELIMINARY (PSAR) INFORMATION
        Not applicable to an FSE.
   1.6 IDENTIFICATION OF DOE ORDERS
        List DOE Orders in effect and in use.
   1.7 IDENTIFICATION OF AGENTS AND CONTRACTORS
        Provide information specified in DOE guidance.

CHAPTER 2.   SUMMARY SAFETY ANALYSIS

   2.1 NORMAL OPERATIONS
        Not applicable to an FSE.
   2.2 ABNORMAL OPERATIONS
        Not applicable to an FSE.
   2.3 ACCIDENTS
        Summarize accident analysis results and conclusions as specified in DOE guidance.
   2.4 CONCLUSIONS
        Present risk conclusions in terms of frequency and consequences of predominant
        contributors. Specify final hazard level recommendation for the facility (or levels
        if facility segments are used). Make no statement of acceptability of risk.

CHAPTER 3.   SITE CHARACTERISTICS
        Not applicable to an FSE.

CHAPTER 4.   PRINCIPAL DESIGN BASES AND CRITERIA
        Not applicable to an FSE.

CHAPTER 5.   FACILITY DESCRIPTION - DESIGN

   5.1 SUMMARY DESCRIPTION
        Describe principal features of the facility.
   5.2 PROCESS BUILDING
        Describe structural design features to the extent required to provide proper
        context for system descriptions. Provide site map with process building

Figure 3. FSE Chapters and Contents

location indicated. If the facility occupies less than the whole building or includes parts of more than one building, show the building wings, rooms, floors and columns bounding the facility addressed in the FSE.

5.3 SAFETY CLASS DESIGN FEATURES

Not applicable to an FSE. If such features exist, discuss in Sects. 6.2 and 7.1

5.4 DESCRIPTION OF SERVICE AND UTILITY SYSTEMS

Generally not applicable to an FSE. Include only to the extent necessary to the development of the accident analysis.

5.5 MISCELLANEOUS FACILITIES

Generally not applicable to an FSE. Include only if necessary to support the development of the accident analysis.

5.6 ENGINEERING DRAWINGS

Not applicable to an FSE.

5.7 CHANGES FROM THE PSAR (FSAR only)

Not applicable to an FSE.

CHAPTER 6.   PROCESS / OPERATION SYSTEMS

6.1 PROCESS/OPERATIONS DESCRIPTION

Provide a detailed narrative description and supporting flow sheets as specified in DOE/OR-901 (ref. 1). Include primary and auxiliary processes and equipment functions, material compositions, quantities, and physical characteristics.

6.2 SAFETY ANALYSIS CONSIDERATIONS

Description of any special design features, administrative controls, chemical hazards, effluent monitoring, operating and shutdown modes (or other items pertinent to the accident analysis) will be included in Sect. 6.1. State <u>presented in Sect. 6.1</u>.

6.3 PRIMARY PROCESS SYSTEMS

Describe all primary process mechanical, chemical, electrical and fluid systems in sufficient detail to support the accident analyses in Chapter 11, using format specified in DOE guidance.

6.4 AUXILIARY PROCESS SYSTEMS

Not applicable to an FSE - include only if these systems are involved in an accident scenario analyzed in Chapter 11.

6.5 CONTROL ROOM AND/OR CONTROL AREAS

Include control area information in Sect. 6.1 to the extent needed to support the accident analysis. State <u>presented in Sect. 6.1</u>.

6.6 ANALYTICAL SAMPLING

Not applicable to an FSE.

6.7 CHANGES FROM THE PSAR

Not applicable to an FSE.

CHAPTER 7.   SAFETY CLASS EQUIPMENT AND INSTRUMENTATION

7.1 SUMMARY OF SAFETY CLASS EQUIPMENT AND INSTRUMENTATION

Provide a list or table of equipment, systems, components or facility features that have been identified in the accident analysis as being safety class. If no SCIs are required for the facility, that fact should be documented in this section.

Figure 3. (Continued)

7.2 SYSTEM SELECTION CRITERIA

Describe the rationale for selecting SCIs. Reference the Energy Systems application guides, the DOE General Design Criteria definition, and the DOE guidance Appendix C guidance (as appropriate) in describing this method and rationale.

7.3 SAFETY CLASS EQUIPMENT AND INSTRUMENTATION DESCRIPTIONS

Provide the information specified in DOE guidance either in this section or by reference to the appropriate text in Sect. 6.1 or in Chapter 11 or 13.

CHAPTER 8.   WASTE CONFINEMENT AND MANAGEMENT

Not applicable to an FSE.

CHAPTER 9.   FACILITY SAFETY PROGRAMS

Not applicable to an FSE.

CHAPTER 10.   NORMAL OPERATIONS

Not applicable to an FSE.

CHAPTER 11.   ACCIDENT ANALYSIS

For the accidents of interest in Phase II, present the accident analysis as specified in DOE guidance Sects. 11.1 through 11.7. Include graphs to illustrate the risk components (frequency and consequences) of the analyzed accidents but make no statement of the acceptability of risk. Do not make insupportable absolute statements (such as "no health effects"). Include clear references to the TSRs in Chapter 13 that document the requirements to maintain the limits or conditions assumed in the accident analysis.

CHAPTER 12.   CONDUCT OF OPERATIONS

Not applicable to an FSE.

CHAPTER 13.   OPERATIONAL SAFETY REQUIREMENTS

This chapter will constitute a contractual agreement between DOE and Energy Systems management regarding the safe operation of the facility. Express the TSRs in the content and format specified in DOE guidance. Reference the specific Chapter 11 accident analysis assumption or condition from which the TSR is derived.

CHAPTER 14.   QUALITY ASSURANCE

Not applicable to an FSE.

Figure 3 (Continued)

any safety functions that utility or service must serve, and discuss the effect of loss of power, air, water, etc., or refer to the accident analysis section that examines this impact.

After the accident analysis is complete, Chapter 7 will be prepared to discuss safety class item selection criteria, to identify SCI for the facility, or in cases where none are required, to explain why.

Design drawings, calculations, and previous safety documentation from which descriptive information was derived should be recorded in a reference section at the end of the chapter.

The guidance provided in DOE guidance is to be closely followed in the performance and documentation of the accident analysis. The accident analysis text in Chapter 11 must convince the FSE reader that each of the hazards has been thoroughly investigated and that the consequences of accident situations have been analyzed to conclusion.

Chapter 2 (which will be written <u>after</u> the descriptions and the analysis chapters are complete) will provide a concise summary of the results and conclusions of the accident analysis through text and tables which identify the bounding consequence accidents and their frequency of occurrence. The facility hazard level classification and its basis will be discussed in this summary chapter.

The analysis in Chapter 11 must identify assumptions made in key areas of equipment failure, hazardous material release, dispersal of material, operator action, response times, and credit taken for safety class systems. The information in this chapter must include specific reference to and provide the detailed basis for the conditions, limitations, and administrative controls invoked by the TSRs.

The Phase II analyses will include the accident sequences for which significant human health consequences were predicted during Phase I. Additional natural phenomena evaluations will be completed during Phase II (in accordance with the <u>Phase II Natural Phenomena Applications Guide</u> and will include such factors as ice and snow loading in addition to the seismic, wind and flood effects which were qualitatively examined during Phase I. Consideration of local flooding (i.e., local exterior-vicinity flooding) will be addressed in Phase II to the extent that data are available. Due to the complexity of local flooding analyses (e.g., great topographical variations and large paved areas), limited existing data, the extended permitting process required prior to test drilling to generate new data, and the lack of a simple directly applicable surface water model, such analysis may require a multi-year effort not compatible with the Phase II schedule. The results of any local flood analysis not scheduled for completion within the Phase II time frame will be considered during Phase III. Natural phenomena evaluation teams will identify existing natural phenomena (NP) analyses that are still applicable as well as direct the performance of additional quantitative NP analyses when required.

PHASE III

Phase III is an expansion of Phase II to include evaluation of those facilities and elements forwarded to Phase III from Phase I, as well as the generation of the remainder of the chapters required for complete FSARs for facilities analyzed in Phase II. The complete FSARs will be produced in accordance with DOE guidance in effect at that time. A great deal of the information applicable to the site will be assembled in a site basis document which will then be referenced by each of the facility analyses generated to this point. Examples of such information are site descriptions, facility safety programs, conduct of operations, and quality assurance. Also, the same generic information (such as demographics, natural phenomena, and meteorological data) can be applicable to other sites in the Oak Ridge complex. Accordingly, plans call for an appropriate consolidation of this information into separate generic documents designed to complement facility-specific FSARs.

The SARs produced in Phase III will only reference and will not repeat the contents of these generic documents. This approach should be economically favorable relative to documentation, review, and approval costs, and should eliminate the possible confusion related to multiple reviews and approvals of identical information. Further, if information in these generic documents changes, only the single generic document would have to be revised rather than requiring the revision of all SARs. Specific SARs may require change if the generic document revision affects the accident analysis assumptions (e.g., demographic shifts could affect toxic release consequence predictions).

Process Hazard Screening—A Method for Identifying Batch Pharmaceutical Processes with the Greatest Risk for Catastrophic Accident

Steven J. Schwartz
with

William R. Fassel, Michael P. Roesner, and William M. Walasinski
Eli Lilly and Company, Indianapolis, IN

Abstract

Chemical manufacturing facilities faced with the task of conducting Process Hazards Reviews (PHR's) for all OSHA regulated processes must do so given specific boundaries. Regulated processes must be analyzed in twenty-five percent increments from May 1994 until May 1997. Furthermore, the Process Hazards Reviews must be conducted in a priority order based on the potential consequences which may arise from uncontrolled releases of hazardous materials. Eli Lilly and Company has developed a Process Hazard Screen which facilitates the prioritization mechanism based on objective process data. The goal of this document is to explain the elements of the Process Hazard Screen in detail, examine results compiled from various types of manufacturing related processes, analyze the limitations of the Process Hazard Screen, and to list constructive comments which will be implemented for an upcoming revision of the program.

Introduction

Eli Lilly and Company is a global producer of chemically derived pharmaceutical products. Many of the processes utilized to synthesize such products involve the use of toxic, reactive, and flammable materials. Given the requirements of the OSHA Process Safety Management Standard, CFR 1910.119, Eli Lilly and Company must conduct Process Hazards Reviews (PHR's) for all processes using amounts of highly hazardous materials greater than or equal to OSHA specified threshold limits. A Process Hazard Screen was developed to provide a priority order for all regulated processes based on a quantitative analysis.

Given the resource demands necessary to conduct a large number of Process Hazards Reviews, it is prudent to evaluate those processes first which pose the greatest risk to the employees, facility, and community. Eli Lilly and Company has developed the Process Hazard Screen as an appropriate countermeasure, then standardized the methodology for all global operations which process toxic, reactive, and flammable materials.

The intent of the Process Hazard Screen is to provide a relatively quick and easy approach to quantitatively define which processes present the greatest risk (potential for occurrence and magnitude of severity) in the event of a catastrophic accident. The Process Hazard Screen also allows for subjective data to be analyzed when selecting a specific process to be reviewed. The data required to provide a quantitative risk evaluation for a batch pharmaceutical process includes:

- the process boundaries;
- the hazards associated with the materials used in the process;
- the potential associated with chemical reactions;
- the safe operating parameters of the process;
- an indication of the complexity of the process;
- the potential for accident based on process volume and frequency;
- and subjective data including process history, management considerations, and on-site exposure.

Corporate facilities with operations involving highly hazardous materials were identified, then asked to complete a Process Hazard Screen for each individual process. Although the Process Hazard Screen was used for a wide range of processes - fermentations, separations, environmental operations, and utility operations - the method was based on, and therefore most effective for, screening batch pharmaceutical manufacturing processes.

The Process Hazard Screen is completed in two steps. First, a facility representative is asked to obtain all the pertinent processing data required by the Process Hazard Screen worksheet. Once the data have been collected, it is transferred into a microcomputer based software package which calculates an overall Process Hazard Screen value. For each process screened, the software program will communicate the final screen value plus provide a list of "prioritized" processes as sorted by descending screen values. The completed database and worksheets are retained by the facility Process Safety Coordinator. Data are used by the site management to plan Process Hazards Reviews as well as to estimate corresponding resource requirements. At Eli Lilly and Company, all facilities with a Process Safety Management Program have completed an initial Process Hazard Screen for every identified process.

Materials and Methods

The Process Hazard Screen developed by Eli Lilly and Company is comprised of seven distinct categories. This section, concerning **Materials and Methods**, will describe the data necessary and the calculations performed to determine a Process Hazards Screen score. After a screen score has been determined, it is the responsibility of each facility Process Safety Management Committee to determine a schedule for performing Process Hazards Reviews based on the objective (numerical) analysis as well as any pertinent subjective data.

It is the intent of the Process Safety Department to provide Eli Lilly and Company with a standard method of determining the relative risk of any process. For each process identified on a particular site, the facility Process Safety Management Committee is provided with a Process Hazards Screen packet, complete with tutorial, to be filled out by the process "expert". Once completed, the information is returned to site management and used to plan Process Hazards Reviews and estimate the required resources. The following steps are followed by each process "expert" to screen their respective process.

Process Identification and Boundary Data

The *Process Identification and Boundary Data* section is the cover page of every completed Process Hazard Screen packet. Process identifications and the establishment of process boundaries are documented on the first page of the Process Hazard Screen packet (figure 1). The "process name" entered, on the first line of the first page, may not be a unique identifier. It is often common for the same process to be operated at separate facilities. Process Hazard Screens are to be completed by all processing departments, including environmental and utility services.

Since the same process may be conducted at other sites, the Plant Site facility name is recorded on the *Process Identification and Boundary Data* page. Other pertinent information on the cover page includes: the building name, the operation category, the material produced/processed, and the ultimate end product supported. These data clearly represent the scope of a given process, and provide the facility Process Safety Management Committee a knowledge of which operations represent the greatest potential hazards. In many instances, the "process" is actually an intermediate step of a final product. Other times, the specific utility or environmental services supports many intermediates or end products.

If applicable, there is a space to complete the specific operation number. Many processes can be further stratified into specific operation steps - including

Eli Lilly and Company
Corporate Process Safety Department

Process Hazard Screening
Data Gathering Form

Process Identification and Boundary Data

Process Name : ____

Plant Site : ____

Building : ____

Operation Category : ____

Material Produced/Processed : ______________________________

Ultimate End Product Supported : ___________________________

Operation Number : ____

Department Responsible : ___________________________________

Strategic Business Unit : __________________________________

Date : ____

Process Description :

____
____
____
____
____
____
____
____
____
____
____
____
____
____
____
____
____
____
____
____
____
____
____
____
____
____
____
____
____

Figure 1 - Process Identification and Boundary Data

dryers, centrifuges, and filter presses. If necessary, a department may break the
process into separate operations. The final information supplied for the
Process Identification and Boundary Data section include: the name of the
operating department, the overall "strategic business unit", and room for the
person completing the form to describe the process, operations, and boundaries
thereof in greater detail.

Although none of the information in the *Process Identification and Boundary Data* section is used to actually calculate a final Process Hazard Screen score, it is imperative to collect this information for the purpose of documenting the analysis on the subsequent pages.

Material Data

The *Material Data* page is provided to collect the hazard data for all materials used within the process boundaries (figure 2). This information is collected for each material listed on the Bill of Materials (BOM) for the process being screened. The BOM for a particular process includes the name of the material, the item code (quality assurance number), and the quantity (in either mass or volume) of the material necessary for a specific process step. Utility and environmental services do not use a BOM.

For each item listed on the BOM, the National Fire Protection Association (NFPA) codes for **health, fire**, and **reactivity** are to be listed. The NFPA codes range from zero to four, with a zero indicating the lowest hazard. If the NFPA rating is not available or cannot be located, a default value is provided. For the **health** and **fire** fields, the default value is 2.4; for the **reactivity** field, the default value is 1.3, unless the material can be described as "non-chemical", which would create a default value of zero.

To determine the process **health** score, the highest NFPA **health** rating is selected from the list and recorded at the bottom of the page. Using the same basis, the process **fire** score is determined by selecting the highest NFPA **fire** rating from the list and recording it at the bottom of the page. The **reactivity** score for the Process Screen is generated by adding each **reactivity** field value for each material listed and recording the sum at the bottom of the page. The **reactivity** score is selected using Table 1.

Table 1   Determining the **reactivity** score as a function of the sum of all **reactivity** field values

Reactivity Sum	Reactivity Score
> 9.0	4
5.1 to 9.0	3
2.6 to 5.0	2
> 2.5	1

Process Name : ___________________________   **Item Code :** ___________

Material Data

| Material | Item Code | National Fire Protection Association (NFPA) Ratings | | Reactivity | Accelerated Rate Calorimetry (ARC) | | | | Stability Rating |
		Health	Fire		Self-Heat Rate $^\circ$C/min	Onset Temp. $^\circ$C	Max. Processing Temp. $^\circ$C	Temp. Diff. (Onset - Max.) $^\circ$C	
________	____						______		
________	____						______		
________	____						______		
________	____						______		
________	____						______		
________	____						______		
________	____						______		
________	____						______		
________	____						______		
________	____						______		
________	____						______		
________	____						______		
________	____						______		
________	____						______		
________	____						______		
________	____						______		
________	____						______		
________	____						______		
________	____						______		
________	____						______		
________	____						______		

Health and Fire Scores   *High*   *High*   *Sum*

Reactivity Score . . . . .

Stability Score . . . . . *High*

Figure 2 - Material Data

Material **stability** data is also requested on the *Materials Data* information sheet. **Stability** data is not available for all materials, however the Eli Lilly and Company Chemical Hazards Data Base provides information for many compounds, both intermediates and final bulk products, that are processed in the solid phase. The Chemical Hazards Data Base includes data acquired from an Accelerated Rate Calorimeter (ARC), which is used to determine the material's onset temperature prior to exotherm as well as the material's self-heat rate, in degrees Celsius per minute. The difference between the onset temperature and the maximum processing temperature is calculated, and the **stability** rating is determined according to Table 2. The **stability** value from Table 2 is to be extracted and recorded as the material **stability** rating. The highest **stability** rating is selected from the list, and it is recorded as the process **stability** score.

Reaction Data

The third page of the Process Hazard Screen calculates the relative risk of a runaway reaction hazard for a particular process (Figure 3). The *Reaction Data* calculations produce an exothermic score for the chemical reaction which releases the most heat. In some processes, more than one chemical reaction is expected. In other processes, there are no chemical reactions. Default values are provided for endothermic chemical reactions and systems that are defined as "non-chemical".

The Process Hazard Screen requires a listing of all chemical components and the mass of each component used in the reaction. If the process is continuous, then the calculation should be based on the amount of each component in the reactor per unit time.

Table 2 Determining the **stability** score as a function of self-heat rate and temperature difference

Rate of Self-Heating in deg. C/min.	Temperature Difference (Onset-Max) deg. C		
	< 50	50 -100	100 - 200
< 2	2	2	1
2 to 50	4	3	2
> 50	4	4	3

Eli Lilly and Company
Corporate Process Safety Department

Process Name : ____________________ **Item Code :** ____________

Reaction Data

Component in Reaction Mixture	Item Code:	Mass kg.

Total Mass of Reaction Mixture, kg : - - - - - -

Reaction Info:
☐ Endothermic ☐ No Data ☐ Non Chemical

Reaction Equation:

Limiting Reagent :

Mass of Limiting Reagent (kg):

Molecular Weight of Limiting Reagent, g / gmol :

Heat of Reaction cal / gmol of A :

Component A is :

Total Heat From Chemical Reaction, cal :

Heat Available (Released) per Unit Mass, cal / g : - - - - - -

Exothermic Score : - - - - - -

Default Score : - - - - - -

Final Exothermic Score : - - - - - - - - - -

Figure 3 - Reaction Data

The next step is to calculate the total heat released per unit mass of the exothermic reaction mixture. This is accomplished by listing the applicable reaction data: the stoichiometric chemical equation, the molecular weight of the limiting reagent, the heat of reaction of the limiting reagent, and the total mass of the reaction mixture. The total heat generated by the chemical reaction can be divided by the total mass of the reaction mixture to produce the heat generated by unit mass. This calculation assumes a 100 percent conversion of reactants, however it is possible for a more accurate value to be used. The exothermic score is then derived as a function of the heat released per unit mass, using Table 3. This table can be used more than one time, depending on the number of reactions identified in the process. The reaction which results in the highest heat release per unit mass should be used in Table 3.

The default values which were developed for specific situations include: 1.1 for "non-chemical" processes, 1.0 for endothermic reactions, and 2.2 for exothermic reactions where data are not available.

Process Data

Catastrophic accidents are sometimes attributed to the failure to process hazardous materials within established operating limits. Extreme temperature conditions may lead to thermal expansion or contraction, embrittlement of metals, and hazardous phase changes of process materials. The failure to control the pressure of the process can result in vessel ruptures or implosions, environmental releases, and uncontrolled temperature fluctuations. The rationale for capturing the *Process Data* in the Process Hazard Screen is to quantify the relative probability and severity of the hazards resulting from excessive process conditions (Figure 4).

Table 3   <u>Determining the exothermic score as a function of the heat released per unit mass of reactants</u>

Heat Released per Unit Mass (cal. / g)	Exothermic Score
> 40	4
15 to 40	3
5 to 15	2
< 5	1

**Eli Lilly and Company
Corporate Process Safety Department**

Process Name : ___________________________   **Item Code :** _____________

Process Data   Instructions Data

Process Condition	°C	Score	Document	Number of Pages	Multiplier	Number of Equivalent Pages
Minimum Temperature	______	. . . . .	Ticket	________	1	. . . .
Maximum Temperature	______	. . . . . .	Protocol	________	1	. . . .
			Written Procedures	________	5	. . . .

psig

| Minimum Pressure | ______ | . . . . . |
| Maximum Pressure | ______ | . . . . . |

Total Number of Equivalent Pages . . . .   *Sum*

Complexity Score . . . .

Range of Process Conditions Score . . . . .   *High*

Production Data

Number of lots / year (or "Process being scaled up") : _______________________

Experience Score : ----------------------

Number of lots / week : _______________________________

Volume of material processed per lot (gallons / lot) : _________________________________

Weekly processing rate (gallons / week) : ---

Volume Score : ------------------------------

Figure 4 - Process, Instructions, and Production Data

The following tables (4 through 7) provide individual *Process Data* scores for minimum/maximum pressure and temperature points.   The process value determined should result from normal operation of the entire process within the specified boundaries.   The final *Process Data* score is the highest of the four recorded.

Table 4   Determining the minimum process temperature score

Minimum Temperature (degrees C)	Score
< -80	4
-80 to -41	3
-40 to -16	2
> -15	1

Table 5   Determining the maximum process temperature score

Maximum Temperature (degrees C)	Score
> 250	4
180 to 250	3
130 to 179	2
< 130	1

Table 6   Determining the minimum process pressure score (of two possible)

Minimum Pressure:

psia	psig	in. H20	mmHg	Score
< 14.3	< -0.4	< -10	< 741	2
≥ 14.3	≥ -0.4	≥ -10	≥ 741	1

Table 7   <u>Determining the maximum process pressure score</u>

Maximum Pressure (psig)	Score
> 150	4
50 to 150	3
7 to 50	2
< 7	1

Instructions Data

Eli Lilly and Company produces pharmaceutical, agricultural, and animal health care products. The processes designed to manufacture these products can vary in complexity, depending on the synthesis, separation, and purification steps required. The *Instructions Data* section of the Process Hazard Screen (Figure 4) is a crude method of quantifying the complexity of the process step based on the number of written documents necessary to operate the process. The written documents may come in the form of manufacturing tickets (batch or continuous process recipes), department protocols (specific process procedures utilized to maintain consistent quality), and written procedures (standard operating procedures referenced by either the ticket or protocol).

The *Instructions Data* score is an attempt to calculate the probability of a human error occurrence based on the total number of written documents the operator must follow. The total number of pages of all documents used in the process is factored into the *Instructions Data* score. Tickets and protocols are used specifically for determining the quantity and quality of process inputs,

Table 8   <u>Determining the complexity score as a function of the equivalent number of document pages</u>

Total number of Equivalent Pages	Complexity score
> 36	4
21 to 35	3
11 to 20	2
1 to 10	1

therefore each page is only counted as one. Standard operating procedures are relatively shorter and are not process specific. The total number of operating procedures used is multiplied by five. This analysis was based on an engineering judgment which assumes that the number of operations which occur within a process is related to length of the instructions documents.

Once the total number of equivalent pages (tickets plus protocols plus five times procedures) is calculated, Table 8 may be referenced to determine the overall complexity score.

Production Data

Production Data (Figure 4) is comprised of an Experience Score and a Volume Score. The intent is to quantify the potential for an accident to occur based on the frequency that the process is operated and the quantity of product manufactured.

The only data point required to determine an Experience Score is the number of process "lots" produced in one year. The Experience Score is based on the assumption that risk is inversely proportional to the quantity of material produced. As the number of "lots" produced increases, the experience operating the process increases, and the potential for a catastrophe decreases.

Where utilities, environmental operations, or continuous chemical processes are involved, the Experience score is compensated accordingly. Experience Scores can also be determined qualitatively for processes which are "scaled-up" in the pilot plant. Once the frequency of the process is determined, the Experience Score can be determined using Table 9.

Table 9   Determining the experience score as a function of material processed in one year

Qualitative Processing Frequency	Quantitative Processing Frequency ("lots"/yr.)	Experience Score
Being "Scaled-Up"	-	4
Seldom/Infrequent	< 10	3
Frequent	10 to 50	2
Continuous	> 50	1

The basis of the Volume Score is that *Production Data* can be directly proportional to the potential of a catastrophic accident occurring based on the quantity of material processed through the pipeline. The probability and severity of an accidental release of hazardous material is increased as larger volumes of material are processed.

If the number of "lots" processed on a weekly basis is multiplied by the total volume of material processed per "lot", the weekly processing rate (in gallons per week) can be evaluated to determine a Volume Score. Continuous processes may also be evaluated by calculating the total volume of material processed over one week's time. The Volume Score can then be determined as a function of the weekly processing rate using Table 10. A qualitative column has also been included for processes which operate in the pilot plant.

Table 10 <u>Determining the volume score as a function of the processing rate</u>

Volume (gal/week)	Qualitative Meaning	Batch Size (gallons)	Lots/week	Volume Score
> 60,000	Large	2,000	> 15	4
		4,000	> 30	
36,001 - 60,000		2,000	10 to 15	3.5
		4,000	19 to 30	
12,001 - 36,000	Typical	4,000	4 to 9	3
		2,000	7 to 18	
7,201 - 12,000		4,000	2 to 3	2.5
		2,000	4 to 6	
		500	15 to 24	
2,401 - 7,200	Small	4,000	1	2
		2,000	2 to 3	
		500	5 to 14	
1,451 - 2,400		2,000	1	1.5
		500	3 to 4	
501 - 1,450	Pilot Plant	500	2	1
		300	2 to 4	
< 501		500	1	0.5
		300	1	
		200	1 to 2	
		100	1 to 5	
		30	1 to 16	

Subjective Data

The information referenced in the *Subjective Data* section (Figure 5) is used to identify those potential hazards which were not quantified in the previous five sections. Although the *Subjective Data* is not used in the calculation of the final Process Hazard Screen score, it may influence Plant Management or the Process Safety Management Committee to perform a Process Hazards Review that may not have the highest overall Screen score. This qualitative analysis also provides a basis for prioritizing between two processes that have similar scores on the Process Hazard Screen.

The categories which are analyzed in the *Subjective Data* section include:

On-site exposure, the magnitude and severity of personnel exposure on-site which can be expected during the accidental release of a hazardous material. This category can represent proximity to a building containing people or containing other potentially hazardous materials.

Off-site exposure, the magnitude and severity of personnel exposure off-site which can be expected during the accidental release of a hazardous material. Off-site exposure is concerned with the consequences of releasing a quantity of hazardous material towards an off-site population or natural body of water.

Phase, the analysis of the hazardous materials at processing conditions as well as ambient pressure and temperature. The accidental release of a toxic material processed at elevated temperature and pressure may likely vaporize to form a toxic cloud.

Facility, an evaluation of the facilities in which the hazardous materials are processed. If the facility does not include the best available technologies to mitigate an accidental release, these data are best noted in this paragraph.

Incident History, a reference to the number of injuries and near-misses which have occurred within the process boundaries. Although this number is quantifiable, it serves purpose to highlight those processes which have trends indicating a higher accident rate.

Skills Required, a paragraph to identify any particular operating techniques which must be executed precisely to prevent an accidental release. If specialized training or a specific procedure must be utilized to operate the process, the potential for an accident to occur may exist.

Eli Lilly and Company
Corporate Process Safety Department

Process Name : ________________________   Item Code : __________

Subjective Data

On-Site Exposure :

____
____
____
____

Off-Site Exposure :

____
____
____
____

Phase :

____
____
____
____

Facility :

____
____
____
____

Incident History :

____
____
____
____

Skills Required :

____
____
____
____

Management Considerations :

____
____
____
____
____

Figure 5 - Subjective Data

Management Considerations, any other criteria specified by area management
or the Process Safety Management Committee. Public scrutiny, discontinuation
of a process, or possible allergenic reactions may cause an area manager to
request a Process Hazards Review of a particular process over another, even
though the screen score may be less.

Calculating a Process Hazard Screen Score

The final Process Hazard Screen score is calculated using the equation:

$$\text{Score} = \Sigma\ ((i=1 \text{ to } 3)\ \Sigma\ (j=1 \text{ to } 6)\ E_i \times I_j \times F_{ij})$$

where:
E_i = Extrinsic variable scores (Volume, Experience, Complexity)
I_j = Intrinsic variable scores (Toxicity, Flammability, Reactivity, Exothermic, Process Conditions, Stability)
F_{ij} = Weighting factor for interaction.

The purpose of the weighting factor, F_{ij}, is to quantify the interaction effects between the intrinsic and extrinsic variables. As the Process Hazard Screen tool was developed, it was determined that the extrinsic variables, those which were contingent on the characteristics of every material used in the process, would cause greater risk depending upon the current operating parameters. These current operating parameters, labeled intrinsic variables, were subject to change as "lot" sizes, production schedules, and procedural changes were made. An interaction between the extrinsic and intrinsic variables was utilized to quantify such differences as "large quantities of toxic materials" versus "a relatively simple process with no chemical reactions".

The Process Hazard Screen, computed on a software database, requires the user to only input the intrinsic and extrinsic variables determined in the previous paragraphs. Once these variables are typed into the computer, the program was designed to weight the interaction of the variables using the interaction variables in Table 11. The result was a final Process Hazard Screen score which could be used to rank each process based on the quantities and hazards of materials used, the operating limits of the process, the frequency and knowledge of the process, and the interaction between all the process variables listed.

Table 11 <u>Interaction weighting factor, F, as a function of intrinsic and</u>
<u>extrinsic variables</u>

	Volume	Experience	Complexity
Toxicity	F = 3	1	2
Flammability	3	1	2
Reactivity	3	2	1
Exothermic	3	2	0
Process Cond. Range	2	3	2
Stability	3	2	1

Given the equation:

$$\text{Score} = \Sigma\ ((i=1 \text{ to } 3)\ \Sigma\ (j=1 \text{ to } 6)\ E_i \times I_j \times F_{ij})$$

the minimum Process Hazard Screen score that could be obtained was 24. The maximum obtainable score was 576. All plant site processes were evaluated and scores ranged from the double digits to over 400. An example of a calculation performed at an Eli Lilly and Company site is provided below:

Example 1   Calculation of a Process Hazard Screen score for a batch pharmaceutical process

Volume (2.5)	x Toxicity (3)	x	Weight Factor (3)	=	22.5
Volume (2.5)	x Flammability (4)	x	Weight Factor (3)	=	30.0
Volume (2.5)	x Reactivity (3)	x	Weight Factor (3)	=	22.5
Volume (2.5)	x Exothermic (2.2)	x	Weight Factor (3)	=	16.5
Volume (2.5)	x Proc. Cond. (3)	x	Weight Factor (2)	=	15.0
Volume (2.5)	x Stability (4)	x	Weight Factor (3)	=	30.0
			Volume Factor Calculation	=	136.5

Experience (3)	x	Toxicity (3)	x Weight Factor (1)	=	9.0
Experience (3)	x	Flamm. (4)	x Weight Factor (1)	=	12.0
Experience (3)	x	Reactivity (3)	x Weight Factor (2)	=	18.0
Experience (3)	x	Exoth. (2.2)	x Weight Factor (2)	=	13.2
Experience (3)	x	Proc. Cond. (3)	x Weight Factor (3)	=	27.0
Experience (3)	x	Stability (4)	x Weight Factor (2)	=	24.0
			Experience Factor Calc.	=	103.2

Complexity (2)	x	Toxicity (3)	x Weight Factor (2)	=	12.0
Complexity (2)	x	Flamm. (4)	x Weight Factor (2)	=	16.0
Complexity (2)	x	Reactivity (3)	x Weight Factor (1)	=	6.0
Complexity (2)	x	Exoth. (2.2)	x Weight Factor (0)	=	0.0
Complexity (2)	x	Proc. Cond. (3)	x Weight Factor (2)	=	12.0
Complexity (2)	x	Stability (4)	x Weight Factor (1)	=	8.0
			Complexity Factor Calc.	=	54.0
			Process Hazard Screen	=	**293.7**

This Process Hazard Screen score, along with any subjective data supplied by the department, was used to determine the relative risk. This value was then compared with the other Process Hazard Screen scores for all remaining operations on site. The Process Hazard Reviews for all site operations were prioritized based on the data compiled from the Hazard Screens.

Results/Discussion

Process Hazards Screens were conducted at six Eli Lilly and Company sites, resulting in the screening of over four hundred process operations. The scores from all sites ranged from the 40's to the 440's, and these data were used to propose four-year completion plans for the processes with the highest risks. The operations which produced the highest screen scores were those that operated continuously and processed toxic and flammable materials. Many of the Environmental processes, such as liquid incineration and flammable off-gas incineration scored in the top ten percent at each site. Infrequent operations, such as tank truck unloading of small volume solvents, scored very low.

There was much feedback received from the Plant Site customers who were asked to complete the Process Hazard Screen. Although all processes have been screened once to determine the initial priority order, certain valid requirements were raised by the customers to be included in a "new and improved" version of the Process Hazard Screen. A few of the remarks raised by the sites are captured in the following paragraphs.

Volume scores could falsely alter the final Process Hazard Screen score. Many of the biochemical operations conducted at Eli Lilly and Company are aqueous based, yet use small amounts of hazardous reagents. Unfortunately, a small quantity of hazardous material would be weighted heavily in the overall score since each component is not weighted individually, but weighted based on the overall process volume. Many "friendly" processes, those which utilized large aqueous volumes, calculated artificially high Process Hazard Screen scores due to small volume acid additions. The Process Safety Department intends to revise the method so that the mass of each component processed is rated with the specific hazards common to the particular component.

Accelerated Rate Calorimeter (ARC) data is not available for many of the materials used in processes. ARC data is not available for most chemical solvents. Nor is ARC data available for many raw materials used as reagents or buffers. ARC data is most readily available for bulk intermediate and final product solids which are created from batch pharmaceutical processes. Many of the processes screened at each site were forced to use artificial default values for stability scores since ARC data did not apply in such cases.

Many of the processes screened do not involve a chemical reaction. At all six Eli Lilly and Company sites where the Process Hazard Screen was conducted, every process was evaluated for relative risk. Given all the Utility, Environmental, Biosynthetic, Purification, and Material Transfer operations conducted throughout the corporation, the percentage of processes that utilize a true chemical reaction is very low. Many of the sites were forced to record a default value for the *Exothermic score*. The default value was the same regardless of the operation, and was sometimes very comparable to the *Exothermic score* for a process that actually conducted a chemical reaction. This default value is intended to be changed based on the nature of the process.

The complexity score assumptions do not reflect the written instructions philosophy for all sites. There was much feedback from the customers concerning the validity of the *Complexity score*. The assumption was made that the complexity of a process was directly proportional to the number of pages of written instructions. Two problems surfaced from this assumption. First, some of the continuous utility operations (such as Electricity distribution) did not use a written procedure, resulting in an artificial default value. Second, many of the operations conducted at the bulk production facilities exceeded the 36 page maximum. Most every process screened for a *Complexity score* resulted in a maximum score of 4, or a default value of 2.2. There was a request from many sites to revise the method of measuring the complexity of the process.

The Subjective Data were not quantified, and therefore not incorporated into the final Process Hazard Screen score. Many of the departments which completed the screen felt that the Process Hazard Review schedules were developed strictly from the Hazard Screen score. Furthermore, the *Subjective Data* which was included on every Process Hazard Screen was not transferred from the information package to the spreadsheet that sorted all Hazard Screen scores. Although the intent was to evaluate the final score and the *Subjective Data,* many customers felt as though the exercise to supply this data was not value-adding since it was not reviewed during the prioritization of the processes. Many process engineers felt that knowledge of the *Subjective Data* would be the best information to consider when scheduling a Process Hazards Review.

The original Process Hazard Screen was completed by process engineers and chemists and was recorded in individual information packets. These packets are documented at each site as well as the Corporate Process Safety Department. Although the data captured is very important, the screen packets are not updated as process changes are implemented. Furthermore, as screen packets are not updated, relative risk values remain constant on paper while the actual risk may change in the field. It is the intent of the Eli Lilly and Company Process Safety Department to create a corporate database. Management of change will require engineers to update the database as process changes are implemented, and the Process Hazard Screen value will change electronically to reflect the revised operation. The database will also allow the user to compare "apples to apples", as Utility processes at a particular site can be prioritized while Pharmaceutical syntheses can be prioritized as a separate sub-group.

There were many more points of interested raised by the customers to be included on the next Process Hazard Screen revision. Some customer comments, although valuable, was not consistent with the requirement of providing a quick method of calculating a relative risk. The consensus was, however, that the first approach to Process Hazard Screen was very useful as an analytical technique for developing a prioritized schedule of Process Hazard Reviews.

Summary

The Process Hazard Screen was originally developed by one Eli Lilly and Company bulk production facility as a method of prioritizing the 100-plus processes which were anticipated to operate during the course of the year. The Process Hazard Screen was designed for Bulk Pharmaceutical Processes, and then the use expanded to include Environmental and Utility operations. Finally, the Process Hazard Screen was accepted and standardized at the other Eli Lilly and Company facilities with a Process Safety Management program.

The goal of screening every process in a quick and efficient fashion to determine the relative risk was considered successful. The Process Hazard Screen was not so particular that it provided a distinct score for each process. In fact, processes that were considered similarly were often found to have exact Hazard Screen scores. One other positive point was that processes which were similar yet conducted at different facilities were found to have similar Hazard screen scores.

Given the analysis of material stability, chemical reaction data, and annual production volume, there is no doubt that the Process Hazard Screen is best utilized to determine the relative risk of batch pharmaceutical processes. Fortunately there is the ability and the intent to refine the method and create a tool which better addresses the applicability of other site operations. As improvements are made to the analysis, and processes are screened two and three times, Eli Lilly and Company will continue in its effort to evaluate the hazards and risks associated with the Global Health Care production, and continue to provide operations which protect the integrity of our personnel and environmental resources.

An Approach to Prioritizing Risk-Reducing Projects

Daniel G. Brooks, Allen C. Miller, Gregory L. Hamm,
and Miley W. Merkhofer
Applied Decision Analysis, Inc., 2710 Sand Hill Road, Menlo Park, CA 94025

INTRODUCTION

The standards used to determine spending priorities for risk reduction within many large organizations are often haphazard. Projects developed to address particular risks within an organization may be funded for a variety of reasons: the project is driven by regulatory compliance issues, the project advocates are influential, the budget is affordable, or the type of risk addressed by the project is politically visible. This process leads to much lower risk reduction than could be achieved at the same cost with more effective priorities. Unless an organization clearly defines its objectives for risk management, it will have difficulty allocating scarce resources effectively, documenting that it is using its resources to manage risk responsibly, and communicating its risk management objectives to the public, employees, and regulators.

To evaluate the cost effectiveness of alternative risk-reduction activities, an organization must clearly identify the benefits to be achieved by each activity. The benefits can then be weighed against the cost of achieving them, and the activities evaluated on the basis of benefit returned on investment. Risk-based prioritization is a systematic means of discriminating among alternative activities. It determines which activities utilize resources in the most effective manner. The key to this process is accurately defining and quantifying benefits when: the results of undertaking a risk-reducing activity are uncertain, there are multiple motivations for implementing the activity, parties that will be impacted by the activity have conflicting interests, and the activity requires long-term funding from a variety of sources.

This paper presents a logical and consistent process for solving risk-based prioritization problems. The methodology is technically sound, defensible, and adaptable to a wide range of risk-management environments. This paper describes an application of the process to the prioritization of several hundred activities addressing environmental, safety and health deficiencies at Los Alamos National

Laboratory. As a result of employing this approach, the Laboratory management was able to address a variety of significant concerns, document responsible risk management, and keep the cost of risk reduction projects within budget constraints.

The next section gives some background for the problem of risk-based prioritization and the Los Alamos application. Following sections show how to define and quantify risk reduction benefits, and then combine them with project costs to establish priorities.

BACKGROUND

For several years, the Department of Energy (DOE) has encouraged its facilities to conduct regular and systematic reviews of their operations. The reviews are focused on protecting environmental resources and the health and safety of workers and the public. The DOE sent audit teams to many of its facilities to review the facilities' self assessments. In the fall of 1991, nearly 170 DOE inspectors spent six weeks analyzing the environmental, safety and health activities initiated at Los Alamos National Laboratory as a result of the Laboratory's self assessment. The inspectors compared this analysis to what they found in thorough inspections of the Laboratory's facilities and operations. The result of this audit was a list of over 1000 "findings," each indicating some deficiency in current Laboratory operations relative to DOE and other government regulations.

The audit team's findings were consolidated, and "action plans" were developed to address the findings. This resulted in over 200 action plans with a total estimated cost of almost $1 billion. Since the Laboratory's budget precluded additional expenditures at this level, the Laboratory took the position that meeting these commitments would require a concerted effort to review, plan, and prioritize the action plans. The DOE agreed with the Laboratory.

The Laboratory adopted a prioritization process that attempts to achieve as much risk reduction as possible with the available resources. The process identifies when to stop spending resources because the value of a risk reduction is less than the resources required to achieve it. The process also serves as a communication tool. It facilitates communication between technical experts and policy makers in charge of risk management. It also provides a basis for communicating the logic of prioritization to the public, employees, regulators, and, if necessary, the courts.

Finally, the prioritization process provides a sound, defensible audit trail, so that the basis upon which limited funds were allocated to risk-reducing activities would stand up to technical and legal review. This audit trail shows how the concerns of each stakeholder impacted by the Laboratory's decisions are addressed. Stakeholders include policy makers who determine the relative importance of the various risks, technical specialists who estimate the level of risk, employees and members of the public whose resources are being protected, and regulators who authorize and provide funding for the activities. The audit trail makes explicit the technical judgments supporting risk estimates, the value judgments underlying tradeoffs among risks, major sources of uncertainty in the process, and the way in which consequences were ultimately valued. This gives the stakeholders a framework for participating in the process effectively.

DEFINING AND QUANTIFYING BENEFITS OF RISK REDUCTION

The prioritization process developed at Los Alamos is based on multiattribute utility theory. This is a prescriptive, axiomatic methodology for modeling values when there are multiple objectives to be achieved, multiple stakeholders effected by the decisions, uncertainty about the outcomes of some activities, and the need for coordinating technical inputs from disparate disciplines. The initial developments in the field of single-variable utility theory occurred in the 1940's. The theory was expanded in the 1960's to include multiple evaluation measures. The application of this methodology to prioritization problems has been reviewed by the National Academy of Sciences and by various technical and academic review groups. They concluded that the approach is technically sound and appropriate.

The methodology bases the allocation of limited resources to risk reduction activities on the value judgments of decision makers. The explicitness of the process allows effective communication of these value judgments to employees, regulators, and the public so that differences in judgments can be discussed and understood. The multiattribute utility approach to evaluation is quantitative, using evaluation measures (often called attributes) to measure the degree to which an activity reduces each risk. These quantitative measures are then combined into a single benefit value. The combination rule is based on an investigation of the relationships between the various types of risks and the way in which they are measured.

The remainder of this section describes the overall evaluation process, the evaluation categories that define benefit, the measures that quantify results in each category, how these measures are used to estimate averted risk, and how the averted risks are combined to estimate the overall benefit of implementing an action plan.

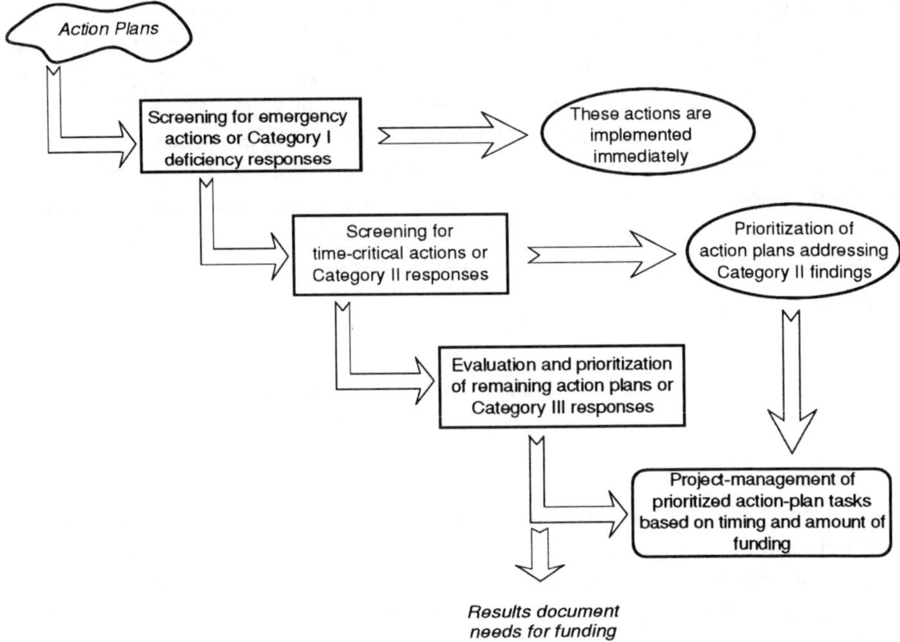

FIGURE 1. Overall Activity Evaluation and Implementation Process

The Evaluation Process

Figure 1 illustrates how the prioritization system is incorporated into an overall process of activity review, screening, and priority-setting. The formal prioritization process deals with activities that remain after a screening process that detects emergency and time-critical activities.

Evaluation Categories

The Laboratory's prioritization system determines the value of an action plan by measuring the degree to which the plan reduces "total risk," as defined by both DOE and the Laboratory. Total risk includes worker and public health and safety, environmental impacts, public concerns and confidence, regulatory and statutory compliance (including the impact of regulations on Laboratory programs and mission), and impacts on efficiency. The total risk is represented by a set of evaluation categories corresponding to these issues. Figure 2 shows the risk

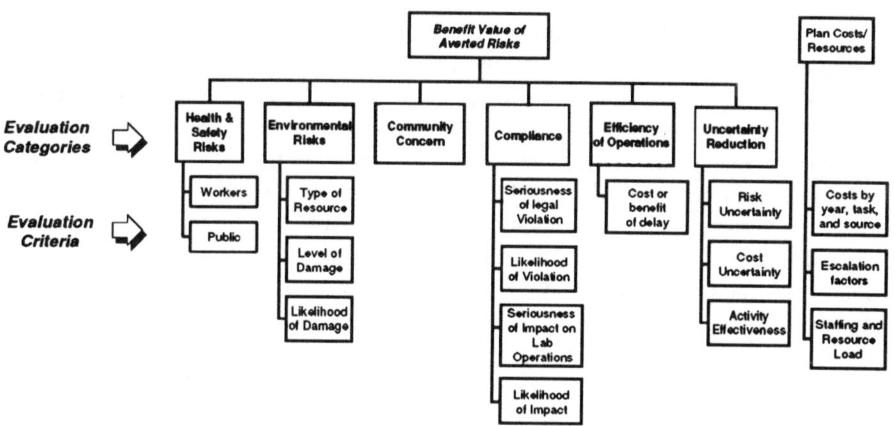

FIGURE 2. Evaluation Categories Defining the Benefit of Averted Risk

categories. Quantitative scales were developed to measure the reduction of risks in these categories.

These categories were developed from a series of interviews with Laboratory senior management. After a first round of interviews, further interaction with special technical groups helped provide more detailed descriptions of the categories. For instance, one Laboratory group, established and funded to serve as a liaison between Laboratory managers and local citizens, was consulted on development of the "Community Concerns" category. Laboratory legal council was consulted on compliance issues; health and safety managers were involved in defining both worker and public health and safety issues.

The risk evaluation categories also correspond to the "risk concerns" identified by the DOE audit team that came to review the Laboratory's self-assessment report. This correspondence was important in showing that while the Laboratory used a formal structure to evaluate activities, this structure was fully consistent with the regulators' concerns.

Measurement Scales and Value Functions

A series of meetings with appropriate Laboratory technical specialists produced a quantitative scale for measuring each evaluation criterion under the categories shown in Figure 2. These scales measure the degree to which an activity addresses the evaluation criteria. Figure 3 shows the scales associated with worker health and safety. Laboratory health and safety professionals helped develop these scales. The product of the scores on these scales estimates the expected number of health impacts, weighted for time urgency.

The scales were reviewed by several teams of experts to ensure that they accurately represented the technical information requested in the scoring process. Laboratory management also reviewed the measurement scales. They then participated in a series of formal elicitation sessions where they responded to tradeoff questions. These sessions produced a value function for each scale in each evaluation category.

For example, Figure 4 shows their value function for the severity of health and safety effects.

Eliciting Risk Scores

The person most familiar with each action plan (usually the person who developed the plan) used the measurement scales to estimate how well the plan reduced risks. This process was called risk scoring. Each plan was scored in a structured elicitation process. The experts provided two sets of scores for each plan. The first set estimated the baseline risk, or the risk level before an activity was implemented. The second set of scores estimated the modified risk, or the level of risk after the plan was implemented. The difference between the baseline and modified risk levels is an estimate of the risk reduction attributable to the plan.

There is always the potential that the person scoring an action plan, while being well informed about the plan, may not score it objectively. To guard against this sort of project advocacy, the scores underwent a series of reviews. The review procedure was designed to control both within-activity variation (internal

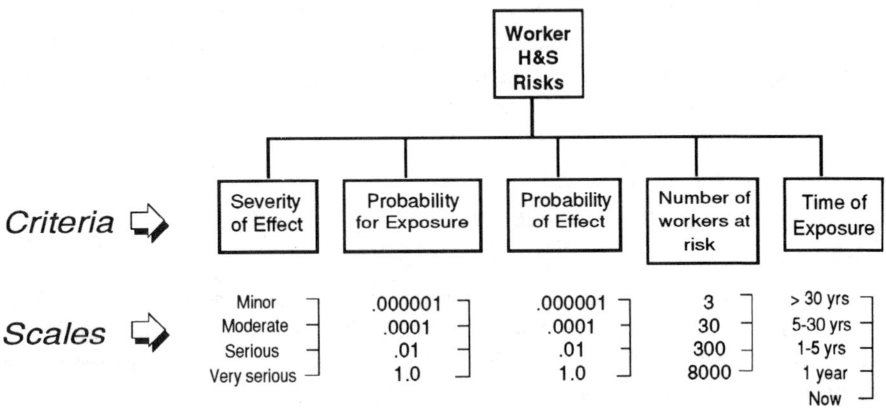

FIGURE 3. Measurement Scales for Worker Health and Safety

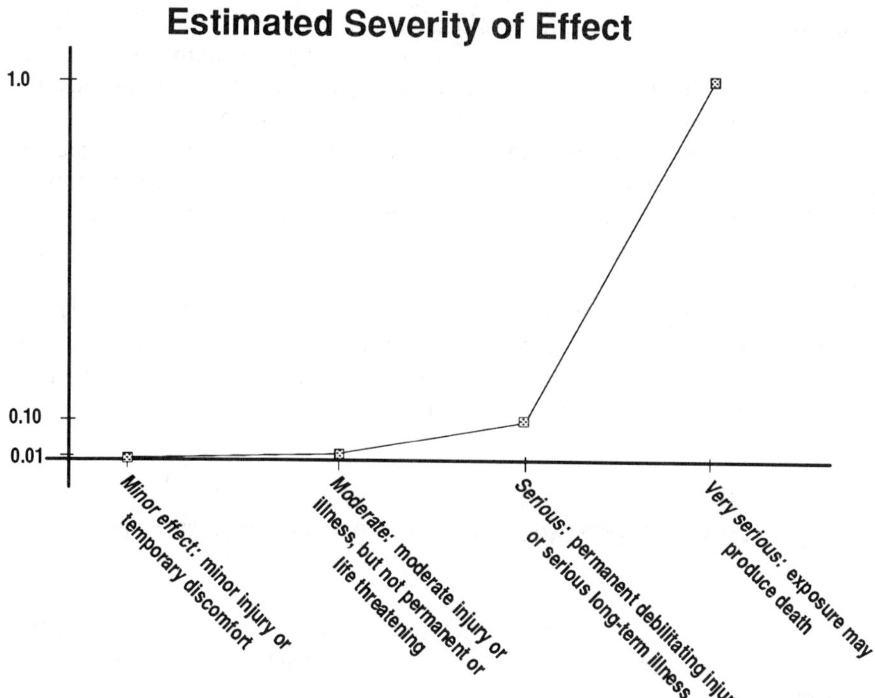

FIGURE 4. Value Function for Severity of Health Effect

consistency) and between-activity variation (consistency across activities). Several interdisciplinary technical and managerial review teams participated in the reviews. They had authority to modify scores that were clearly inconsistent.

Combining Evaluation Measurements

The next step was to combine the scores associated with each action plan into a single measure of benefit. The objective was to represent the benefit derived from implementing an action plan with a single number. Simple models combined the scores within each evaluation category, and tradeoff weights supplied by the Laboratory management combined the risk category scores into an overall benefit estimate.

These weights, or tradeoff coefficients, reflect the fact that the evaluation categories are not of equal importance. For example, worker health and safety is much more important than Laboratory efficiency. A formal elicitation determined the set of weights for the risk categories, reflecting the relative importance management ascribes to each of the categories. This set of weights represents the values of risk managers when they allocate resources to competing risk reduction opportunities. The development of the tradeoff weights required discussion between different groups within the Laboratory so that all management perspectives on relative importance were taken into account.

In addition to determining the relative weights for the evaluation categories, the assessments also produced a rule for combining these weighted measures. In the Laboratory application, a series of tradeoff exercises done with various groups of managers showed that it was valid to use a linear combination of category scores.

Sensitivity analysis showed that changing the tradeoff weights over plausible ranges had only a modest effect on the benefit associated with each activity. Even major changes in the weights (orders of magnitude) had little impact on the overall ranking of activities by total benefit.

RESOURCE ESTIMATES AND PRIORITIZING ACTIVITIES

The first part of the prioritization process estimated the benefit derived from each activity. The second part examined the constraints on resources needed to implement the activity.

Implementing an activity requires a payment stream that may last for years. It is much like acquiring a mortgage. The payment stream may have "balloon payments" in future years. Although it may be attractive to start an activity in the current fiscal year, implementing it now may obligate a large portion of some future year's budget. Further, an activity may require funds from different sources with different budget limitations. It is important that the resource analysis take into account not just the total cost of an activity, but the payment stream and source of funds.

For some activities, the most difficult resources to obtain may not be monetary. An activity may require personnel with rare skills, or specialized (and scarce) equipment. The Laboratory's prioritization system elicited these special resource requirements, and they were entered as resource requirements along with monetary costs.

Prioritizing Activities

Selecting the set of action plans that yields the greatest benefit and can still be funded under the set of resource constraints (by year and source) is an optimization problem. Because resources come from different sources in different years with different limits, it isn't possible to just "follow a list until you run out of money." Instead, the process must find the solution to the following "portfolio" problem. What is the portfolio of activities consistent with all the resource constraints that maximizes the total benefit obtained? This problem was solved in Laboratory's prioritization system using a mixed-integer programming model.

The graph in Figure 5 shows a simplified version of the results. It shows the cumulative benefit of implementing action plans in order of their priority, as a function of their cumulative cost. It is clear from the graph that some activities are well worth implementing, some provide benefit but are expensive and should be re-designed, and some provide little benefit and require substantial resources. The shape of the graph also shows that a major portion of the benefits can be obtained for only a small portion of the total cost. This graph is just a summary of the results available from solving the portfolio problem. The details of the portfolio solution tell managers how to allocate resources over time and within various budget categories.

The prioritization process documented the Laboratory's efforts to manage the risks to workers, the public, and the environment. The process also provided an auditable system for defining risk, measuring its reduction, and effectively managing it with limited budgets.

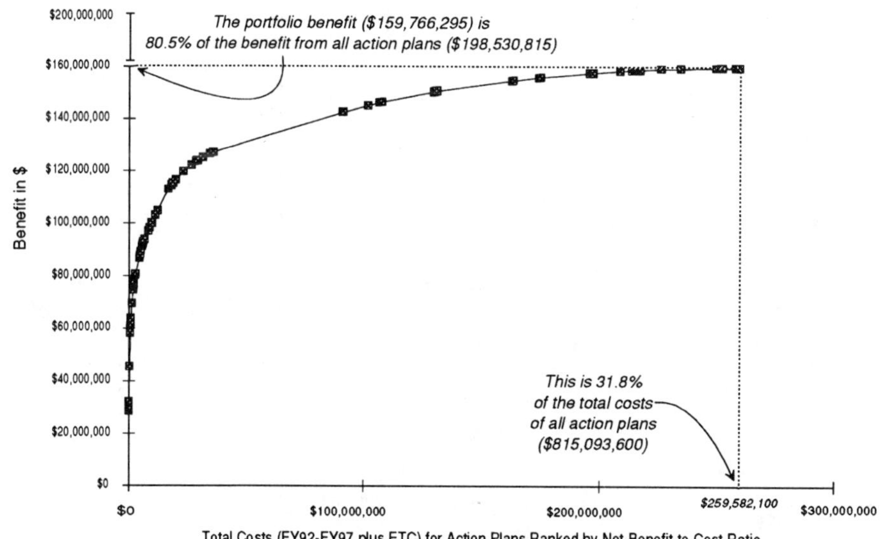

FIGURE 5. Benefit to Cost for a Portfolio

Catastrophes—The Underlying Causes

Robert M. Arnold, Jr.
DNV—Loss Control Management
International Loss Control Institute, Inc.
Loganville, GA

This briefing paper will demonstrate that the causes of catastrophic loss events, regardless of their effects to people, property and the environment, are quite similar to the causes of incidents, in general, and that the consequences of incidents are largely fortuitous. Indeed, once a sequence of events is initiated which could result in loss, it is difficult to interrupt these events. Often, the only thing which can be done is to mitigate the consequences of the incident.

A UNIVERSAL LOSS CAUSATION MODEL AND ITS RELEVANCY TO THE CATASTROPHIC LOSS PRODUCING EVENTS

To understand the impact Loss Control Management (LCM) can have in reducing catastrophic losses, it is vital to understand what is meant by the approach.

For the purpose of this paper, an incident is defined as an undesired event that results in harm to people, environment, property and/or process.

The causes and effects of incidents are illustrated throughout the practical loss causation model illustrated in Figure 1.

Loss

The result of an incident is loss. As reflected in the incident definition, the most obvious losses are harm to people, property, environment or the process. Once the sequence has occurred, the type and degree of loss are somewhat a matter of chance. The effects may range from insignificant to catastrophic. The type and degree of loss depend partly on fortuitous circumstances and partly on the actions taken to minimize losses.

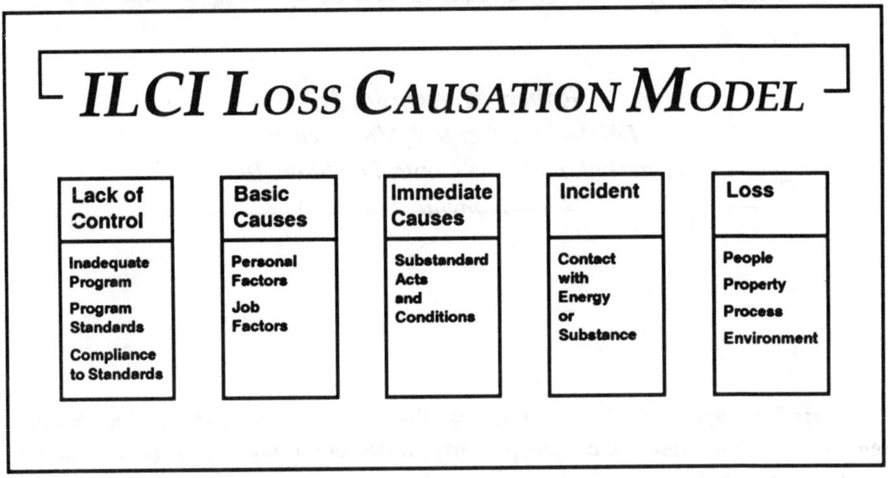

Figure 1. ILCI Loss Causation Model

Incident

This is the event that precedes the "loss" and involves a contact that causes it. When potential causes of incidents are permitted to exist, the way is always open for a contact with a source of energy or substance above the threshold limit of the body, environment or object.

Immediate Causes

The "immediate" causes of incidents are the circumstances that immediately precede the contact. They are substandard acts (behaviors which could permit the occurrence of an incident) and substandard conditions (conditions which could permit the occurrence of an incident). An increasing body of evidence confirms the results from research in quality control which indicates that 80% of the mistakes that people make are the result of factors over which only management has control. This finding calls into question the long-held concept that the majority of incidents result from the unsafe acts or faults of people. Progressive organizations are focusing on how management systems influence behaviors and conditions rather than focusing on "unsafe acts" of people.

Basic Causes

Basic causes are the causes behind the symptoms; the reasons why substandard acts and conditions exist; the factors that, when identified, permit meaningful management control. Just as it is helpful to consider two major categories of immediate causes (substandard practices and substandard conditions), so it is also helpful to think of basic causes in two major categories:

Personal Factors such as:

- Inadequate capability
  - Physical/Physiological
  - Mental/Psychological
- Lack of knowledge
- Lack of skill
- Stress
- Improper motivation

Job Factors (Work Environment) such as:

- Inadequate leadership and/or supervision
- Inadequate engineering
- Inadequate purchasing
- Inadequate maintenance
- Inadequate tools, equipment, materials
- Inadequate work standards
- Wear and tear
- Abuse and misuse

Basic causes are the origins of substandard practices and conditions. However, they are not the beginning of the causes and effect sequence. What starts the sequence, ending in loss, is "lack of control."

Lack of Control

Control is one of the four essential management functions: plan, organize, lead, control. These functions relate to anyone required to manage an activity, regardless of level or title. Whether the function is administration, marketing, production, quality, engineering, purchasing or loss control, the supervisor/leader/manager must plan, organize, lead and control to be effective. Without adequate control, the incident cause and effect sequence is started and, unless corrected in time, leads to losses.

There are three common reasons for lack of control: 1) inadequate program, 2) inadequate program standards, and 3) inadequate compliance with standards.

Inadequate Program

A loss control program may be inadequate because of too few program activities or focusing on improper program activities. While the necessary program activities vary with an organization's scope, nature and type, significant research and experiences of successful companies show the program activities in Figure 2 to be fairly common.

Twenty Elements of an Effective LCM System

1. Leadership and Administration
2. Management Training
3. Planned Inspection
4. Task Analysis and Procedures
5. Accident/Incident Investigations
6. Task Observations
7. Emergency Preparedness
8. Organizational Rules
9. Accident/Incident Analysis
10. Employee Training
11. Personal Protective Equipment
12. Health Control
13. Program Evaluation System
14. Engineering Controls
15. Personal Communications
16. Group Meetings
17. General Promotion
18. Hiring and Placement
19. Purchasing Controls
20. Off-the-Job Safety

Figure 2. Twenty Elements of an Effective LCM System

Inadequate Program Standards

A common cause of incidental loss is standards which are not specific enough, clear enough or challenging enough. Adequate standards are essential to controlling proper losses.

Inadequate Compliance with Standards

Lack of compliance with existing standards is quite off a reason for lack of control.

The following case studies of catastrophes from around the world reflect findings of official inquiries and actual investigation reports. These findings can be traced directly back to the lack of control stage of the loss causation model.

CASE 1
EVENT DESCRIPTION

Three Separate Incidents in A Large Refinery
(for proprietary reasons, the organization is not mentioned)

First Event. A crew consisting of two fitters, a rigger and a crane operator were assigned to break a flange and remove a cross-over valve on a flare line. A supervisor performed the necessary checks to satisfy work permit procedures and determined that the system involved was free of liquid and had an acceptably low level of residual gas. As the flange was broken it became apparent to the crew that there was an unacceptable level of both liquid and gas present and they requested a re-check of the area and some "spark proof" tools. Again, a "responsible person" checked and approved the resumption of work. As the last bolts were removed from the flange, the crane suddenly pulled the sections apart and a large amount of liquid poured out. This quickly created a gas cloud which reached an ignition source. The resulting fire killed two of the crew and seriously injured the other two.

Second Event. The hydrocracker was being recommissioned following shutdown. During the process, the plant tripped causing a number of anomalies in the system. The hydrocracker appeared to be satisfactory and was placed on stand-by along with various other system components, pending approval of the area supervisor to continue the recommissioning. Before the supervisor returned to the area, there was a massive explosion and a fire which burned for several hours.

Third Event. A crude oil storage tank had been prepared for cleaning. A significant amount of sludge had built up on the floor of the tank and was to be removed by a crew of contractor personnel. Smoking by at least one of the crew ignited the sludge and the resulting flames and smoke killed one of the crew.

Investigation Findings and Recommendations

Following each of the three events, the appropriate government agency investigated. The results of these investigations clearly identify the need for a systematic, multi-faceted approach to loss control. Among the findings:

1. Management should have carried out more detailed analysis of work and systematic identification of hazards.

2. More checks and balances should have been built into the work permit process.

3. Proper procedures based on task analysis should have been in place with appropriate training.

4. Inspections of equipment were deficient.

5. Skill practice and training were inadequate.

6. Near-miss reporting and investigation of previous events was seriously lacking.

7. Safety trips were routinely bypassed with the supervisor's knowledge and permission.

8. The design of controls led to a high probability of confusion and mistakes.

9. Testing of critical system components was not carried out routinely or with proper follow-up.

10. Changes to plant were made without proper review or loss exposure identification.

11. Contractor controls were inadequate in terms of training requirements, routine inspection of work sites, rules compliance and communications.

12. Emergency procedures and planning were inadequate.

Loss Control Program Elements Indicated

1. Leadership and Supervision
2. Purchasing of Contractor Services
3. Organizational Rules and the Work Permit System
4. Workplace Inspections
5. Incident/Incident Reporting and Investigation

6. Task Analysis and Procedures
7. Emergency Preparedness
8. Training
9. Communications
10. Engineering Controls

CASE 2
EVENT DESCRIPTION

The Kings Cross Fire - London Underground

A fire in an underground public transit station killed and injured over 50 passengers and employees of the London Underground. An investigation concluded that the fire started in the mechanism of a wooden escalator when a cigarette dropped on the stairs came in contact with dust and grease which had accumulated in the mechanism. The fire apparently smoldered for quite a while before igniting the wooden stairs themselves, trapping people underground and sending smoke and intense heat through main public areas.

Investigation Findings and Recommendations

The official government investigation into the tragedy concluded:

1. Low overall performance in safety management including low standards in training, preparation of proper work procedures, engineering controls and the behavioral aspects of safety.

2. An emphasis on reactive measures rather than a proactive approach.

3. A segmented program should be replaced with a systematic, integrated approach.

4. Emergency preparation and planning were deficient.

5. Company rules were lacking and inadequately enforced.

6. Safety committees and safety representatives were not organized according to legislated requirements.

7. No management audit or program assessment was being carried out on a regular basis to identify strengths and weaknesses in the program.

8. Managers were not trained in appropriate safety management techniques.

9. There was no systematic review of tasks to identify training needs of employees.

10. A review of all stations should be carried out to identify the critical parts/items which might need regular inspection.

11. Loss investigation and analysis techniques should be greatly improved. (Similar incidents had occurred with some regularity but had not been reported or adequately investigated.)

12. Engineering controls should be improved with safety reviews at the conception, design and commissioning stages of all new construction.

13. Regular group meetings of personnel should be held to discuss matters of safety.

14. A system of planned inspections should be implemented to ensure proper coverage of all property according to loss potential and organization requirements.

Loss Control Program Elements Indicated

1. Leadership and Supervision
2. Workplace Inspections
3. Engineering Controls
4. Communications
5. Emergency Preparedness
6. Task Analysis and Procedures
7. Incident/Incident Reporting and Investigation
8. Program Auditing
9. Safety Management Training for Company Managers
10. Training Needs Analysis and Development for Personnel
11. Organizational Rules
12. Safety Promotion
13. Task Observation

CASE 3
EVENT DESCRIPTION

The Sinking of the Ocean Ranger

The world's largest and most technologically advanced semi-submersible drilling rig capsized and sank off the coast of Newfoundland killing 84 crewmen. The rig was considered to be almost unsinkable, but fell victim to a series of errors compounded by design flaws and inadequate emergency procedures. During a severe North Atlantic storm, ballast procedures which were supposed to keep the rig stable resulted instead in its loss.

Investigation Findings and Recommendations

1. A misunderstanding of ballast procedures resulted in fatal and irreversible errors.

2. The design of ballast controls led to confusion.

3. Inadequate emergency procedures and equipment contributed to delays and improper operation resulting in the deaths of those who escaped the rig itself.

4. Evacuation procedures were not properly practiced or treated with appropriate concern.

5. Arguments over reporting of incidents and inconsistency in the way critical situations were handled contributed to the loss of the rig.

6. The ballast control room was improperly designed in light of its possible exposure to sea water and its electrically-operated, automated ballast equipment.

7. There was inadequate identification of loss exposures at the design and construction stages of the vessel.

8. The back-up, manual ballast control system was inadequate.

9. Tasks on the rig were "learned by doing." There were no adequate written procedures.

Loss Control Program Elements Indicated

1. Leadership and Supervision
2. Engineering Controls and Design Review
3. Task Analysis and Procedures
4. Communications
5. Emergency Preparedness
6. Training
7. Incident/Incident Reporting and Investigation

CASE 4
EVENT DESCRIPTION

The Explosion of Piper Alpha

On 6 July 1988, a major explosion and fire destroyed the production platform Piper Alpha in the North Sea resulting in the loss of 165 lives. The sequence of events leading to the tragedy began when a cloud of gas condensate accumulated in an area where maintenance work on pumps and valves was in progress. The initial explosion led to a large crude oil fire which eventually touched off a second major explosion when a gas pipeline riser ruptured. Emergency systems were rendered almost inoperative due to the extreme nature of the situation.

Investigation Findings and Recommendations

An official government inquiry into the disaster was undertaken and resulted in a two-volume report of findings and recommendations. The following is a summary of major conclusions.

    1. There was a failure in the work permit system which allowed variations from approved procedures resulting in unsafe actions.

    2. There were deficiencies in the standard of information handed over at the time of shift change.

    3. The fire water system was in substandard condition and changes recommended by a previous audit had not been implemented.

    4. Induction or orientation to emergency procedures was cursory and company-required muster and evacuation drills were not conducted as prescribed.

    5. Safety policies and procedures were in place, but their practice was deficient.

    6. Management assumed that organization requirements in terms of the work permit system were being followed, but had no way of evaluating or measuring compliance.

    7. Management failed to provide proper training in the application of the work permit process.

    8. Management adopted a superficial attitude toward the assessment of risk of a major hazard.

    9. Inspections performed on the platform were severely inadequate and failed to identify major deficiencies.

    10. No action was taken on lessons learned from previous incidents on the platform.

11. There was no systematic assessment of risk in the safety process off-shore.

12. There was no systematic management of safety in the process.

13. There was no regular auditing program being applied which would identify for management critical areas of deficiency in the process itself.

Loss Control Program Elements Indicated

1. Leadership and Supervision
2. Communications
3. Organizational Rules and the Work Permit System
4. Program Evaluation Systems (Program Audits)
5. Workplace Inspections
6. Task Analysis and Procedures
7. Training
8. Emergency Preparedness
9. Systematic Loss Exposure Identification and Evaluation
10. Incident/Incident Investigation and Follow-Up
11. Engineering Controls
12. Contractor Control

CASE 5
EVENT DESCRIPTION

The Investigation of the Space Shuttle "Challenger" Explosion

The space program of the United States was brought to a sudden and tragic halt when the space shuttle "Challenger" exploded in flight killing all seven astronauts aboard.

Investigation Findings and Recommendations

1. There was too much reliance and confidence in past performance and inadequate attention to warning signs.

2. A high productivity demand resulted in the elimination or subrogation of proven safety guidelines.

3. Inadequate controls on contract labor resulted in unskilled workers installing heat tiles on the orbiters.

4. There was inadequate reporting and follow-up of critical problems with O-Ring seals on 10 out of 23 previous flights.

5. Crews were grossly overworked.

6. Back-up systems were ignored or used as primary with no additional redundant capability.

7. Unauthorized tools and equipment were used in shuttle preparation.

8. Changes in materials and processes were not adequately evaluated.

9. Substandard conditions were not reported.

10. The decision-making process was flawed.

11. Strict launch criteria were compromised.

Loss Control Program Elements Indicated

1. Leadership and Supervision
2. Incident/Incident Reporting and Investigation
3. Communications
4. Workplace Inspections
5. Contractor Controls
6. Training
7. Task Analysis and Procedures
8. Loss Exposure Identification and Evaluation
9. Engineering Controls

Analyzing the results of these and other major catastrophes reveals that a pattern of inadequate management controls exists in the following elements:

I. Leadership and Administration
    A. Lack of Management Performance Standards
    B. Lack of Specific Program Objectives (Other Than Statistical Goals)
II. Management Training in Loss Prevention Concepts and Techniques
III. Planned Workplace Inspections
    A. Lack of A Systematic Approach
    B. Inadequate Follow-Up
IV. Task Analysis and Procedures
    A. Inadequate Loss Exposure Identification
    B. Inadequate Procedures
    C. Exceptions and Variations Permitted w/o Adequate Evaluation
    D. Changes to Subsystems (People-Equipment-Material-Environment)

   Not Captured and/or Accommodated in the Task Analysis Program
 V. Accident/Incident Reporting and Investigation
   A. Inadequate Near-Miss Reporting and Follow-Up
   B. Inadequate Cause Analysis
   C. Inadequate Follow-Up and Close-Out of Remedial Actions
 VI. Emergency Preparedness
   A. Inadequate Evaluation of Emergency Response Needs
   B. Inadequate Orientation to Emergency Procedures
   C. Lack of Proper Drills/Practice
   D. Inadequate Inspection and Maintenance of Emergency Equipment
 VII. Organizational Rules and the Work Permit System
   A. Inadequate Training in the Work Permit System
   B. Exceptions and Variations Permitted w/o Proper Risk Evaluation or Authority
   C. Lack of Auditing/Feedback System to Identify Non-Compliance Problems for Management Attention
 VIII. Employee Training
   A. Inadequate Identification of Training Needs
   B. Inadequate Evaluation of Training Program Effectiveness
   C. Lack of Appropriate Refresher Training and/or Testing
 IX. Program Evaluation System
   A. No Comprehensive Audit Systems Being Applied to Identify Strengths and Weaknesses in the Safety/Loss Control Program
   B. Lack of Proper Follow-Up on Audit Results
 X. Engineering Controls
   A. Inadequate Reviews and Loss Exposure Assessment at the Concept, Design and Commissioning Stages of Plant and/or Process
   B. Engineering Changes Made in the Field w/o Proper Review, Evaluation and Authority
 XI. Communications
   A. Inadequate Job-Specific Orientation
   B. Poor Task Instruction
 XII. Purchasing of Contractor Services
   A. Inadequate Evaluation of Potential Contractors
   B. Inadequate Training and/or Qualification of Contract Workers
   C. Inadequate Daily Liaison and Control of Contractors

The pattern is reinforced in the recent article in the November 1990 issue of Professional Safety: "Preventing Incidents at Oil and Chemical Plants" by Najmedin Meshkati.

General Findings Following Major Incidents

1. System defects and inadequate management control activities are the major contributors to incidents.

2. The following are critical components in a proper, systematic approach to safety:

- proper workstation design and ergonomics
- proper work procedures and instructions
- operator and management training
- structured supervisory systems
- responsive feedback and communications systems
- an examination of human needs, capabilities and limitations
- emergency response systems
- genuine management commitment
- an integrated, systematic approach

It has been established that the most effective approach to controlling catastrophic losses is through focusing on management systems, not by focusing on unsafe acts or conditions; and that similar elements, which are effective in controlling injuries, illnesses, property damages and environmental spills, directly relate to catastrophes as well. A question regarding the relative importance of these elements to various industry groupings and for controlling catastrophic events versus more "routine" losses can be asked.

It would be interesting to determine if there is a significant, perceived difference in element importance when considering its effect on reducing just major or catastrophic events. To answer this question a study was commissioned to determine if a difference existed and if one did exist, how significant it was.

The method used to conduct the study was to sample 85 organizations, 35 % of these from the process industry, using a statistical ranking method called "Paired Comparison." The results of the study are shown on the following page.

Dividing the twenty elements into thirds allows us to draw some general conclusions concerning similarities between the rankings. The top third of the elements can be considered to be highly important as opposed to a moderate and low relative importance.

The study indicated that two elements, engineering controls and task analysis and procedures, were included in the top third for catastrophic events. Engineering controls could certainly be equated to plant integrity; however, task analysis and procedures could be usually associated with personal safety, property, environment and process issues. Personal protective equipment and health controls which are primarily "personal safety type of elements" were found in the bottom third of elements important for controlling catastrophic losses. This fairly minor shifting of elements suggests that, in general, the relative importance of elements is fairly similar in relation to their effect on controlling a catastrophic event as well as on controlling routine harm to people.

Paired Comparison Ranking of the
Loss Control Management Program Elements

ELEMENT RANKINGS OF RELATIVE IMPORTANCE FOR CONTROLLING HARM TO PEOPLE, EQUIPMENT, PROPERTY AND ENVIRONMENT.

1. Leadership and Administration
2. Emergency Preparedness
3. Planned Inspections
4. Management Training
5. Accident/Incident Investigation
6. Employee Training
7. Health Controls
8. Organizational Rules/Permits
9. Engineering Controls
10. Task Analysis and Procedures
11. Accident/Incident Analysis
12. Personal Protective Equipment
13. Personal Communication
14. Program Evaluation/Audits
15. Purchasing Controls
16. Group Communications
17. General Promotions
18. Hiring and Placement
19. Task Observation
20. Off the Job Safety

RANKINGS BASED ON "PAIRED COMPARISON" STUDY TO DETERMINE THE IMPORTANCE OF EACH ELEMENT IN CONTROLLING CATASTROPHIC LOSS

1. Leadership and Administration
2. Employee Training
3. Engineering Controls
4. Management Training
5. Task Analysis and Procedures
6. Planned Inspections
7. Emergency Preparedness
8. Accident/Incident Investigation
9. Organizational Rules/Permits
10. Personal Communications
11. Task Observation
12. Accident/Incident Analysis
13. Program Evaluation/Audits
14. Hiring and Placement
15. Purchasing Controls
16. Group Communications
17. Health Controls
18. Personal Protective Equipment
19. General Promotion
20. Off the Job Safety

Following close review of numerous catastrophic events, it has been established that management system failures are the basic causes of harm to people, environment, property and processes. As reflected in the results of inquiries from 5 case studies, it has been established that the same program elements seem to be relevant as well. The relative importance of each element in controlling losses is generally constant for minor as well as major loss events was evidenced in the paired comparison study.

Since Loss Control Management Systems are proven effective at reducing both minor and major incidents, it is logical to conclude that the frequency and consequences of catastrophic events will be reduced by focusing attention on management systems for controlling losses, not on management systems for controlling CATASTROPHES.

History and Development of a Safety Management System Audit for Incorporation into Quantitative Risk Assessment

Linda J. Bellamy and Michael S. Wright
Four Elements, Ltd., Greencoat House, Francis Street, London SW1 1DH, UK

Nick W. Hurst
Health & Safety Executive, Sheffield, S3 7HQ, U.K.

1. HISTORICAL BACKGROUND

Interest in the influence of organisation and management on safety began in the UK in the early 1980s. A study was commissioned by the UK Atomic Energy Authority to investigate whether any data existed to quantify the effects of these influences on human reliability (Bellamy 1983a). At this time the methods of quantifying human error were being reviewed and developed in the UK and a peer group was formed which eventually published a report evaluating and illustrating the various methods available (Safety and Reliability Directorate 1988).

It had been hoped that the study on organisation and management would provide data for input into human reliability assessment (HRA) methods. None of the data used in HRA had been validated, and discussion on methods concentrated on data modelling for tasks, such as following a procedure, or control room response in an emergency.

The results of the study by Bellamy (1983a) indicated that the organisation and management variables neglected in HRA were important determinants in accident causation. But methods of modelling for data collection on these effects were not available. Initially, therefore, it was important to highlight the management failures which appeared to have the greatest contribution to accident causation (Bellamy 1983b). In particular, there were four major areas which recurred in major system failures. The most prevalent was communication failures across organisational boundaries. The other themes were failures caused by pressures (eg. time, peer group, workload, uncertainty), equipment and people resources problems (eg. insufficient means of communication, lack of required skills, organisational overlap in use of resources), and organisational rigidity in response to change (eg. failure to upgrade standards and personnel awareness). Typically then, an organisation prone to accidents might be expected to exhibit many of the following features:

- Poor control of communication and coordination:

  - between shifts.
  - upward from front line personnel to higher management in the organisational hierarchy and downward in terms of implementing safety policy and standards throughout the line of management (particularly in a many-tiered organisation).
  - between different functional groups (eg. between operations and maintenance, between mechanical and electrical).
  - between geographically separated groups.
  - in inter-organisational grouping (particularly where roles and responsibilities overlap) such as in the use of sub-contractors, or in an operation which requires the coordination of multiple groups within the same operational "space".
  - in heeding warnings (which is one of the important manifestations of the above where the indicators of latent failures within an organisation become lost or buried)

- Inadequate control of pressures:

  - in minimising group or social pressures
  - in controlling the influence of workload and time pressures
  - of production
  - of conflicting objectives (eg. causing diversion of effort away from safety considerations)

- Inadequacies in control of human and equipment resources:

  - where there is sharing of resources (where different groups operate on the same equipment), coupled with communication problems. Eg. Lack of a permit-to-work system.
  - where personnel competencies are inadequate for the job or there is a shortage of staff
  - particularly where means of communication are inadequate
  - where equipment and information (eg. at the man-machine or in support documentation) is inadequate to do the job

- Rigidity in system norms such that systems do not exist to:

- adequately assess the effects and requirements of change (eg. a novel situation arises, new equipment is introduced)
- upgrade and implement procedures in the event of change
- ensure that the correct procedures are being implemented and followed
- intervene when assumptions made by front line personnel are at odds with the status of the system
- control the informal learning processes which maintain organisational rigidity

When many of these organisation and management characteristics appear together, the effects can be disastrous if the latent failures within the system are not recovered, particularly when warning indicators (eg. near misses) start to appear. Major accidents in the UK which exhibited organisational and management failures of the types described above include the Flixborough chemical works explosion in 1974 (Parker, 1975) and the Piper Alpha offshore disaster in 1988 (Cullen, 1990). The application of an analysis of the influence of safety management can be found in Bellamy (1986) in an analysis of the Bhopal disaster.

The undertaking of this work on management failure provided a valuable starting point for addressing the question of the influence of management on risk. In particular, the question arose as to the importance of taking account of the management factor when assessing the risk to the public arising from accidents on hazardous installations. This question was stimulated by industry (outside the UK) in the mid-80s.

What followed was the development of an audit technique called MANAGER (MANagement Assessment Guidelines for the Evaluation of Risk) which provided a question set for reviewing the quality of management in the area of the four themes described above (Bellamy and Geyer, 1991). With the questions was a set of "anchor points" which very broadly indicated the characteristics of the good and bad management extremes for each of the four themes. The purpose of the anchor points was to assist auditors in making judgements, and help to provide some consistency in judgement. However, in practice, auditors tended to use their own standards, based on the repertoire of their own experience in the chemical industry of what was average practice.

The technique provided a means of translating the auditor's judgements into a numerical factor. This "Management Factor" was used to modify the generic failure rates of pipework, in-line equipment, and vessels used in Quantitative Risk Assessment. The auditors' judgements were based on using a 3 point scale of good, average and bad. The sum of the good, average and bad attributions for all the subsets of questions were expressed as proportions of 1 (eg. a plant might be rated as 0.2 good, 0.7 average and 0.1 bad).

It was considered that the range of failure rates between good and badly managed plant was around 3 orders of magnitude. This was based on looking at the ranges of failure rates reported in risk assessment. As a result, a plant which would score average on all questions was given a management factor of 1, a good plant 0.1, and a plant which would score bad on all questions 100. A triangular graph of proportions of good, bad or average against Management Factors was constructed. This Management Factor Triangle was produced by expert judgement. It incorporated the concept that it was much harder to be better than the generic average than to be better than bad. Thus, for any site it was possible to generate a Management Factor on a continuum from 0.1 to 100. To date, this MANAGER method has probably been applied to about 30 sites.

Following these developments a new programme of research was started, originally funded by the UK Health and Safety Executive, and later by the Dutch Ministry of Housing, Physical Planning and Environment (VROM), with industry involvement from Norsk Hydro and BP. The current status of this research is that it is now the subject of a collaborative CEC project which will investigate safety management system auditing in a number of European countries. This research programme is described later in the paper.

2. AN ACCIDENT CAUSATION MODEL

The Human Factor in accident causation was becoming an increasingly "hot" topic, as evidenced by the many human failures identified in major accidents occurring during the last decade such as Bhopal and Chernobyl. One area where this aroused particular interest was in Quantitative Risk Assessment (QRA). In 1988 the Health and Safety Executive funded a project to examine the underlying causes of failure in loss of containment accidents on chemical and petrochemical plants. Originally, the intention had been to separate the human from the mechanical causes of failure. Quantitative Risk Assessment in the chemical industry typically uses historical failure data in order to quantify equipment failure rates and evaluate the risks of consequences of failure such as multiple offsite fatalities.

Therefore, human factors are, in a general sense, implicitly included in QRA when generic failure data are applied to details of vessel and pipe sizes and process conditions on a specific site. The question of causes of failure arise when appropriate historical data do not exist. Then, methods such as fault tree analysis are used. The nuclear industry typically uses fault tree analysis in Probabilistic Safety Assessment (PSA) whereas in the chemical industry the use of generic failure rate data is prevalent. Causes of failure must also be considered when identifying means of reducing dominant risks.

The HSE funded research led to the analysis of around 1200 loss of containment accidents of pipework, in-line equipment and vessels (Bellamy, Geyer and Astley, 1989; Bellamy and Geyer 1991). As this research progressed, it soon became apparent that there was a difference between human error as a direct cause of failure and as an underlying deeper cause of equipment failure and human error triggers.

In order to capture this human component at both levels, a 3-dimensional classification scheme was developed. At best, previous classification schemes have only looked at two ways of classifying failure (eg. Blything and Parry, 1988). One is the direct cause such as corrosion or human error (eg. opening a wrong valve). The other looks at the operation taking place within which the failure occurs eg. maintenance.

The 3-D model, however, added the extra dimension of management failure. These were failures to either prevent the unsafe conditions arising, or failures to recover unsafe conditions. The accidents were classified on this dimension by considering what management preventive or recovery mechanisms would have worked in that particular case. This avoided the impossible task of having to obtain data on the management characteristics which were present in the organisation at the time of the accident.

The results were very interesting. For example, in the study of pipework failures (Bellamy, Geyer and Astley 1989; Hurst, Bellamy, Geyer and Astley, 1991), 24.5% of all pipework failures had an underlying management failure contribution in hazard review of design (the biggest single cause), 14.5% in human factors review of maintenance, 13% in supervision of successful completion of maintenance tasks, 11% in human factors review of operations. Maintenance was the biggest single origin of the cause of a pipework failure (38.7% of all origins of causes).

The 3-D scheme showed clearly why previous causal analyses gave such comparatively varied results. The main reason was the overlapping nature of cause descriptions, particularly in one-dimensional schemes, which mix the three dimensions of failure. On the other hand, this does not mean that 3-dimensions captures all the useful data. To date, nothing has been analysed, for example, along the dimension of the organisational aspects of safety. This may well prove to be the next step as links between organisational structures and safety are increasingly investigated (eg. Timmerhuis and Koehorst 1990).

At the time of carrying out this research, the HSE were very interested in modifying generic failure rates in a valid way to allow the quality of safety management influences on the potential for loss of containment to be quantified (Hurst, Nussey and Pape, 1989). The 3-D classification provided a valuable data model of the underlying and direct causes associated with generic failure rates used in QRA. Subsequent analysis of MARS data (reportable accidents under the Seveso directive) (Hurst, 1992) revealed a pattern of causes virtually identical to that for vessel failures.

In parallel to the data analysis, a theoretical model was developed - the

Sociotechnical Pyramid - to link the influence of management to direct causes of failure. This model presents management influences as layers in a chain of more remote (base of pyramid) to most immediate (top of pyramid) causes of failure. This concept of a hierarchical scheme of accident causation is not a new one. However, the scheme takes into account:

- the climate within which a company operates (regulatory, economic, know-how),
- the company organisation and standards,
- the control, communication and coordination processes,
- monitoring of and feedback on the effectiveness of management control, and
- front line personnel competence and task support (interface, tools, procedures etc)

3. GENERATION OF AUDIT QUESTIONS

The pyramid of causes, from climate level up to direct engineering and human reliability causes, was combined with the 3-D data model to provide a question generation mechanism. The idea was that the 3-D data indicated areas in which an assessment of management was important (i.e. prevalent underlying causes of failure), such as hazard review of design. The sociotechnical pyramid indicated an appropriate audit trail. Taken together this would provide a set of questions which would globally cover the management influences assumed to underlie generic failure rates and provide a means of quantifying the importance of the questions, using the data model.

To assist in question generation, the pyramid was divided vertically along the lines of the 4 themes mentioned earlier in the historical background. The structure of the question set was based on separating questions along these four themes within a particular underlying cause category, such as failure in Human Factors Review of Maintenance. Within each theme, the questions were divided according to levels of the pyramid. So, for example, one of the questions from the theme of Pressures in the area of Human Factors of Operations, at the Operator Reliability level in the pyramid, is:

What evidence is there of:

- scheduled training being deferred for operational reasons?
- a backlog of required procedural modifications?

- a significant number of temporary plant modifications to controls and equipment which may increase the chance of operator error?
- a significant number of long-standing temporary modifications to operating procedures still in use?

In much the same way as the MANAGER technique, the application of the questions results in quality ratings by the auditors, whose judgements are translated into a Management Factor. The difference is that where MANAGER gives equal weightings to questions, the research audit weights question areas according to the proportions of underlying causes in the 3-D model. The combination of ratings and weightings then allow recommendations for improvement to be prioritised. Experience of applying the audit indicates that these recommendations tend to reflect common mode causes of problems, such as the need for a training policy, for example, as well as the "smaller" specifics.

4. FURTHER DEVELOPMENTS OF THE AUDIT

The research on a prototype audit is continuing with development of anchor points for each subset of judgements, and each level of rating (currently 5 levels from well above average to well below average). This is to assist in consistency of auditor judgement.

Secondly, work has been completed on analysing the underlying causes of loss of containment mitigation failure as they affect factors such as ignition, release duration and fire fighting, for example. This work (briefly reported in Hurst, Bellamy and Wright, 1992) also needs to be translated into a set of questions.

The important link between generic failure rates and safety performance indicators, such as Lost Time Injury rates, is also being investigated. One company has supplied all its loss of containment data and LTI data, on a plant specific basis, to support this task. It is important to search for data which tests the validity of the assumption of the link between the quality of management of safety and failure rates.

The audit has also been applied in the industrial context to investigate its strengths and weaknesses and hence to enable it to be improved. This has shown to be particularly important in the design of anchor points, in planning out the audit, and in assisting the auditor to navigate the system. The paper by K.B. Ratcliffe (this conference) describes how the audit has been adapted and developed for use by HSE's field inspectors.

The work sponsored by the Commission of the European Communities is beginning. This aims at further development of the audit and a cross-national comparison and validation exercise. The overall objective is to transfer the research development into a valid and acceptable tool within the European context, and to provide a complete audit methodology which achieves the purposes for which it has been designed.

To assist in ensuring the acceptability and usability of this safety management audit tool, an Expert Group has been formed with representatives from industry and regulators. The group will discuss such issues as the structure of the audit, relative importance of individual questions, key issues in each weighted area, anchor points, and auditor training, for example.

Given that the research audit ultimately will be taken and modified by different groups in different countries, as is its purpose, it is important to ensure that sufficiency of structure, content and guidance enables this to be done without compromising the intention behind the audit scheme.

5. REFERENCES

Bellamy, L.J. (1983a) A Literature Survey of Effects of Individual, Social and Organisational Influences on System Reliability. Part 1: Accident Survey. Ergonomics Development Unit, University of Aston in Birmingham, UK. Unpublished report to the NCSR, UK Atomic Energy Authority, 1983.

Bellamy, L.J. (1983b) Neglected individual, social and organisational factors in human reliability assessment. Reliability '83, Proceedings of the 4th National Reliability Conference, Birmingham, UK, Vol.1 pp. 2B/5/1-2B/5/11.

Bellamy, L.J. (1986) The Safety Management Factor: An Analysis of the Human Error Aspects of the Bhopal Disaster. Safety and Reliability Society Symposium, 25 September 1986, Southport, UK.

Bellamy, L.J., Geyer, T.A.W., and Astley, J.A.A. (1989) Evaluation of the human contribution to pipework and in-line equipment failure frequencies. HSE Contract Research Report No. 89/15.

Bellamy, L.J. and Geyer, T.A.W. (1991) Organisational, Management and Human Factors in Quantified Risk Assessment. HSE Contract Research Report 33/1991.

Blything, K.W. and Parry, S.T. (1988) Pipework Failures - A Review of Historical Incidents. Safety and Reliability Report SRD R 411, UK Atomic Energy Authority, Jan 1988.

Cullen, the Hon. Lord (1990) The Public Enquiry into the Piper Alpha Disaster. Department of Energy Cm 1310 2 Vols. London: HMSO. November 1990.

Hurst, N.W. (1992) Personal communication on analysis of MARS data. Hazard Section, UK Health and Safety Executive, Broad Lane, Sheffield S3 7HQ.

Hurst, N.W., Bellamy, L.J. and Geyer, T.A.W.(1991) A classification scheme for pipework failures to include human and sociotechnical errors and their contribution to pipework failure frequencies. J. Hazardous Materials, 26 (1991) 159-186.

Hurst, N.W., Bellamy, L.J., and Wright, M.S. (1992) Research models of safety management of onshore major hazards and their possible application to offshore safety. pp. 129-148 in Proceedings of I.Chem E. Symposium Series No.130, "Onshore and Offshore", Manchester, UK.

Hurst, N.W., Nussey, C., and Pape, R.P (1989) Development and application of a risk assessment tool (RISKAT) in the Health and Safety Executive. Chem. Eng. Res. Des., 67 (1989) 362-372.

Parker, R.J. (Chairman) (1975).
The Flixborough Disaster. Report of the Court of Inquiry. Department of Employment. London: HMSO.

Ratcliffe, K.B. (1993) The development and application of an audit technique (STATAS) for the assessment of safety management systems at major hazard sites. (This conference).

Safety and Reliability Directorate (1988) Human Reliability Assessors Guide. SRD Report RTS 88/95Q, P. Humphreys (Ed), UK Atomic Energy Authority.

Timmerhuis, V. and Koehorst, L. (1990) Organisatie en veiligheid in de chemische industrie. Report to the Ministerie van Sociale Zaken en Werkgelegenheid, Postbus 90804, 2509 LV Den Haag, the Netherlands. ISBN 90 5307 1202.

STATAS: Development of an HSE Audit Scheme for Loss of Containment Incidents

K. B. Ratcliffe
Health & Safety Executive, U.K.

PART 1. A LOSS OF CONTAINMENT MODEL.

1.   INTRODUCTION

HSE has, for a number of years, been concerned about the role of Human Factors, that is human error, ergonomic factors and the influence of management systems in major incidents and have published a number of booklets on these subjects (Ref: 1, 2, 3, 4, 5) .

Original research work has been commissioned by the HSE from consultants Technica and Four Elements (Ref 6-12) to determine how Management andOrganisational factors affect the potential for loss of containment of hazardous substances . HSE's Research and Laboratories Services Division (RLSD) have managed the projects and together with the Hazardous Installations National Interest Group of the Field Operations Division (FOD) have developed a structured audit technique from the research and have recently finished the first of a series of trial audits. The system as developed for FOD is known as STATAS; (Structured Audit Technique for the Assessment of Safety Management Systems) and in this paper I shall be describing its empirical basis, a Management related Loss Causation Model and the development and application of audit question sets derived from them.

2.   BACKGROUND

In the UK, premises storing or using quantities of hazardous substances sufficient to raise concerns about the potential for major accidents and the extent of any residual offsite risk are identified by the Notification of Installations Handling Hazardous Substances Regulations 1982 (NIHHS) and the Control of Industrial Major Accident Hazards Regulations 1984 (CIMAH). Major incidents or events that could escalate into major incidents occur when the hazardous substances involved escape their containment systems, whether these are storage tanks, process vessels or transfer systems including pipework and in-line equipment such as pumps and valves. The first task of the researchers was therefore to analyse and classify data relating to vessel and pipework failures from a number of national and international data bases. A total of 921 pipework and in line equipment failures and 230 vessel failures were examined and an incident causation model developed.

3. DIRECT CAUSES OF LOSS OF CONTAINMENT

At the initial analysis stage it could be seen that most incidents had an obvious or direct cause of failure. Examples include overpressurisation, corrosion or failure due to external impact. These direct causes arose either from a failure of hardware or were the result of human error. A pipe may corrode and leak or a fitter may simply break into a line without adequate isolations. The two, human and hardware failings, may well be linked. A mistake on the part of an operator could lead to overpressurisation in a system and conversely equipment or instrument failure, by demanding human intervention, can lead to an error which causes or compounds the release. Twelve categories of Direct Causes were identified and used to classify the incidents under study. These are as follows:-

Corrosion	Overpressure	Human Error
Erosion	Vibration	Defective Equipment
External loading	Temperature	Other
Impact	Wrong Equipment	Unknown

4. ORIGINS OF FAILURE

Failure from these causes may occur as an immediate and direct consequence of an action or condition or may be latent within the system for some time before manifesting itself. There can exist a mismatch between the company's belief as to the state of the system and its actual condition. A valve for example, may have been installed to the wrong specification but may operate satisfactorily for some time before failing. The origins of the failure lies somewhere else in the lifecycle of the plant than the point at which it occurs. The original design may have been wrong, there may have been failures during manufacture or during the construction of the plant. The plant may have been incorrectly modified or an error may have been made during maintenance. Most of the incidents analysed could be classified in this way as having their origins in one of the major lifecycle phases of the plant, the exceptions being a few cases where the causes were due to external factors such as natural events, sabotage or the domino effects of failure of an adjacent piece of equipment or plant. Nine classifications for the origins of failure were identified and are listed below.

Design	Normal Operations	Manufacture/Assembly
Maintenance	Construction/Installation	Domino
Natural Causes	Sabotage	Unknown

5.   PREVENTION AND RECOVERY MECHANISMS

It is better to prevent an unsafe condition in the first place than to try and recover it once it exists. However, both possibilities, prevention and recovery, exist and have important parts to play in a strategy to minimise failures. Preventative measures are most appropriate in the design of plant and plant modifications while recovery mechanisms are important in the field of routine inspection, testing and maintenance to cope with the normal ageing processes that occur on any plant. These mechanisms can be either in the field of hardware control or equally in the modification of human behaviour. Taken together they represent an over-view of available procedures that can minimise the overall risk of catastrophic failure. Not all failures may be recoverable however, and in some of the incidents studied the recovery mechanism remained unknown. Excluding the unrecoverable and the unknown four broad classes of Prevention/Recovery mechanisms were identified:

Hazard Identification and Assessment	Human Factors Review
Routine Checking and Testing	Task Checking

6.   INCIDENT CLASSIFICATION SCHEME

Each of the pipework, equipment or vessel failures examined, after being classified as to the direct cause of failure and its origin in the lifecycle was further classified as to which of these four potential prevention or recovery mechanisms either failed or had the potential for identifying and preventing the failure occurring. Each incident could therefore be classified in three ways and the method is illustrated in Fig 1. Where more than one mechanism appeared to be involved expert judgement was used to determine contributions to the overall failure. Each incident can be located within the 3-dimensional matrix illustrated. For example, an <u>overpressure</u> failure due to <u>wrong design</u> and not prevented during the initial <u>hazard and operability studies</u> would be located in the cell defined by the 3 coordinates. As incidents are classified data accumulates within the cells and are open to analysis. Figs 2 and 3, for

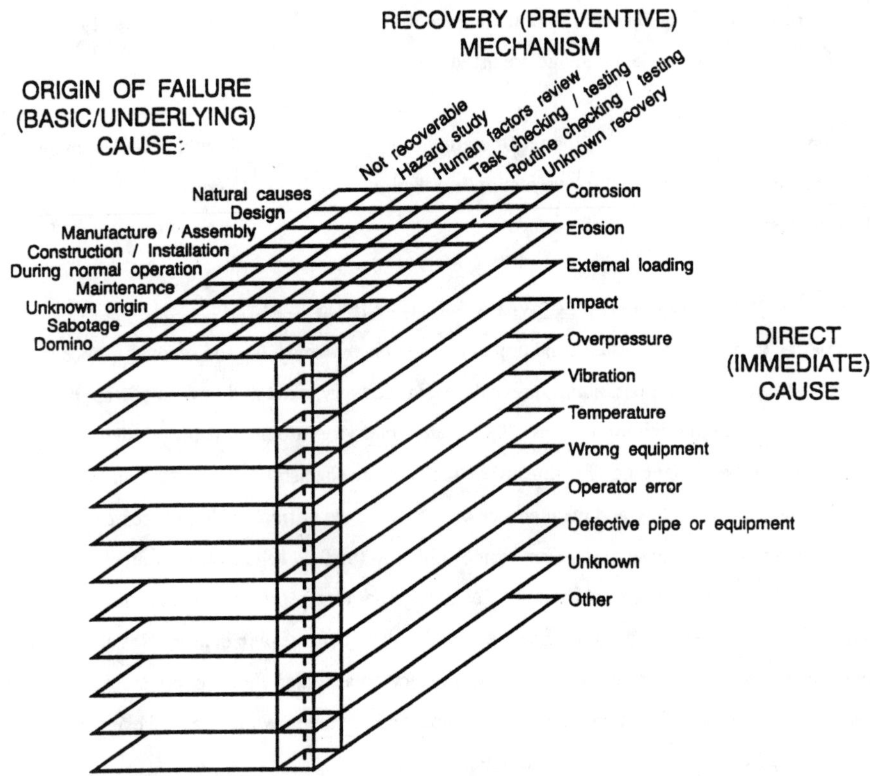

FIG. 1.   A 3-DIMENSIONAL CLASSIFICATION SYSTEM

example, show the contribution of known direct causes to overall failure rates for pipework and vessels. The substantial contribution from operator error, overpressure, corrosion and temperature is illustrated. This in itself is useful information but is much more powerful when illustrated in relation to the origins of the failure and the potential preventative or recovery mechanisms. This reveals in a hierarchical way the areas of activity within each part of the plant's lifecycle which constributes most to failure and where the potential for recovery or prevention lies. This is illustrated in Figs 4 and 5 and represents collapsing the data as it accumulated in the 3-dimensional matrix into "tower blocks". Each tower block consists of the sum of direct causes. Looking at the data for vessels the diagram illustrates that 29% of the overall failure rate was due to faults of design that were not recovered or prevented during hazard identification studies. Similarly 24% of the failure rate had its origins in normal operations where human factors review could have detected and prevented the faults. Figures for pipework and vessels are different, illustrating the relative probability of different failure

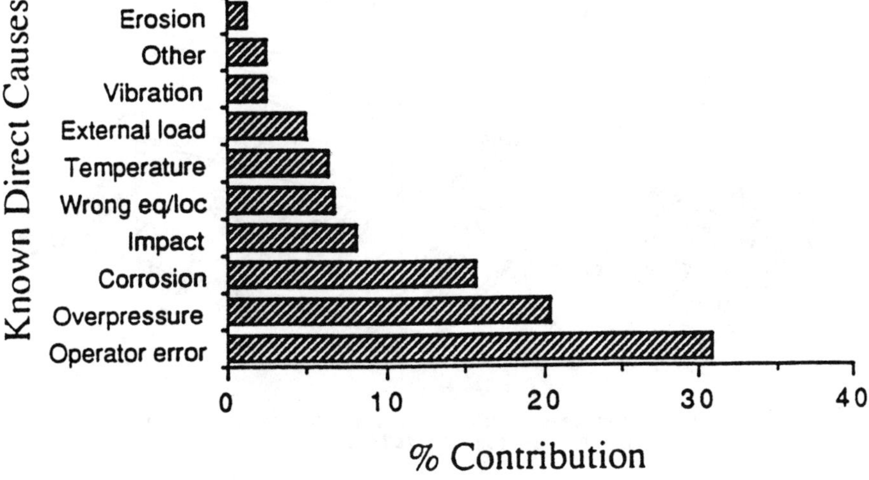

FIG. 2.   PIPEWORK AND IN-LINE EQUIPMENT

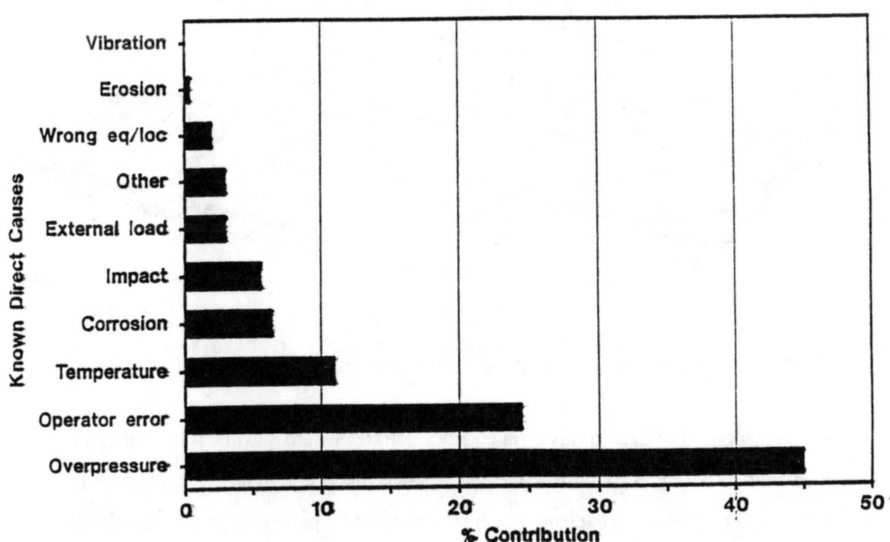

FIG. 3.   VESSELS

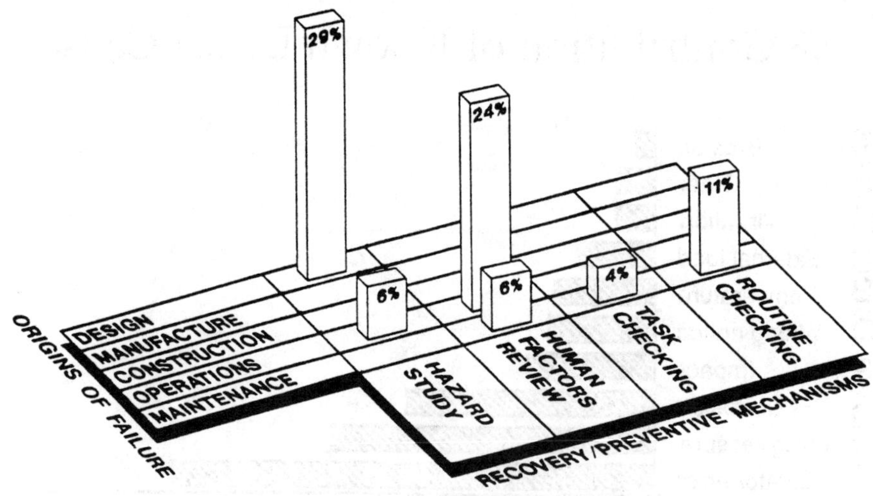

FIG 4

CLASSIFICATION OF VESSEL FAILURES BY ORIGIN OF FAILURE
AND PREVENTION/RECOVERY MECHANISM

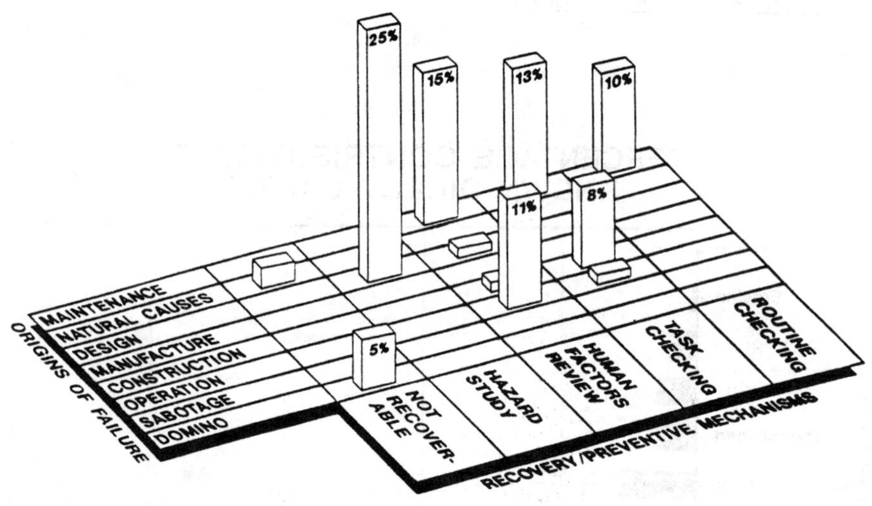

FIG 5

CLASSIFICATION OF PIPEWORK FAILURES BY ORIGIN
OF FAILURE AND PREVENTION/RECOVERY MECHANISM

modes. Presentation of data in this form clearly illustrates where maximum benefit can be obtained from investing in sound management systems that are designed to implement the prevention and recovery mechanisms which are relevant to a particular part of the plant's lifecycle.

These mechanisms include engineering based prevention through hazard identification and hazard assessment and recovery through routine inspection checking and testing. Human factors prevention mechanisms include task analysis, competencies, communications etc while recovery mechanisms include task checking and testing. In the next part of the paper I shall look at how the overall engineering and human reliability of these systems can be linked to a management model, and how this lends itself to the development of a structured audit scheme.

PART 2. MANAGEMENT AND ORGANISATIONAL FACTORS: A SOCIO-TECHNICAL MODEL OF ACCIDENT CAUSATION.

In this part of the paper I will describe the factors which have an effect on the frequency of loss of containment incidents using a Management and Organisation Model, the Socio-Technical Pyramid, and link this to HSE's guidance, "Successful Health and Safety Management", HS(G)65. (Ref 1).

2.   THE SOCIO-TECHNICAL PYRAMID

This model (Fig 6) has been developed from case studies of major accidents including Three Mile Island and the Challenger Space Shuttle Disaster (Ref 6) and a review of existing management models. Models have the advantage of rendering a complex organisation more easily understandable so that it can be looked at as a whole or examined in part in a simplified way yet retain the essential characteristics of the whole.

The Socio-Technical Pyramid represents, progressing from Level 1 to Level 5, increasingly remote factors that, nonetheless, have a casual link to the frequency of undesired events on site, specifically the loss of containment of hazardous substances. Fig 7 gives an indication of factors operating at the various levels.

3.   LEVEL 5 - THE SYSTEM CLIMATE

Factors exist which are generally outside the control of the company under review. These factors include the state of knowledge within the industry, the effects of historical incidents both on and off site and industry norms particularly in the field of hardware standards and maintenance procedures. The existence or absence of specific legislation and the activity of the regulatory body also have an influence as do official guidance and Approved Codes of Practice, on the standards adopted on site. There may be political and financial pressures

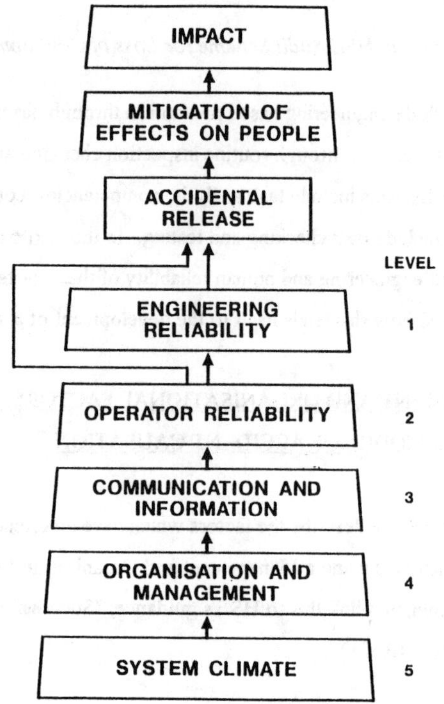

FIG 6

THE SOCIO-TECHNICAL PYRAMID

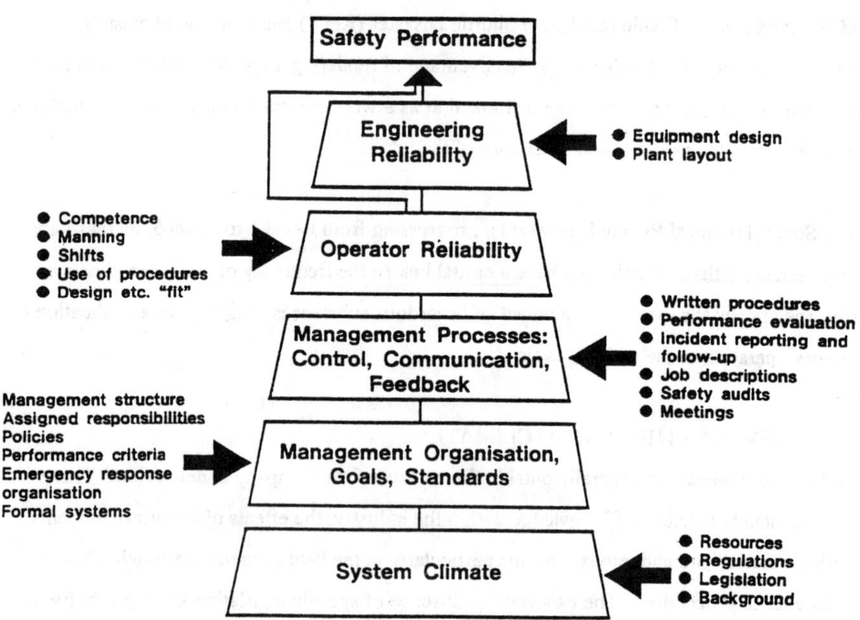

FIG 7

THE SOCIO-TECHNICAL PYRAMID
FACTORS AFFECTING PERFORMANCE

540

operating as well as local public opinion. The scale of the hazard, the availability of relevant resources, the activities of the emergency services, the location of the plant are all factors which will influence the way in which site management commits itself to the desired outcome of the lowest practicable incident rates.

4.   LEVEL 4 - ORGANISATION AND MANAGEMENT

This is the level within the management model where organisational and management structures are determined, policy decided and arrangements made. Of particular importance are:

a)   Policy. Corporate goals should be set via policy documents. There should be clear leadership at the highest level on targets and priorities.

b)   The Organisational Structure. This is the mechanics of how the company organises itself to achieve its desired goals. There may be discreet operational, maintenance, safety and training structures or these may be integrated in an infinite variety of ways.

c)   Defined Responsibilities. Whatever the actual organisation it is important that roles and responsibilities are clearly defined and methods exist for those with health and safety responsibilities to be held accountable. Mechanisms for control, eg review and revision of objectives should be specified, and individual scope for decision making defined.

d)   Site Standards. The arrangements made for the selection and setting of appropriate site standards and the mechanism for maintaining and improving these. Standards should take due note of Regulatory requirements, industry norms and relevant national and international standards. They should also be routinely reviewed and revised.

e)   Resources. The allocation of resources, setting of budgets and deadlines and ground rules for control of contractors and use of specialist equipment.

f)   Information. Good data management assist all these activities. It is important to have a policy on the use of data from monitoring, auditing and inspections to close the management control loop and develop a climate of continuous improvement.

g)   Reward and Punishments Systems. These systems should form a constructive part of the whole system, aiming for a no-blame culture.

This level (Level 4) defines the "Mission". This is achieved by the implementation of good management practice through planning, control, leadership and organisation. Activity at this level will change and evolve through influence from lower levels as well as in response to changes in system climate.

5.   LEVEL 3 - COMMUNICATION AND FEEDBACK

At this level management processes are in operation. It is here that systems must demonstrate the elements of control, communication, coordination and cooperation and the essential monitoring, programme review and development necessary to the management control loop. Issues include:

a)   Formal and informal communications. Written communication includes the development and dissemination of instructions, use of log books and handover documentation, and the development and review of standard operating procedures. Verbal communications include formal meetings, eg safety committee meetings and informal meetings at all levels whereby the company's policy is spread and reinforced.

b)   Documentation. There should be a policy covering the generation, dissemination and updating of documents. These include records of inspection, incident investigation reports, standard operating procedures, P and I diagrams, hazard and operability studies, permit to work systems, design of documents, eg accident report forms and the availability of standards and data sheets.

c)   Man machine interface systems (MMI). Control room operability can be effected by the design of data presentation on visual displays and instruments.

d)   Communications hardware. The reliability of hardware systems ensuring communication particularly during emergencies.

e)   The activities of supervisors in the role of task checking and inspection.

f)   Monitoring, feedback and programme evaluation and review. The role of auditing.
       Auditing is a necessary part of the system whereby a 'third party' can give an
       independent and structured report on the health of the controlling Safety Management
       Systems.

g)   Data presentation and acquisition and the use of performance measures. Traditional
       measures can be too coarse - loss time accident data alone rarely gives sufficient
       insight into the underlying causes of incidents or illuminates early trends. Appropriate
       performance measures need to be identified concentrating on acquisition and use of
       near-miss information and other measures that can be built-in to system such as
       progress with Safety Committee items, response time to deviation reports etc.

6.   **LEVEL 2 - OPERATOR RELIABILITY**

At this level the overall safety of the system depends upon the reliability and competence of
the operator. This is in turn effected by activity at Level 3 through Performance Shaping
Factors. These include:

a)   The demands and design of tasks.

b)   Operator selection and, through training and acquired experience, levels of skills,
       knowledge and overall competency.

c)   Man machine interfaces. Stresses and pressures, peer group influence, influences of
       the environment.

7.   **LEVEL 1 - ENGINEERING RELIABILITY**

At this level the integrity of system relies upon the overall control of plant design and site
modification. Issues include:

a)   Use of failure rates and failure analysis.

b)   Instrumentation and controls. Hardware design and maintenance.

c)   Programmable electronic systems, potential failure modes and ergonomic weak-
       nesses.

A hardware failure can not only be a manifestation of a failure within the controlling management system but could be as a direct result of human failure. Operator error could result in part of the system being subject to conditions outside its operating criteria, eg overpressure. Conversely a hardware failure requiring human intervention gives rise to the possibility of the operator making a wrong decision. This may be particularly so if emergency action is needed with the operator required to follow infrequently used procedures under pressure.

8.   LEVEL 0 - LOSS OF CONTAINMENT AND MITIGATION

The actual impact of an event can be effectively mitigated by action taken once the alarm is raised. Mitigation covers all the activities that should be contained in a good on-site emergency plan. Issues include:

a)   Bunding arrangements and post release action, eg foaming.

b)   Control of ignition sources.

c)   Arrangements for fire fighting from onsite resources and attendance by Fire Authority.

d)   Use of water sprays, scrubbing systems and venting systems. These systems are normally included in initial design considerations and appropriate standards.

e)   On and offsite emergency planning including arrangements for escape, site evacuation, temporary refuges, use of PPE during evacuation.

f)   Effectiveness of emergency shut down systems, automatic shut off valves, remotely operated shut off valves or manual isolation valves. Time taken to control a release is typically quoted as one minute for an ASOV, 5 minutes for a ROSOV and 20 minutes for manual intervention. (Bellamy and Geyer 1990 TECHNICA). Mitigation of impact is also influenced by plant layout, design and accessibility of controls, initial hazard assessment and awareness, eg appreciation of the likely consequences of a release and its development with time, gas detection systems and containment systems ncluding secondary containment.

9.   MANAGEMENT CONTROL

The Socio-Technical Pyramid as a model of a management structure is consistent with the management model and control mechanisms illustrated in the HSE's Accident Prevention Advisory Unit's guidance "Successful Health and Safety Management" HS(G)65. (Ref 1) since the Socio-Technical Pyramid is derived from analysis of loss of containment scenarios and explores increasingly remote systems failure through Level 1 and 2 (hardware and operator reliability) to Level 5 (system climate) it is inverted in relation to the APAU model (Fig 8) which starts with management level commitment to a policy for health and safety. The Socio-Technical Pyramid is a loss causation model with casual links between each level. The

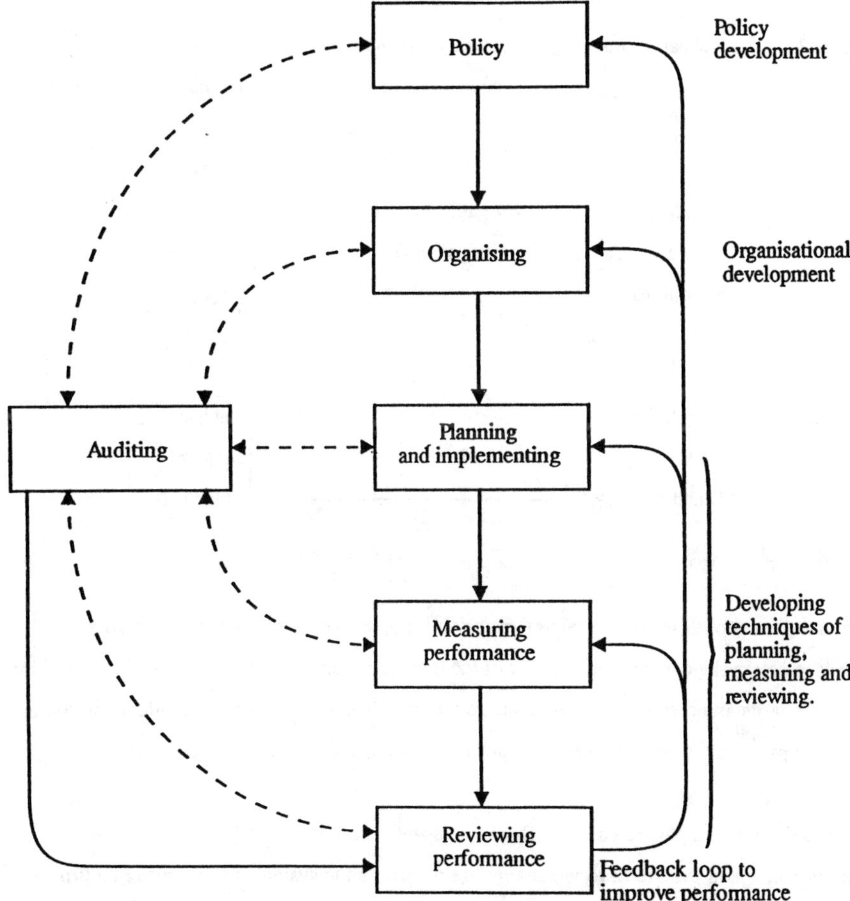

Fig 8   - Key elements of successful health and safety management
Source:Health and Safety Series booklet HS(G)65.

APAU model is a diagrammatic representation of the key elements of a successful health and safety management system incorporating the essential feedback loops through monitoring and auditing to ensure the continued improvement and development of the system.

The complimentary nature of the models can be illustrated by considering a systems analysis for training needs (Fig 9).

CONTROL LOOP - TRAINING

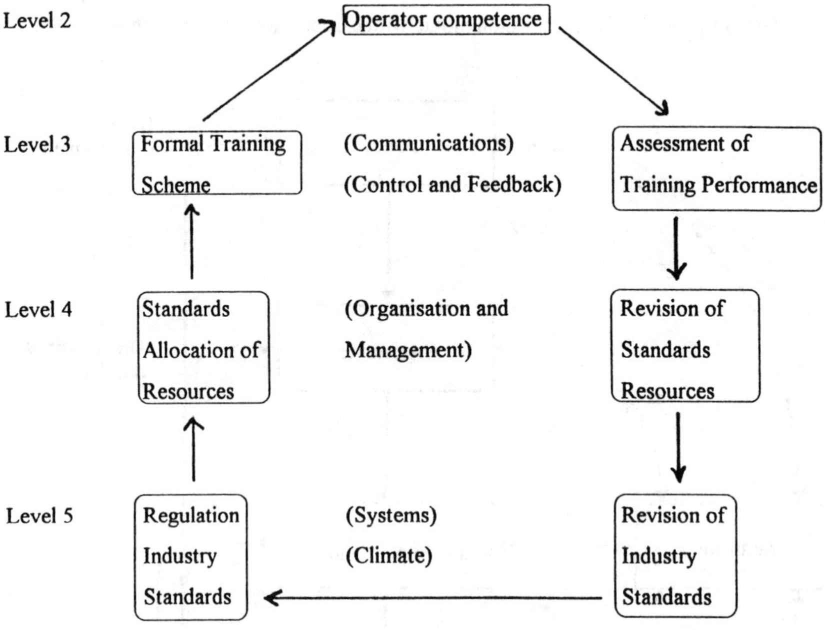

Information about operator reliability from routine assessment or evaluation of errors is fed back to assess the content and delivery of the training scheme. If the scheme is found wanting the standards and resource allocation are revised and through external communication routes the company's experience may be fed back into a revision of an industry standard.

The Key Elements illustrated in Fig 8 are the essential components of a successful health and safety policy and the extent to which they are present in the system is a measure of a firms progress towards a self- improving health and safety culture. The Key Elements are summarised below.

a)   Policy.

The lead for setting a policy which recognises the hazards associated with a firm's activity and lays down an overall strategy for evaluation, elimination and control must come from the top of the management structure. To be successful the policy should:

i.   Be comprehensive.

ii.   Recognise the letter and the spirit of legal requirements.

iii.   Recognise the contribution that a commitment to high standards of health and safety makes to the overall performance of the company.

iv.   Demonstrate real commitment and involvement from the highest levels of management.

v.   Demonstrates the will and ability to develop the necessary structures, systems and safety culture.

vi.   Commit the necessary resources to achieving identifiable goals.

vii.   Be systematic in the identification of hazards, the assessment of risk and the application of control measures.

viii.   Recognise the essential value of measuring output from systems as a necessary step in reviewing and developing standards.

ix.   Recognise the value of regular independent auditing of systems.

b)   Organising.

Organising for successful health and safety management requires building upon the essential areas of developing controls, cooperation between functions, communication between individuals and groups and promoting high levels of competence for all individuals within these functions. Elements of control include:

i.   Defining responsibilities allocated to managers in different functions.

ii.   Comprehensive job descriptions for line managers and defining the role of safety adviser.

iii. Enabling individuals with authority, competence, time and resources to carry out duties effectively.

iv.   Developing systems with accountability and motivation through target setting.

v.   Providing adequate supervision, instruction and guidance.

Cooperation is developed by:

i.   Good communications between functions.

ii.   Involving employees and safety representatives in developing and monitoring performance indicators.

iii. Participative systems for recognising achievements.

Good communications are developed by:

i.   The visible involvement of management in control activities.

ii.   Provision of clear, comprehensive supporting documentation.

iii. An efficient system of formal and informal meetings.

Competence is assured by:

i.   A sound policy on recruitment and selection and promotion to key jobs.

ii.   Standards of training with monitoring of the effectiveness of the results.

iii.   The availability of specialist advice to decision makers.

c)   Planning and Implementing.

A successful safety management system must be seen to be drawing up short and long term plans, setting relevant performance standards aimed at eliminating and controlling risks and to be putting those plans in to practice. These objectives are achieved by:

i.   Identifying objectives and setting targets to be achieved within specified time limits.

ii.   Setting appropriate standards to achieve, maintain and improve a positive health and safety culture.

iii.   Using hazard identification and risk assessment as a basis for the control of risk.

iv.   Having as a prime aim the elimination of risk where possible and otherwise the minimisation and control of risk through engineering reliability and operator competence.

v.   Establishing priorities.

vi.   Establishing a system for the development, dissemination and review of documentation.

d)   Measuring Performance.

Measurement is an essential part of the managerial control loop. Some systems have simple performance indicators, eg lost time injury rates, other require judgements as to their success or a combination of both. For example, the effectiveness of a training programme can be judged both through scores in tests and by an evaluation of operator competency via line management. Measuring systems should include:

i.   Active monitoring systems as routine comparing planned attainment of objectives and standards within time limits against actual performance.

ii.   Reactive systems for collection and analysis of data relating to failures in the system and its use to remedy underlying causes. These include causes of injuries and ill health, dangerous occurrences and near misses.

iii.   The evaluation of information at the appropriate management levels.

e)   Reviewing and Auditing.

Information relating to the success or failure of safety management systems must be collected routinely and used to develop and improve the system. These systems should:

i.   Test the health of the whole system.

ii.   Identify remedial action where the system is deficient or no standards have been set.

iii.   Measure the degree of compliance of the systems against stated policy.

iv.   Allow for third party audit for unbiased judgement and to enable comparisons to be made.

PART 3 . CONSTRUCTING AN AUDIT SCHEME

1.   INTRODUCTION

In the earlier parts of this paper I have described the development of an incident model which classified releases of dangerous substances in terms of a direct cause, an origin of failure (within the plant's life cycle) and a potential prevention or recovery mechanism. The system failures underlying a release were related to a management loss causation model, the Socio-Technical Pyramid and good management practice as outlined in the HSE Guidance Booklet HS(G)65 "Successful Health and Safety Management". In this part I will show how these models can be combined to form the structure of an audit scheme and how STATAS, a scheme developed for possible use by HSE's Field Operations Division evolved.

2.   MAJOR CONTRIBUTIONS TO FAILURES

The purpose of an audit is to provide an independent assessment of management's planning, organisation and control systems. It should support all other activities on site by providing information about the relevance of standards and the extent to which plans and systems are properly implemented and effective. Any audit scheme needs to be comprehensive in its examination of the relevant safety management systems and the extent to which the key elements of the management control loop are in place.

The empirical basis for STATAS starts with the research work described earlier which analysed and classified pipework and vessel failures. In particular with the quantitative data, the "tower blocks" (Figs 4 and 5) which represents the sum of all direct causes located at their origin in one of five identified phases of the plant's life cycle and according to what potential prevention or recovery mechanism either failed or was available to identify and correct the error. The 5 phases of the plant's life cycle were identified as:

a)   Design; (Des), including the choice of design, and the design process. This also covers modifications.

b)   Manufacture; (Manf), including off-site manufacture of components and pre-construction assembly.

c)   Construction; (Con), including installation and commissioning.

d)   Operation; (Op), covers all normal process operations including start-up, shut downs and emergencies.

e)   Maintenance; (Maint), includes all planned inspection and testing as well as routine replacement and emergency maintenance..

The 4 prevention/recovery mechanisms are divided between assessment of hardware or human behaviour and whether the activity is designed to prevent faults and errors or detect any that occur. The categories therefore are, for hardware:

1.   Hazard identification and assessment techniques (HAZ)

2. Routine inspection, testing and sampling (ROUT)

Similarly for human activities:

3. Human factors review (HF),

4. Task checking (TCHECK).

Different forms of these activities are relevant at different parts of the plant's life cycle. The information represented graphically in Fig 4 and 5 is shown numerically in Table 1. Although much of this paper has been about theoretical models it must be remembered that the exercise starts with real incidents and the prevention/recovery mechanisms refer to real activity on sites. For example when designing plant or plant modifications all firms, to a greater or lesser extent, will use a variety of hazard identification and assessment techniques. Precisely what may be done depends upon many factors but would include Preliminary Hazardous Assessment, Hazard and Operability Studies, Event and Fault Tree Analysis, Check

TABLE 1

ORIGINS OF FAILURE/PREVENTATIVE MECHANISMS COMBINATIONS CONTRIBUTIONS TOWARDS OVERALL FAILURE RATES

Combination	Contribution		Cumulative Totals	
	Pipework	Vessels	Pipework	Vessels
Des/HAZ	25	29	25	29
Maint/HF	15	6	40	35
Maint/TCHECK	13	4	53	39
Maint/ROUT	10	11	63	50
Op/HF	11	24	74	74
Con/TCHECK	8	2	82	76
OP/HAZ	–	5	82	82
Op/TCHECK	2	2	84	83
=========================	=========================				Cut off
Des/HF	2	–	86	83
Manf/TCHECK	2	–	88	83
Maint/HAZ	–	2	88	85
Con/HF	2	–	90	85
etc (all other known)	2	2	92	87
Unknown/Not recoverable	8	13	100	100

All contributions > 1%, figures rounded to nearest whole number.

Lists etc. Similarly for other activities at other parts of the life cycle. It is these activities that are examined by the audit and the percentage contribution towards overall failure rates suggests a hierarchical approach.

3.   STRUCTURE OF THE STATAS AUDIT

There are 20 possible combinations of origins of failure (5) and prevention/recovery mechanisms (4). It can be seen from Table 1 that some of the combinations contribute little to the overall failure rates. For STATAS it was decided to concentrate on the 8 areas of activity above the cut-off point. These 8 areas cover 84% of the total contribution towards pipework failures and 83% towards vessel failures. This is compared with the maximum recovery potential of 92% for pipework and 87% for vessels, the balance of the contribution comes from combinations that were not recoverable or unknown. The audit therefore consists of 8 question sets, one covering the use of hazard identification techniques in the design and modification of plant, one covering task checking during and after construction work, 3 linked to maintenance activities and covering human factors, routine inspection and testing and task checking, and 3 linked to normal operations covering human factors, hazard identification and task checking.

Each identifiable activity which has a direct causal link to a human or hardware failure constitutes in whole, or in part, a safety management system. The audit probes the systems at each relevant level of the Socio-Technical Pyramid, tailoring questions to establish operator competency at Level 2, the effectiveness of a company's communication control and feedback mechanisms at Level 3 and establishing details of the management organisation at Level 4. Level 5, the systems climate, has been left out of the formal audit structure as these are matters which are outside the direct control of site management. It is, however, important to understand the environment within which the company operates and STATAS uses guidelines to assist the auditor in assessing the system climate. This may be particularly useful where it becomes apparent at a multi-site manufacturer that site practices and powers of decision making are controlled or restricted by corporate policies. In this case it is necessary to influence the highest decision maker, by direct approach at corporate level.

4.   THEMES

At each level of management questions are grouped into 4 job related themes. These are:

Theme A   - Structures, systems and procedures.

Theme B   - Standards and criteria.

Theme C   - Mitigation of pressures.

Theme D   - Availability and use of resources.

A fuller description of these themes is given in the next section but it can be seen that the
principles now described totally define the structure of the question set and the role and
purpose of individual questions. The structure is illustrated at Fig 10 with the questions
themselves designed to show the extent to which the management system under review can
demonstrate the existence of the key elements of successful health and safety management,
particularly the management control loops. Indeed there should be many embedded control
loops at each level of the system as well as a major one covering the systems main output, eg
training. The smallest sub set of questions is a "cell" which may consist of between 2 and 10
questions, exploring one of the themes at a particular level within the management structure

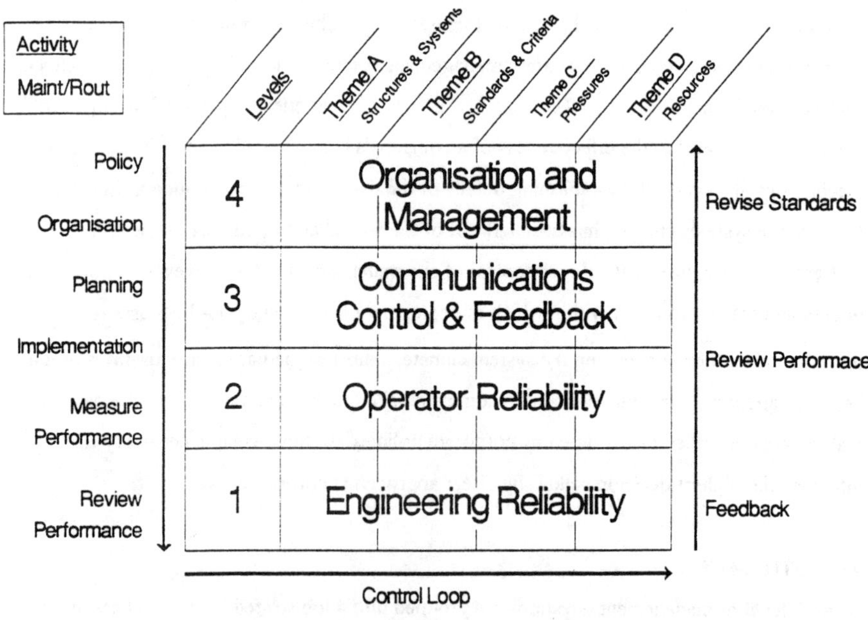

Fig 10   Arrangement of Question Set Cells

relating to one area of activity. At the moment the questions are a mixture of the general and the specific. General questions are designed to let the company demonstrate its range of activities in a particular area and specific questions are on those systems which would seem to be essential at any site, for example permit to work procedures. The questions are not designed to be followed slavishly. Once the auditor has established the extent of the activity and the quality of its controlling management system by "key elements" questions then he/she can move on. Neither is it necessary to use all of the audit at once. The question sets can be used in any combination, for example concentrating on maintenance activities or using 2 of the sets to look at human factors across maintenance and normal operations. This flexibility enables the auditor to judge and tailor the input of resource to individual companies.

5.   JOB RELATED THEMES

The themes in more detail, are as follows.

Theme A - Structures, systems and procedures. These are the arrangements that are in place to ensure that the system operates correctly. This includes the way rules and responsibilities are specified and how people are held accountable, the development of procedures and means of communication between individuals and groups.

Theme B - Standards and criteria. These are the means that ensure that proper use is made and notice taken of statutory requirements, approved codes of practice, guidance from the regulatory authority, industry standards and American, British or other International standards. It also includes the way in which these are incorporated into site standards and how any deviations from accepted norms are justified. Arrangements should also be made for monitoring and reviewing the performance of these standards with the objective of routine revision and improvement.

Theme C - Mitigation of pressures. These questions explore the extent to which economic, operational and other pressures may interfere with the achievement of policy objectives. Pressures may arise from schedules of work, customer demand etc and arrangements should exist for recognising the potential for pressure and ensuring that safety considerations are given proper regard.

Theme D - Resources. Questions seek to show that proper resources have been allocated to various functions, including properly competent personnel, information and equipment.

Each question set consists of between 40 and 80 questions and is fronted by a summary sheet (Fig 11) showing the percentage contribution to overall failure rate to this particular activity and summarising the key issues. This provides a "map" for the auditor to establish his/her position in the audit structure.

6.   THE ASSESSMENT PROCESS

The way in which the audit is conducted is based upon current best practice. The extent of the audit is determined from operational needs and an assessment of information already held about the firm. In the UK sites subject to the top tier requirements of the Control of Industrial Major Accident Hazards Regulations must submit Written Reports which include descriptions of their management arrangements. This report may point to activities which would be appropriate for an audit approach. The firm is contacted at highest site level with a preliminary visit to explain the methodology and agree on personnel to be interviewed. Such a visit is also useful to establish the system climate (Level 5).

The audit does not attempt to interview everyone on site but aims to sample "horizontal" and "vertical" slices, a degree of redundancy is built into these slices so that perceptions are obtained from more than one person and, to support information obtained from interview relevant documentation is examined and formal inspection carried out of procedures. The horizontal slice includes managers with influence and duties at Levels 4 and 3 and therefore with involvement in the process of policy and standard setting. The vertical slice looks at those involved with the delivery of engineering reliability and operator competence as appropriate. The 3 "legs" of the audit are mutually supporting. Information from formal interviews directs the auditor to particular documentation and claims made about systems are tested at operator and maintenance fitter level. The audit is conducted in a series of visits as close together as possible with 2 auditors mutually assisting and supporting each other. In practice it is found that interviews of about one hour are sufficient although this can be less for minor players.

S T A T A S

HAZARD IDENTIFICATION IN THE DESIGN OF PLANT
(Des/HAZ)

Origins of Failure	Prevention/ Recovery Mechanism	Weightings %	
		Pipework	Vessels
Design	Hazard studies	25	29

KEY ISSUES: Design criteria, materials, venting, fire precaution, process parameters.
Modification procedures.
Formal hazard studies (HAZAN, HAZOP etc).
Consequential changes (P and I diagrams), Standard Operating Procedures.
Multi disciplinary team composition; leadership and duties.
Programmable electronic systems and ergonomic factors.
Instrumentation and controls.
Location of hazardous plant in relation to control rooms and office locations.
Lessons from previous incidents locally, nationally and worldwide.

TESTED AT:
LEVEL 4 Organisation and management structures.
Roles, responsibilities and resources.
Priorities and standards; setting and review.
Data management.

LEVEL 3: Procedures for monitoring, feedback and auditing.
Formal, informal and written communications and meetings.
Documentation control.
Team membership and job specifications.
Training, experience and qualifications of team members.

LEVEL 2: Operator awareness and competency.
Active participation in change procedures.

LEVEL 1: Evidence of implementation of procedures.
Appropriate shutdown systems, pressure relief settings etc.

ASK	OBSERVE	CHECK

FIG 11

EXAMPLE OF SUMMARY SHEET

After all the evidence is assembled together with supporting documentation and the experience of inspection the auditors seek to make judgements about site performance in each theme of the activity. Ratings are made on a 5 point scale with "average" as the centre point. In areas of activity where clear standards exist then judgement can be made directly against these standards. In many areas, however, there are no such standards and judgement is made against the evidence of management activity. For example, at an average site it would be expected that there would be some management activity in most areas although procedures would not be formalised and there would be little feedback through monitoring. A site marked well below would have no measurable control activities and indeed, hazardous conditions would be likely to exist. For an assessment of "well above average" it should be seen that programmes exist with properly chosen standards, that management control principles are understood and applied and control loops built into the systems are working and generating system improvements.

The choice of "anchor points" is an important one to any audit scheme. In the development of STATAS and parallel work this is an area where further research is continuing and more robust points are being produced. Once these judgements have been made the results are assembled and can be presented to the company for an Improvement Plan to be agreed. In the UK it is seen as important that firms safety representatives and union officials are included in these procedures. It is almost axiomatic that a company well on the path of good management and safety performance would have a high degree of involvement from the workforce.

Priorities can be assigned to recommendations by considering the overall contribution of that activity to failure rates. Experience has shown that this is not quite as straight forward as it seems and auditor judgement is usually mixed with company pragmatism to prioritise the recommendations.

7.　　CONCLUSIONS AND FURTHER WORK

The development team have now concluded a number of audits using STATAS developing and refining the system as a consequence. Although the Field Operations Division of the Health and Safety Executive is not committed to adopting any such system at this time it is very actively looking at the role of audits and how a system such as STATAS could be used to augment more traditional inspection techniques. The main qualities of STATAS are its sound theoretical and empirical basis and the fusion of management and loss causation models to produce a series of judgement based question sets that, taken together with reviews of documents and traditional inspection, produce a robust assessment of site activities designed to control risks of major accidents, and a framework for improvements.

Further work is continuing on the development of anchor points and methodology and a further module covering mitigation.

REFERENCES

1.   HS(G)65 "Successful Health and Safety Management"
      HSE: Accident Prevention Advisory Unit, HMSO ISBN 0 11 885988 9

2.   HS(G)48 "Human Factors in Industrial Safety"
      HSE, HMSO ISBN 0 11 885486 0

3.   "Guidance on Permit to Work Systems in the Petroleum Industry"
      OIAC, HMSO ISBN 0 11 885688 X

4.   "Guidance on Multi-Skilling in the Petroleum Industry"
      OIAC, HMSO ISBN 0 11 886319 3

5.   "Dangerous Maintenance: A Study of Maintenance Accidents in the Chemical Industry
      and how to prevent them"
      HSE, HMSO ISBN 0 11 883957 8

6.   L J Bellamy, T A W Geyer and J A Astley
      Evaluation of the Human Contribution to Pipework and In-Line Equipment Failure
      Frequencies. Contract Research Report No 89/15. HSE. ISBN 0717603245

7.   N W Hurst, L J Bellamy, T A W Geyer and J A Astley.
      A Classification Scheme for Pipework Failures to Include Human and Sociotechnical
      Errors and their Contribution to Pipework Failure Frequencies. J Haz Mat 26 (1991)
      159-186

8.   L J Bellamy, T A W Geyer, M S Wright and N W Hurst.
      The Development in the UK of Techniques to Take Account of Management,
      Organisational and Human Factors in the Modification of Risk Estimates. AiChemE.
      Spring National Meeting, Orlando 1990

9.   T A W Geyer and L J Bellamy.
      Pipework Failure, Failure Cases and the Management Factor. I Mech Eng 1991

10.   N W Hurst, L J Bellamy and T A W Geyer.
      Organisational, Management and Human Factors in Quantified Risk Assessment. A
      Theoretical and Empirical Basis for Modification of Risk Estimates. Safety and
      Reliability in the 90's (SARRS '90). Ed Water and Cox. Elsevier Applied Science

11.   L J Bellamy and T A W Geyer (ed J C Williams).
       Organisational, Management and Human Factors in Quantified Risk Assessment. HSE
       Contract Research Report 33/1991

12.   P I Harrison and J C Williams.
       Organisational Management and Human Factors in Qualified Risk Assessment. HSE
       Contract Research Report 34/1991